European Consortium for
Mathematics in Industry

E. Eich-Soellner and C. Führer

Numerical Methods
in Multibody Dynamics

# European Consortium for Mathematics in Industry

Edited by
Leif Arkeryd, Göteborg
Heinz Engl, Linz
Antonio Fasano, Firenze
Robert M. M. Mattheij, Eindhoven
Pekka Neittaanmäki, Jyväskylä
Helmut Neunzert, Kaiserslautern

Within Europe a number of academic groups have accepted their responsibility towards European industry and have proposed to found a European Consortium for Mathematics in Industry (ECMI) as an expression of this responsibility.

One of the activities of ECMI is the publication of books, which reflect its general philosophy; the texts of the series will help in promoting the use of mathematics in industry and in educating mathematicians for industry. They will consider different fields of applications, present casestudies, introduce new mathematical concepts in their relation to practical applications. They shall also represent the variety of the European mathematical traditions, for example practical asymptotics and differential equations in Britian, sophisticated numerical analysis from France, powerful computation in Germany, novel discrete mathematics in Holland, elegant real analysis from Italy. They will demonstrate that all these branches of mathematics are applicable to real problems, and industry and universities in any country can clearly benefit from the skills of the complete range of European applied mathematics.

# Numerical Methods in Multibody Dynamics

Edda Eich-Soellner

Fachbereich Informatik / Mathematik
Fachhochschule München, Germany

Claus Führer

Department of Computer Science and
Numerical Analysis
Lund University, Sweden

Springer Fachmedien Wiesbaden
GmbH 1998

Prof. Dr. rer. nat. Edda Eich-Soellner
Fachbereich Informatik/Mathematik
Fachhochschule München
Lothstraße 34
D-80335 München Germany
e-mail: edda.eich@informatik.fh-muenchen.de
WWW: http://www.informatik.fh-muenchen.de/professoren/EddaEich-Soellner/


Docent Dr. rer. nat. Claus Führer
Department of Computer Science and Numerical Analysis
Lund University
Box 118
S-22100 Lund Sweden
e-mail: claus@dna.lth.se
WWW: http://www.dna.lth.se/home/Claus_Fuhrer/

Die Deutsche Bibliothek – CIP-Einheitsaufnahme

**Eich-Soellner, Edda:**
Numerical methods in multibody dynamics / Edda Eich-Soellner; Claus Führer.
 (European Consortium for Mathematics in Industry)
  ISBN 978-3-663-09830-0     ISBN 978-3-663-09828-7 (eBook)
  DOI 10.1007/978-3-663-09828-7

© Copyright 1998 by Springer Fachmedien Wiesbaden
Originally published by B. G. Teubner Stuttgart in 1998

Cover design by Peter Pfitz Stuttgart.

# Preface

Numerical Analysis is an interdisciplinary topic which develops its strength only when viewed in close connection with applications.

Nowadays, mechanical engineers having computer simulation as a daily engineering tool have to learn more and more techniques from that field. Mathematicians, on the other hand, are increasingly confronted with the need for developing special purpose methods and codes. This requires a broad interdisciplinary understanding and a sense for model-method interactions.

With this monograph we give an introduction to selected topics of Numerical Analysis based on these facts. We dedicate our presentations to an interesting discipline in computational engineering: multibody dynamics. Though the basic ideas and methods apply to other engineering fields too, we emphasize on having one homogeneous class of applications.

Both authors worked through many years in teams developing multibody codes. Interdisciplinary work also includes transferring ideas from one field to the other and a big amount of teaching - and that was the idea of this book.

This book is intended for students of mathematics, engineering and computer science, as well as for people already concerned with the solution of related topics in university and industry.

After a short introduction to multibody systems and the mathematical formulation of the equations of motion, different numerical methods used to solve simulation tasks are presented. The presentation is supported by a simple model of a truck. This truck model will follow the reader from the title page to the appendix in various versions, specially adapted to the topics.

The models used in this book are not intended to be real-life models. They are constructed to demonstrate typical effects and properties occurring in practical simulations.

The methods presented include linear algebra methods (linearization, stability analysis of the linear system, constrained linear systems, computation of nominal interaction forces), nonlinear methods (Newton and continuation methods for the computation of equilibrium states), simulation methods (solution of discontinuous ordinary differential and differential algebraic equations) and solution methods for inverse problems (parameter identification). Whenever possible, a more general presentation of the methods is followed by a special view, taking the structure of multibody equations into consideration.

Each chapter is divided into sections. Some of the material can be skipped during a first reading. An asterisk (*) in the section title is indicating these parts.

Nearly all methods and examples are computed using MATLAB programs and nearly all examples are related to the truck model. Those MATLAB programs which describe the truck itself are given in the appendix for supporting the description of the model. Others are given as fragments in the text, where MATLAB is used as a piece of meta language to describe an algorithm.
Some of the examples had been used in universitary and post universitary courses. These can be obtained among other information related to this book via the book's homepage[1].

We want to thank everyone who has helped us to write this book: our teachers, colleagues, friends and families.

January 1998                                    Edda Eich-Soellner and Claus Führer

---

[1] http://www.teubner.de/cgi-bin/teubner-anzeige.sh?buch_no=12

# Contents

# 1 Multibody Systems

## 1.1 What is a Multibody System?

Let us assume somebody in a truck manufacturing company has the idea to improve the driver's comfort by damping the oscillations of the driver's cabin. For this end he plans to introduce a new type of dampers governed by a sophisticated control law and a complicated mechanical coupling device. The best argument for convincing the decision makers in his company would be a demonstration of his idea. The classical "demonstration" of such an idea would require the construction of the new device and a lot of field tests. And this costs money and takes time and might be too expensive just for checking an idea.

The modern way to check ideas and to convince others is computer simulation. There, a computer model of the system under investigation is established and all kinds of studies and "measurements" are made on this model. The art of transforming a complex mechanical system into a computer model is covered by the rather young discipline *computational mechanics*.

There are different tools to transform a mechanical system into a computer model. We will consider here the technique to model a mechanical system as a so-called *multibody system*. A multibody system (MBS) is a result of describing a mechanical system by simple components specially designed for mechanical systems. These consist of *bodies*, which have masses and torques and can be rigid or elastic. They are linked together by mass-less *interconnections* which can be *force elements*, like springs and dampers, or by *joints* reducing the degrees of freedom, see Fig. 1.1. Typical joints are constraining the motion in one or more directions, they may also force the bodies to follow a particular curve.

This description of the mechanical system is used by special multibody programs to automatically transform it into a set of mathematical equations, mostly ordinary differential equations (ODEs).

Often the engineer using such a tool is not interested in the mathematical description of his problem. He asks for a machinery inside the multibody program which handles these equations in a numerically appropriate way and gives answers to his questions. In the above example the critical speed of the vehicle might be such an answer.

However, it cannot be expected that the numerical methods implemented in multibody programs run satisfactory for *any* multibody problem. A particular model might cause problems such as the fascinating error message "corrector could not

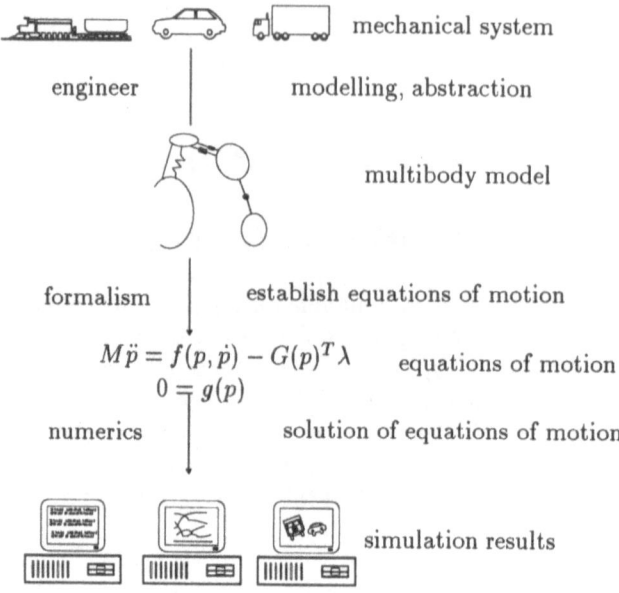

Figure 1.1: Typical steps for simulating a multibody system

converge". In other cases the performance indicates that there might be problems with the model.

Then it becomes necessary for the engineer to know something about the numerical "machinery" inside multibody programs. There are usually a lot of varieties how to model a system. The choice of the model and the mathematical formulation have a large impact on the simulation performance. Let us take a simulation of a truck as an example: Which coordinates are taken for the description? What about the friction models?

These are questions which can only be answered when based on a deep knowledge about the numerical algorithms.

The other way round, the numerical analyst working together with engineers is asked to understand their models and the influence of the models on the behavior of the numerical algorithms.

This book is not a course about multibody systems, nor a description of a particular multibody program. This can be found in various other places, e.g. [Hau89, Sch90]. Our goal is to give insights into the numerical methods inside multibody programs and the relationship between model formulation and numerical methods to be used.

In this first chapter we discuss the basic mathematical tasks and the typical form of equations of motion of multibody systems. We present a simple planar model of a truck used throughout this book for demonstration purposes. The chapter is concluded by the linearization of the equations of motion and the computation of nominal interaction forces.

## 1.2 Basic Mathematical Tasks in Multibody Dynamics

Multibody system analysis includes a number of different mathematical tasks and questions:

- *Kinematic analysis*
  Is the model reasonable? Do the bodies penetrate each other?
  Given a robot and its hand position. How must the joint angles be chosen?
  These questions lead to systems of nonlinear equations, which are normally solved by Newton's method, see Ch. 3.

- *Static equilibrium position*
  How must the forces be chosen such that the system is in equilibrium?
  Mathematically, a nonlinear system of equations must be solved. This is done by using Newton's method, possibly together with a continuation method, see Ch. 3.

- *Dynamic simulation*
  How does the system behave dynamically?
  This requires the numerical solution of explicit ordinary differential equations or differential-algebraic equations, see Chs. 4, 5, 6.

- *Linear system analysis*
  Is the system stable? How is its input/output behavior?
  This requires an eigenvalue analysis of the linearized system and the numerical computation of its frequency response, see Ch. 2.

- *Design, optimal control*
  How can the behavior of the system be optimized?
  This questions includes parameter optimization, optimization of the structure, optimal control problems and *parameter identification problems*. The latter problem arises if some of the system parameters are unknown and measurements are taken to determine them.
  This leads to optimization problems with explicit ordinary differential or differential-algebraic equations as constraints, see Ch. 7.

## 1.3  Basic Form of Equations of Motion of Multi-body Systems

We discuss in this section the basic forms of these equations, and we refer the reader interested in the way how to derive them efficiently to the literature on multibody formalisms given e.g. in [RS88].

There are various classes of MBS having differently structured equations of motion:

- unconstrained MBS;

- constrained systems with rheonomic, time dependent, or scleronomic, time independent constraints. In these systems a special class of forces, the so-called constraint forces must be taken into account;

- constrained systems with friction elements, where the friction force depends on the constraint forces;

- MBS with special additional dynamic force laws resulting from control devices etc.

Moreover, the type of the resulting equations depends also on the type of variables used. In some cases it may be advantageous to use redundant variables. This may be the case when considering constrained systems but also when describing large rotations even in unconstrained systems.

### 1.3.1  Unconstrained Planar Multibody Systems

Let us start with the simplest case, the *unconstrained*, planar system: With $p$ describing the $n_p$ position coordinates of a given MBS the equations of motion have the form

$$M\ddot{p} = f_a(t, p, \dot{p}). \qquad (1.3.1)$$

This is just *Newton's law* saying that mass times acceleration equals force. Herein, $M$ is a positive definite $n_p \times n_p$ *mass matrix* and $f_a(t, p, \dot{p})$ are the applied forces. $t$ denotes the time, $\dot{p}$ the velocity and $\ddot{p}$ the acceleration.

**Example 1.3.1** *Let us consider the planar model of the truck shown in Fig. 1.2 to demonstrate the individual terms of Eq. (1.3.1). The set of position coordinates $p_i, i = 1, \ldots, n_p$ is related to the degrees of freedom of the individual bodies* *composing the truck:*

$p_1$   *Vertical motion of the rear wheel (body 1)*
$p_2$   *Vertical motion of the truck chassis (body 2)*
$p_3$   *Rotation of the truck chassis (body 2)*
$p_4$   *Vertical motion of the front wheel (body 3)*
$p_5$   *Vertical motion of the driver cabin (body 4)*
$p_6$   *Rotation of the driver cabin (body 4)*
$p_7$   *Horizontal motion of the loading area (body 5)*
$p_8$   *Vertical motion of the loading area (body 5)*
$p_9$   *Rotation of the loading area (body 5)*

*All coordinates are defined with respect to the inertial system.*
*The equations of motion can be derived by simply summing up all forces acting on the individual bodies:*

$$
\begin{aligned}
m_1\ddot{p}_1 &= -f_{10_2} + f_{12_2} - m_1 g_{gr} \\
m_2\ddot{p}_2 &= -f_{12_2} - f_{23_2} + f_{24_2} + f_{42_2} + f_{25_2} + f_{1d_2} + f_{2d_2} - m_2 g_{gr} \\
l_2\ddot{p}_3 &= (-a_{23}f_{23_2} - a_{12}f_{12_2} - h_1(f_{23_1} + f_{12_1})) \cos p_3 - \\
&\quad (-a_{23}f_{23_1} - a_{12}f_{12_1} - h_1(f_{23_2} + f_{12_2})) \sin p_3 + \\
&\quad (a_{25}f_{25_2} + a_{52}(f_{1d_2} + f_{2d_2}) + h_2(f_{25_1} + f_{1d_1} + f_{2d_1})) \cos p_3 - \\
&\quad (a_{25}f_{25_1} + a_{52}(f_{1d_1} + f_{2d_1}) + h_2(f_{25_2} + f_{1d_2} + f_{2d_2})) \sin p_3 - \\
&\quad (a_{24}f_{24_2} + a_{42}f_{42_2} + h_2(f_{24_1} + f_{42_1})) \cos p_3 - \\
&\quad (a_{24}f_{24_1} + a_{42}f_{42_1} + h_2(f_{24_2} + f_{42_2})) \sin p_3 \\
m_3\ddot{p}_4 &= -f_{30_2} + f_{23_2} - m_3 g_{gr} \\
m_4\ddot{p}_5 &= -f_{42_2} - f_{24_2} - m_4 g_{gr} \\
l_4\ddot{p}_6 &= (-b_{24}f_{24_2} - b_{42}f_{42_2} - h_3(f_{24_1} + f_{42_1})) \cos p_6 - \\
&\quad (-b_{24}f_{24_1} - b_{42}f_{42_1} - h_3(f_{24_2} + f_{42_2})) \sin p_6 \\
m_5\ddot{p}_7 &= -f_{25_1} - f_{1d_1} - f_{2d_1} \\
m_5\ddot{p}_8 &= -f_{25_2} - f_{1d_2} - f_{2d_2} - m_5 g_{gr} \\
l_5\ddot{p}_9 &= (-c_{25}f_{25_2} - c_{d1}f_{1d_2} - c_{d2}f_{2d_2} - h_3(f_{25_1} + f_{1d_1} + f_{2d_1})) \cos p_9 - \\
&\quad (-c_{25}f_{25_1} - c_{d1}f_{1d_1} - c_{d2}f_{2d_1} - h_3(f_{25_2} + f_{1d_2} + f_{2d_2})) \sin p_9
\end{aligned}
$$

*Herein $m_i, l_i$ are the masses and torques of body $i$, $g_{gr}$ is the gravitational accelera-tion constant and the other symbols denote geometric constants and forces defined in Figs. 1.3, 1.4 and Tab. A.3. All forces induced by springs and dampers in parallel are given by simple linear relations. Thus, $f_{10_2}$ is given as*

$$
f_{10_2} = k_{10}\,(p_1 - u(t)) + d_{10}\,(\dot{p}_1 - \dot{u}(t)) + f_{10}^0.
$$

*The time dependent function $u$ describes the excitation of the truck due to the road roughness. The road roughness normally is described by a function $\xi$ having the distance from a particular point of the road to a fixed starting point as its*

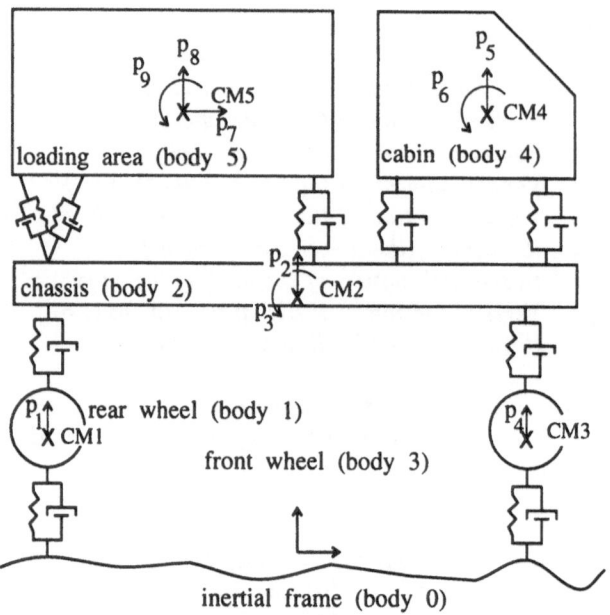

Figure 1.2: Planar unconstrained MBS model of a truck. "x" denotes the center of mass (CM) of the individual bodies.

*independent variable. If we assume the truck having a constant forward speed $\nu$ we get the relation*

$$u(t) := \xi(\nu \cdot t).$$

*This road roughness function is acting in the same way but with a delay $T_1 - T_2$ on the front and rear wheels. These delays are given by the relations $T_1 := \nu \cdot a_{12}$ and $T_2 := -\nu \cdot a_{23}$.*
*Another example is the force $f_{25}$:*

$$
\begin{aligned}
f_{25}(t) &= k_{25}\|\rho_{25}(t)\| + d_{25}\tfrac{d}{dt}\|\rho_{25}(t)\| + f_{25}^0 \text{ and} \\
(f_{25_1}(t), f_{25_2}(t))^T &= f_{25}(t)\tfrac{\rho_{25}(t)}{\|\rho_{25}(t)\|}
\end{aligned}
$$

*with $\rho_{ij}$ being the vector between the points of attachment of bodies $i$ and $j$:*

$$\rho_{25} = \begin{pmatrix} p_7 \\ p_8 \end{pmatrix} + S(p_9) \begin{pmatrix} c_{25} \\ -h_3 \end{pmatrix} - \left[ \begin{pmatrix} 0 \\ p_2 \end{pmatrix} + S(p_3) \begin{pmatrix} a_{25} \\ h_2 \end{pmatrix} \right] \qquad (1.3.2)$$

*with the 2D-rotation matrix*

$$S(\alpha) := \begin{pmatrix} \cos\alpha & -\sin\alpha \\ \sin\alpha & \cos\alpha \end{pmatrix}. \qquad (1.3.3)$$

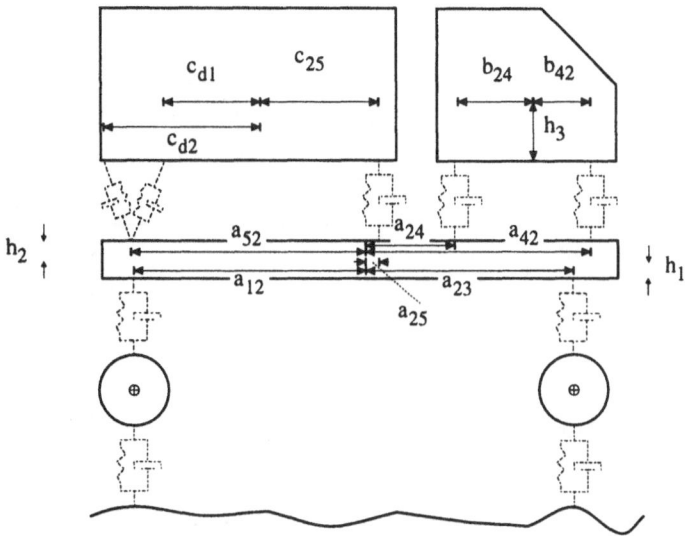

Figure 1.3: Geometry of the unconstrained truck

*We also make the assumption that the points of attachment are chosen in such a way that $\|\rho_{ij}\| \neq 0$ for all motions of interest.*

*Note that the computation of $\dot{\rho}_{ij}$ involves the computation of the time derivative of the rotation matrix $S$. We will come back to that point later in Sec. 1.3.5.*

*Introducing now the geometric and inertia data given in Tables A.3-A.4 in the appendix and reformulating the equations in vector and matrix notation we end up with a set of equations in a form like Eq. (1.3.1). There are still unknown quantities in the equations, the nominal forces $f_{10}^0, \ldots, f_{52}^0$. These forces are in general unknown. We will compute these forces by requiring that the system is initially in a stable equilibrium position. The way this task can be solved in general will be discussed in Sec. 1.6.*

## 1.3.2 Constrained Planar Multibody Systems

The more general case is the *constrained*, planar system. A constraint is a relation between a subset of the coordinates, which must hold for every possible system configuration. A typical constraint is a *joint*, which allows motions parallel to certain axes only or restricts certain rotations, e.g. a Kardan joint. More complicated joints are constraints which permit motions only along a prescribed curve or surface. Typical technical examples for this type of joints can be found in contact mechanics, see Sec. 5.5.

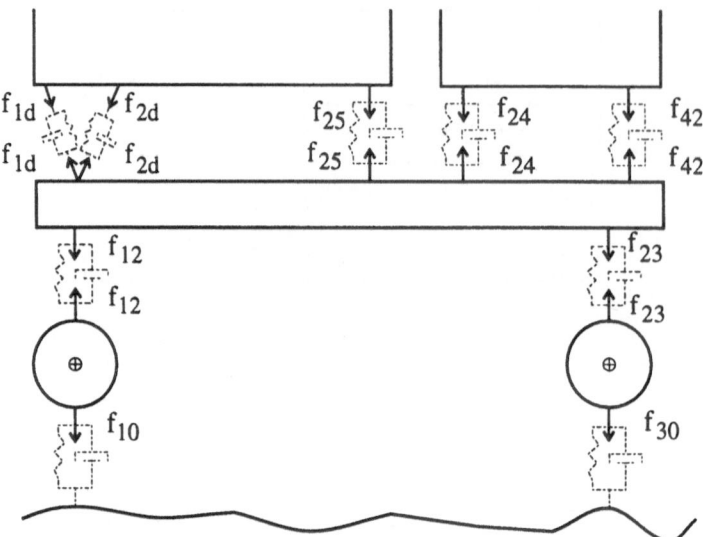

Figure 1.4: Forces acting on the unconstrained truck

The equations of motion in the constrained case have the form

$$M\ddot{p} = f_a(t, p, \dot{p}) - f_c(p, \lambda)$$
$$0 = g(p)$$

where $g(p)$ is a vector valued function describing the $n_c$ constraints; $f_c$ describes additional forces acting on the system. These so-called *generalized constraint forces* are responsible for the constraint to be satisfied. The constraint defines a manifold of free motion. By basic physical principles, like *d'Alembert's principle*, it can be shown that constraint forces are orthogonal to this manifold. This leads to

$$f_c(p, \lambda) = G(p)^T \lambda$$

with the *constraint matrix* $G(p) := \frac{d}{dp}g(p)$ and $n_\lambda$ unknown parameters $\lambda$, the so-called *Lagrange multipliers*.

Thus, the equations of motion for a holonomically constrained system have the structure

$$M\ddot{p} = f_a(t, p, \dot{p}) - G(p)^T \lambda \qquad (1.3.4a)$$
$$0 = g(p). \qquad (1.3.4b)$$

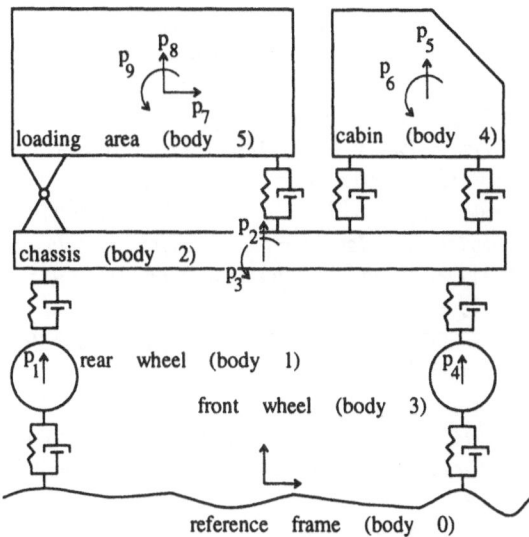

Figure 1.5: MBS model of the planar truck model with a constraint

**Example 1.3.2** *We modify now the truck given in Example 1.3.1 by replacing the spring and damper at the rear end of the loading area by a joint, which allows the loading area to be tilted, cf. Fig. 1.5. The constraint says that the distance between chassis and loading area at the joint's rotation axis must be zero. We must therefore compute the relative vector $\rho_{52}$ between the points of attachment of the joint 52*

$$\rho_{52} = \begin{pmatrix} p_7 \\ p_8 \end{pmatrix} + S(p_9) \begin{pmatrix} -c_{c_1} \\ -c_{c_2} \end{pmatrix} - \left[ \begin{pmatrix} 0 \\ p_2 \end{pmatrix} + S(p_3) \begin{pmatrix} -a_{c_1} \\ a_{c_2} \end{pmatrix} \right]. \qquad (1.3.5)$$

*Thus, we have*

$$g(p) = \rho_{52}(p) = 0.$$

*This defines the $n_p \times n_\lambda$ constraint matrix*

$$G(p)^T = \begin{pmatrix} 0 & 0 \\ 0 & -1 \\ -a_{c_1} \sin p_3 + a_{c_2} \cos p_3 & a_{c_1} \cos p_3 + a_{c_2} \sin p_3 \\ 0 & 0 \\ 0 & 0 \\ 0 & 0 \\ 1 & 0 \\ 0 & 1 \\ c_{c_1} \sin p_9 + c_{c_2} \cos p_9 & -c_{c_1} \cos p_9 + c_{c_2} \sin p_9 \end{pmatrix}$$

*with the geometric constants $a_{c_1}, \ldots c_{c_2}$ given in Tab. A.3.*
*Altogether the equations of motion of the constrained truck are given by*

$$
\begin{aligned}
m_1 \ddot{p}_1 &= -f_{10_2} + f_{12_2} - m_1 g_{gr} \\
m_2 \ddot{p}_2 &= -f_{12_2} - f_{23_2} + f_{24_2} + f_{42_2} + f_{25_2} - m_2 g_{gr} + \lambda_2 \\
l_2 \ddot{p}_3 &= (-a_{23} f_{23_2} - a_{12} f_{12_2} - h_1(f_{23_1} + f_{12_1})) \cos p_3 - \\
&\quad (-a_{23} f_{23_1} - a_{12} f_{12_1} - h_1(f_{23_2} + f_{12_2})) \sin p_3 + \\
&\quad (a_{25} f_{25_2} + h_2 f_{25_1}) \cos p_3 - (a_{25} f_{25_1} + h_2 f_{25_2}) \sin p_3 - \\
&\quad (a_{24} f_{24_2} + a_{42} f_{42_2} + h_2(f_{24_1} + f_{42_1})) \cos p_3 - \\
&\quad (a_{24} f_{24_1} + a_{42} f_{42_1} + h_2(f_{24_2} + f_{42_2})) \sin p_3 - \\
&\quad (-a_{c_1} \sin p_3 + a_{c_2} \cos p_3) \lambda_1 - (a_{c_1} \cos p_3 + a_{c_2} \sin p_3) \lambda_2 \\
m_3 \ddot{p}_4 &= -f_{30_2} + f_{23_2} - m_3 g_{gr} \\
m_4 \ddot{p}_5 &= -f_{42_2} - f_{24_2} - m_4 g_{gr} \\
l_4 \ddot{p}_6 &= (-b_{24} f_{24_2} - b_{42} f_{42_2} - h_3(f_{24_1} + f_{42_1})) \cos p_6 - \\
&\quad (-b_{24} f_{24_1} - b_{42} f_{42_1} - h_3(f_{24_2} + f_{42_2})) \sin p_6 \\
m_5 \ddot{p}_7 &= -f_{25_1} - \lambda_1 \\
m_5 \ddot{p}_8 &= -f_{25_2} - m_5 g_{gr} - \lambda_2 \\
l_5 \ddot{p}_9 &= (-c_{25} f_{25_2} - h_3 f_{25_1}) \cos p_9 - (-c_{25} f_{25_1} - h_3 f_{25_2}) \sin p_9 - \\
&\quad (c_{c_1} \sin p_9 + c_{c_2} \cos p_9) \lambda_1 - (-c_{c_1} \cos p_9 + c_{c_2} \sin p_9) \lambda_2 \\
0 &= p_7 - c_{c_1} \cos p_9 + c_{c_2} \sin p_9 + a_{c_1} \cos p_3 + a_{c_2} \sin p_3 \\
0 &= p_8 - c_{c_1} \sin p_9 - c_{c_2} \cos p_9 - p_2 + a_{c_1} \sin p_3 - a_{c_2} \cos p_3
\end{aligned}
$$

Condition (1.3.4b) not only constrains the position coordinates $p$ but also the corresponding velocities $\dot{p}$. This can be seen by forming the first derivative with respect to time, which must also vanish along the solution $p(t)$

$$G(p)\dot{p} = 0. \qquad (1.3.6)$$

This is a formulation of the *constraint on velocity level*.

### 1.3.3  Multibody Systems with Nonholonomic Constraints*

One might confound constraints given on velocity level with nonholonomic constraints. Though we will not treat nonholonomic constraints in this book we will give a definition to show the difference:

**Definition 1.3.3** *A constraint[1] $G(p)\dot{p} = 0$ is called* holonomic, *iff there exists a function $g(p)$ with $\frac{d}{dp}g(p) = G(p)$. Otherwise the constraint is called nonholonomic.*

Thus, in order to be a holonomic constraint every row of $G(p)$ must be a gradient of a function. A necessary condition for a vector to be a gradient is that its derivative is a symmetric matrix, e.g. [OR70]. Thus, the $i$-th constraint is nonholonomic, iff

---

[1] As in mechanics there are only constraints known which are linear in the velocities, we formulate the definition only for that class.

the derivative with respect to $p$ of the $i$-th column of $G(p)^T$ is a non symmetric $n_p \times n_p$ matrix.

Nonholonomic constraints occur in mechanics for example when "pure" rolling of two surfaces is modeled, see [RS88]. In practice "pure" rolling is an idealization, if creepage effects can be neglected.

### 1.3.4 System with Dynamical Force Elements*

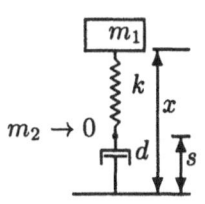

Figure 1.6: Spring and damper in series

The equations of motion presented so far are second order ODEs with or without constraints. In some cases there might be also additional *first order differential equations*. Consider, for example, the system shown in Fig. 1.6. If the mass $m_2$ is taken to be zero the mass $m_1$ is suspended by an interconnection consisting of a *spring and a damper in series*.

To establish the equations of motion of that system under the influence of gravitation we will first give $m_2$ a nonzero mass. This yields

$$m_1\ddot{x} = m_1 g_{gr} - k(x - s)$$
$$m_2\ddot{s} = m_2 g_{gr} - d\dot{s} + k(x - s).$$

If in the second equation $m_2$ is set to zero, we obtain a mixed system consisting in a second and a first order ODE

$$m_1\ddot{x} = m_1 g_{gr} - k(x - s)$$
$$\dot{s} = \frac{k}{d}(x - s).$$

Mixed systems of this type arise also when mechanical systems are influenced by non mechanical dynamic forces, like electro-magnetic forces in a magnetically levitated (maglev) vehicle. Also, multibody systems with additional control devices, hydraulic components etc. may result in such a mixed form.

### 1.3.5 General Systems

In the 3D-case the equations of unconstrained motion are given in the form

$$M\dot{v} = f_a(t, p, v), \qquad (1.3.7)$$

where $v$ is a velocity vector composed of translational velocities and *angular velocities*. In the 3D-case angular velocities are not just the derivatives of the corresponding angles. They are related to position coordinates $p$ in a nonlinear way described by the *kinematical differential equations*

$$\dot{p} = Z(p)v, \qquad (1.3.8)$$

which we will derive now.

A 3D-rotation can be expressed in the form[2]

$$S(\alpha, \beta, \gamma) = S(0, 0, \gamma)S(0, \beta, 0)S(\alpha, 0, 0)$$

with the elementary rotation matrices

$$S(\alpha, 0, 0) = \begin{pmatrix} 1 & 0 & 0 \\ 0 & \cos\alpha & -\sin\alpha \\ 0 & \sin\alpha & \cos\alpha \end{pmatrix}$$

$$S(0, \beta, 0) = \begin{pmatrix} \cos\beta & 0 & -\sin\beta \\ 0 & 1 & 0 \\ \sin\beta & 0 & \cos\beta \end{pmatrix}$$

$$S(0, 0, \gamma) = \begin{pmatrix} \cos\gamma & -\sin\gamma & 0 \\ \sin\gamma & \cos\gamma & 0 \\ 0 & 0 & 1 \end{pmatrix}.$$

Herein $\alpha$ is the angle of the rotation about the $x$-axis, $\beta$ the angle of rotation about the (new) $y$-axis and $\gamma$ the corresponding one about the (new) $z$-axis.

From that, it can be easily shown that there exists a matrix $\tilde{\omega}$ of the form

$$\tilde{\omega} := \begin{pmatrix} 0 & -\omega_3 & \omega_2 \\ \omega_3 & 0 & -\omega_1 \\ -\omega_2 & \omega_1 & 0 \end{pmatrix}$$

fulfilling the so-called *Poisson equation*

$$\dot{S}(\alpha, \beta, \gamma) = \tilde{\omega} S(\alpha, \beta, \gamma). \tag{1.3.9}$$

This equation defines the *angular velocities* $\omega_1, \omega_2, \omega_3$. From (1.3.9) the transformation between angular velocities and $\dot{\alpha}, \dot{\beta}, \dot{\gamma}$ can be established:

$$\begin{pmatrix} \dot{\alpha} \\ \dot{\beta} \\ \dot{\gamma} \end{pmatrix} = \begin{pmatrix} \cos\gamma/\cos\beta & \sin\gamma/\cos\beta & 0 \\ \sin\gamma & -\cos\gamma & 0 \\ -\cos\gamma\tan\beta & -\sin\gamma\tan\beta & 1 \end{pmatrix} \begin{pmatrix} \omega_1 \\ \omega_2 \\ \omega_3 \end{pmatrix}. \tag{1.3.10}$$

For an entire system this equation defines in an obvious way the regular transformation matrix $Z(p)$ in (1.3.8).

In the above truck example a 2D-model is described and derivatives of angles and angular velocities are identical. Thus, in that case $Z(p)$ is the identity matrix.

---

[2] This parameterization of the rotation matrix is called the Kardan angle or Tait-Bryan angle parameterization. For a detailed discussion of various parameterizations of the rotation matrix, see [RS88].

The general first order form of the equation of motion of a constrained system then reads

$$
\begin{aligned}
\dot{p} &= Z(p)v & \text{(1.3.11a)} \\
M\dot{v} &= f_a(t, p, v, s) - Z(p)^T G(p)^T \lambda & \text{(1.3.11b)} \\
\dot{s} &= f_s(t, p, v, s) & \text{(1.3.11c)} \\
0 &= g(p). & \text{(1.3.11d)}
\end{aligned}
$$

Note that this equation cannot be transformed into a second order differential equation in the standard way. This transformation can be performed after first introducing velocity variables $\bar{v} := Z(p)v$.
A transformation into this set of variables leads to

$$
\begin{aligned}
\dot{p} &= \bar{v} \\
\bar{M}(p)\dot{\bar{v}} &= Z(p)^{-T}\left(f_a(t, p, Z(p)^{-1}\bar{v}, s) - M\dot{Z}(p)^{-1}\bar{v}\right) - G(p)^T\lambda \\
\dot{s} &= f_s(t, p, Z(p)^{-1}\bar{v}, s) \\
0 &= g(p),
\end{aligned}
$$

with a $p$-dependent transformed mass matrix $\bar{M}(p) := Z(p)^{-T}MZ(p)^{-1}$.
It should be noted that the transformation matrix becomes unbounded for $\beta \to \pi/2$. This is the reason for taking other parameterizations of the rotation matrix if $\beta$ tends towards $\pi/2$. Such a reparameterization introduces discontinuities which can be avoided when using a redundant set of rotation coordinates. One typically uses *quaternions* often also called *Euler parameters*. These are four coordinates instead of the three angles and one additional normalizing equation, see Ex. 5.1.10. This normalizing equation describes a property of the motion, a so-called *solution invariant*. Differential equations with invariants will be discussed in Sec. 5.3.
For the description of most of the methods in this book the particular choice of $v$ and the presence of equations of the type (1.3.11c) plays no role. We will thus assume $Z(p) = I$ and the dimension of the vector $s$ being zero.

## 1.4   Relative and Absolute Coordinates*

In this section we will see how the choice of coordinates affects the structure of the equations of motion (1.3.11).
We saw from the example above that if absolute coordinates were taken, we get equations of motion having two important structural properties:

1. The mass matrix is constant and block-diagonal. This makes it easy to convert Eq. (1.3.11b) into an explicit equation.

2. There are algebraic equations describing the constraints, so that the system (1.3.11) is a system of differential-algebraic equations (DAE). These systems are principally different from ODEs and require special numerical methods for their solution.

Unless there are robust numerical methods available to treat DAE-systems, the last property turns out to be rather disadvantageous, while the first property simplifies the numerical treatment a lot.

By taking relative coordinates the system gets just the opposite features:

1. The mass matrix is dense and now position dependent.

2. At least for tree structured systems the constraints can be easily eliminated.

This will be explained now by considering the constrained truck model again. For this end we will assume that there exists at most one interconnection between bodies.

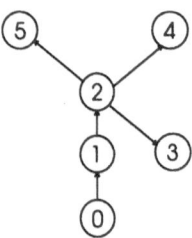

Figure 1.7: Kinematical Graph of the Constrained Truck

If, like in the example, there is more than one, it is possible to replace them by a single idealized interconnection having the same mechanical effect. This is always possible for connections between rigid bodies. Then, the system's topology can be represented by a labeled graph (see Fig. 1.7) by representing a body by a labeled node and an interconnection by a labeled edge of the graph. If the graph consists of a connected tree it is possible to select a single node as the root of the tree. There is a parent - child relation which can be assigned to a tree structured graph. A parent of a node is a node nearer to the root node connected by an edge to the child. Children are labeled with higher numbers than parents and edges linking parents to children are labeled with the child's number. The motion of a body can then be described relatively to the motion of its parents. This leads to a set of *relative coordinates* or *joint coordinates*.[3]

There are different ways to select joint coordinates. We take between two nodes of the spanning tree the interconnection with the smallest number of degrees of freedom and take these degrees of freedom as joint coordinates.

This will be demonstrated for the truck model and its spanning tree given in Fig. 1.7. For the constrained truck we take the following interconnections to describe the spanning tree: $10, 12, 23, 24, 52$, cf. Fig. 1.4. Interconnections $10, 23$ have

---

[3]Note that the spanning tree gives one possibility to compute the relative coordinates. Thus the front wheel can be described using other relative coordinates in the tree and it is not necessary to add an edge between bodies "0" and 4.

one translatory degree of freedom, interconnections $21, 24$ have a translatory and a rotatory and interconnection 52 has a single rotatory degree of freedom. Thus, the system has seven degrees of freedom, which can be described by the following joint coordinates:

$$
\begin{aligned}
q_1 &= \|\rho_{10}\| \\
q_2 &= \|\rho_{12}\| \\
q_3 &= p_3 \\
q_4 &= \|\rho_{23}\| \\
q_5 &= \|\rho_{24}\| \\
q_6 &= p_6 - p_3 \\
q_7 &= p_9 - p_3 \, .
\end{aligned}
\tag{1.4.1}
$$

Herein, we denote by $\rho_{ij}$ the relative vector along joint $ij^4$. The set of joint co-ordinates of a constrained system is smaller than the set of absolute coordinates and the number of joint coordinates corresponds for tree structured systems to the number of degrees of freedom (DOF). In the constrained truck example we have $q = (q_1, \ldots, q_7)^T$.

There is a nonlinear relationship between relative coordinates $q$ and absolute coordinates $p$ given by the $n_q$ equations

$$
0 = \tilde{g}(p, q),
\tag{1.4.2}
$$

which can be obtained directly from the definition (1.4.1) of the relative coordinates. On the other hand, in order to obtain the absolute coordinates from the relative coordinates we need in addition to the relation (1.4.2) the position constraints

$$
0 = \begin{pmatrix} g(p) \\ \tilde{g}(p, q) \end{pmatrix} =: \gamma(p, q).
\tag{1.4.3}
$$

These are $n_p$ equations for the $n_p$ unknowns $p$, given $q$. If the coordinates have been chosen correctly, these equations must be solvable with respect to $p$. We can then assume that the Jacobian $\Gamma_p(p, q) = \frac{\partial}{\partial p} \gamma(p, q)$ is regular.

By taking in (1.4.3) the first derivative with respect to time we get

$$
0 = \underbrace{\frac{\partial}{\partial p} \gamma(p, q)}_{=: \Gamma_p(p,q)} \dot{p} + \underbrace{\frac{\partial}{\partial q} \gamma(p, q)}_{=: \Gamma_q(p,q)} \dot{q},
\tag{1.4.4}
$$

and

$$
\dot{p} = \underbrace{-\Gamma_p(p, q)^{-1} \Gamma_q(p, q)}_{=: V(p,q)} \dot{q},
\tag{1.4.5}
$$

---

[4]Note that due to the assumption made on the choice of the points of attachments there is no loss of information by taking only the norms of the relative vectors.

a relation between the absolute and relative velocities.

From that we obtain by taking another time derivative a relation between the corresponding accelerations

$$\ddot{p} = V(p,q)\ddot{q} + \zeta(p,q,\dot{p},\dot{q}).$$

with the function $\zeta$ collecting all terms with lower derivatives of $p$ and $q$. It is important to note that $\dot{p}$ defined by Eq. (1.4.5) satisfies by construction the constraint equations on velocity level, i.e. Eq. (1.3.6). This implies the property

$$G(p)V(p,q) = 0.$$

Using this and the expression for the accelerations, we can transform the equations of motion from absolute to relative coordinates and get after a premultiplication with $V^T$

$$\underbrace{V(p,q)^T M V(p,q)}_{=:\widetilde{M}(p,q)} \ddot{q} = V(p,q)^T (f_a(p,\dot{p}) - M\zeta(p,q,\dot{p},\dot{q})). \qquad (1.4.6)$$

This together with (1.4.2) and (1.4.5) is the formulation of the equations of motion in a minimal set of coordinates. Constraint forces and constraints have been eliminated by this transformation. In case of tree structured systems this is a transformation to a *state space form*. The general way to transform the equations of motion into a state space form and a definition of these forms will be given in Ch. 2.1.

We note that by this transformation the mass matrix looses its simple structure and even becomes dependent on $p$ and $q$.

## 1.4.1  Mixed Coordinate Formulation for Tree Structured Systems*

At least for tree structured systems, there is a formulation of the equations of motion, which has the advantages of both the relative and the absolute coordinate formulations without suffering from their disadvantages. This method is often referred to as the $\mathcal{O}(n)$-method of establishing the equations of motion. We saw that by using relative coordinates the price for having a minimal coordinate formulation must be paid by the need of inverting a state dependent, dense mass matrix $\widetilde{M}$ for obtaining an explicit form[5]. This step requires a number of operations which grows like $n^3$, where $n$ is the number of bodies in the system. If instead, both coordinates, the absolute coordinates $p$ and the relative coordinates $q$ are taken, the explicit form can be obtained by a number of operations increasing only linearly

---

[5]We will see in Ch. 5.2.4.1, that depending on the solution method, it might not always be necessary to transform the equations into their explicit form.

with the number of bodies. This type of methods has been suggested independently by various authors, cf. [Fea83, BJO86]. We will explain this approach following the more linear algebra oriented report [LENP95]:

For this end we write the equations of motion in a somehow artificial manner: We take the equations of motion of the unconstrained system in absolute coordinates, together with the trivial equation $0\ddot{q} = 0$ and impose to this set of equations the equation $\gamma(p, q) = 0$ as a constraint by introducing Lagrange multipliers $\mu$ in the usual way.

Because of $\Gamma_q^T \mu = 0$ we obtain

$$\begin{pmatrix} M & 0 \\ 0 & 0 \end{pmatrix} \begin{pmatrix} \ddot{p} \\ \ddot{q} \end{pmatrix} = \begin{pmatrix} f_a(t, p, \dot{p}) \\ 0 \end{pmatrix} - \begin{pmatrix} \Gamma_p(p, q)^T \\ \Gamma_q(p, q)^T \end{pmatrix} \mu \qquad (1.4.7a)$$

$$0 = \gamma(p, q). \qquad (1.4.7b)$$

By construction is

$$\mu = \begin{pmatrix} \lambda \\ 0 \end{pmatrix} \text{ and } \Gamma_p(p, q)^T = \left( G(p)^T, \frac{\partial}{\partial p} \tilde{g}(p, q)^T \right)$$

with $G$ being the constraint matrix as above.

Differentiating (1.4.7b) twice with respect to time, we obtain

$$\begin{pmatrix} M & 0 & \Gamma_p^T \\ 0 & 0 & \Gamma_q^T \\ \Gamma_p & \Gamma_q & 0 \end{pmatrix} \begin{pmatrix} \ddot{p} \\ \ddot{q} \\ \mu \end{pmatrix} = \begin{pmatrix} f_a(t, p, \dot{p}) \\ 0 \\ z(p, q) \end{pmatrix} \qquad (1.4.8)$$

with $z(p, q, \dot{p}, \dot{q}) := -\frac{\partial}{\partial(p,q)} \left( \Gamma_p(p, q)\dot{p} + \Gamma_q(p, q)\dot{q} \right) \begin{pmatrix} \dot{p} \\ \dot{q} \end{pmatrix}$.

Solving this equation starting with the last line, we first get

$$\ddot{p} = \underbrace{-\Gamma_p^{-1}\Gamma_q \ddot{q}}_{=:V} + \underbrace{\Gamma_p^{-1} z}_{=:\zeta}. \qquad (1.4.9)$$

Inserting this result into the first line and premultiplying by $V^T$ gives

$$V^T M V \ddot{q} - \Gamma_q^T \mu = V^T (f_a - M\zeta)$$

because of $V^T \Gamma_p = \Gamma_q$.

From this we finally get

$$\ddot{q} = (V^T M V)^{-1} V^T (f_a - M\zeta) \qquad (1.4.10)$$

(cf. (1.4.6)).

By reordering the coordinates, these equations can be solved in an efficient way.

We will demonstrate this on a MBS having a *chain structure*.

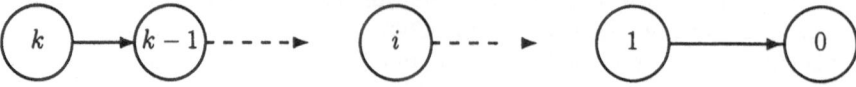

Let us denote by $p^i, q^i$ the coordinates of body $i$ and by $\mu^i$ the multiplier corresponding to the $i$-th interconnection, which is the one linking body $i$ to body $i-1$. Then by arranging the coordinates in an order reflecting the topological structure of the MBS

$$x = (\dots , (\ddot{p}^i)^T, (\ddot{q}^i)^T, (\mu^i)^T, (\ddot{p}^{i-1})^T, (\ddot{q}^{i-1})^T, (\mu^{i-1})^T, \dots )^T$$

the leading matrix in (1.4.8) gets a structure which facilitates the solution process. In order to demonstrate this, it is important to note that the constraint $\gamma^i$ corresponding to the $i$-th interconnection has the form

$$\gamma^i(p, q) = \gamma^i(p^i, p^{i-1}, q^i).$$

Thus, the $i$-th row block of the matrices $\Gamma_p$ and $\Gamma_q$ have the form

$$\Gamma^i_p = ( \ \dots, \quad 0, \quad \Gamma_{p_{i,i}}, \quad \Gamma_{p_{i,i-1}}, \quad 0, \quad \dots \ )$$

and

$$\Gamma^i_q = ( \ \dots, \quad 0, \quad \Gamma_{q_{i,i}}, \quad 0, \quad 0, \quad \dots \ )$$

with $\Gamma_{p_{i,j}} = \frac{\partial}{\partial p_j}\gamma^i$ and $\Gamma_{q_{i,j}} = \frac{\partial}{\partial q_j}\gamma^i$.
By this, the linear system (1.4.8) has the block structure

$$
\begin{pmatrix}
A_k & C_k^T & & & & & & \\
C_k & A_{k-1} & C_{k-1}^T & & & & & \\
& C_{k-1} & A_{k-2} & C_{k-2}^T & & & & \\
& & C_{k-2} & & & & & \\
& & & \ddots & & & & \\
& & & & C_3^T & & & \\
& & & & C_3 & A_2 & C_2^T & \\
& & & & & C_2 & A_1 & C_1^T \\
& & & & & & C_1 & A_0
\end{pmatrix}
\begin{pmatrix}
x^k \\ x^{k-1} \\ x^{k-2} \\ \vdots \\ \vdots \\ x^2 \\ x^1 \\ x^0
\end{pmatrix}
=
\begin{pmatrix}
b^k \\ b^{k-1} \\ b^{k-2} \\ \vdots \\ \vdots \\ b^2 \\ b^1 \\ b^0
\end{pmatrix}
\qquad (1.4.11)
$$

with $x^i = ((\ddot{p}^i)^T, (\ddot{q}^i)^T, (\mu^i)^T)^T$ and $b^i = ((f_a^i)^T, 0, (z^i)^T)^T$. The $A_i$ have the same form as the leading matrix in (1.4.8)

$$A_i := \begin{pmatrix} M_i & 0 & \Gamma_{p_{i,i}}^T \\ 0 & 0 & \Gamma_{q_{i,i}}^T \\ \Gamma_{p_{i,i}} & \Gamma_{q_{i,i}} & 0 \end{pmatrix}. \qquad (1.4.12)$$

Note that $A_i$ is a square matrix having a maximal dimension of $18 = 3 \times 6$, which is achieved if the $i$-th body is not constrained with respect to its parent. $C_i$ has the form

$$C_i^T := \begin{pmatrix} 0 & 0 & 0 \\ 0 & 0 & 0 \\ \Gamma_{p_{i,i-1}} & 0 & 0 \end{pmatrix}.$$

The system (1.4.11) can be solved by the recursions

a.) Downward recursion

- Initialization:

$$\begin{aligned} \hat{A}_k &:= A_k \\ \hat{b}^k &:= b^k. \end{aligned} \qquad (1.4.13)$$

- Recursion:
  for $i = k - 1, \ldots, 0$

$$\begin{aligned} \hat{A}_i &:= A_i - C_{i+1}\hat{A}_{i+1}^{-1}C_{i+1}^T \\ \hat{b}^i &:= b^i - C_i\hat{A}_i^{-1}\hat{b}^{i+1}. \end{aligned} \qquad (1.4.14)$$

b.) Upward recursion:
   Solve for $x^0$ and then for $x^i$ with $i = 1, \ldots, k$:

$$\begin{aligned} \hat{A}_0 x^0 &= \hat{b}^0 \\ \hat{A}_i x^i &= \hat{b}^i - C_i^T x^{i-1} \qquad \text{with} \qquad i = 1, \ldots, k. \end{aligned} \qquad (1.4.15)$$

It is important to note that $\hat{A}_i$ has the same structure as $A_i$ shown in (1.4.12). Especially its submatrix $\hat{M}_i$ is positive definite, [LENP95]. By this, the linear systems to be solved at every step have all the same structure. This can be exploited by a numerical method.

Thus, the linear system (1.4.8) can be solved by solving $t+1$ small systems (1.4.15) with a maximal dimension $18 \times 18$ and the effort grows linearly when increasing the numbers of bodies in the chain.

We described this algorithm for chain structured systems and leave it to the reader to extend it to tree structured systems.

The shorter the individual branches of the tree the smaller is the gain in efficiency of this formulation. For a system having a kinematical tree with a relative large number of branches the direct method using Eq. (1.4.10) may be faster.

To summarize, in general both types of coordinates, absolute and relative ones, are required to evaluate the equations of motion. This section described an effective way to obtain relative accelerations $\ddot{q}$ given the relative coordinates $q, \dot{q}$ at time $t$:

- Compute $p$ and $\dot{p}$ by recursing upwards in the tree and by computing $p^i$ from $p^{i-1}$ and $q^i$.

- Evaluate the forces $f_a(t, p, \dot{p})$ and $z(p, q, \dot{p}, \dot{q})$ to build the components $b^i$ of the right-hand side vector in system (1.4.11).

- Solve for $\ddot{q}$ by the above down- and upward recursion, cf. (1.4.14), (1.4.15).

From that $q$ and $\dot{q}$ can be obtained at a later time point by means of performing one step of numerical integration. This will be discussed in Ch. 4.

## 1.5   Linearization of Equations of Motion

In general, the equations of motion are nonlinear. Making special assumptions on the MBS under consideration some multibody formalisms result in partially linear equations. For example the kinematics can be assumed to be linear in some applications of vehicle dynamics: If the motion of a railway vehicle on a track is considered, the deviation of the vehicle from the nominal motion defined by the track can be assumed to be so small that one can linearize around that nominal motion. If in addition also the forces are linear, the resulting equations of motion are linear. For a straight line track and linear, time independent force laws the equations are linear with constant coefficients. For details on multibody formalisms which establish linear or partially linear equations, the reader is referred to [Wal89].

In this section we will assume that the equations of motion are given in a nonlinear form and we will show how the linearized equations look like. In subsequent chapters we will refer then to the linear equations when discussing linear system analysis methods.

For the linearization we consider time dependent functions $f_a, g$. We will assume for the following that the time dependency of the applied forces can be described in terms of an input function $u(t)$, such that the applied forces are $f_a(p, v, u(t))$. $u(t)$ plays in the sequel the role of an *input* or *forcing function*.

Let the time dependency of the constraint have the form of an additive *kinematic excitation*

$$g(t, p) = g(p) - z(t). \tag{1.5.1}$$

We assume for both $u(t)$ and $z(t)$ that they are small deviations from their nominal values, which are assumed to be zero

$$u_N(t) = 0 \qquad\qquad z_N(t) = 0.$$

We consider a nominal solution, $p_N(t), v_N(t)$ and $\lambda_N(t)$, of the system

$$\dot{p} \;=\; v \tag{1.5.2a}$$

$$M(p)\dot{v} \;=\; f_a(p, v, u(t)) - \frac{d}{dp} g(p)^T \lambda \tag{1.5.2b}$$

$$0 \;=\; g(p) - z(t) \tag{1.5.2c}$$

corresponding to the initial values $p_N(0), v_N(0)$ and $\lambda_N(0)$. Let us set

$$p(t) := p_N(t) + \Delta p(t), \quad v(t) := v_N(t) + \Delta v(t), \quad \lambda(t) := \lambda_N(t) + \Delta \lambda(t)$$

and expand all functions in Eq. (1.5.2) in a Taylor series around the nominal solution and around the nominal input function and nominal kinematic excitation. If we assume that squares and products of the increment variables $\Delta p(t), \Delta v(t), \Delta \lambda(t)$ with each other and $\Delta p(t)\Delta \dot{v}(t)$ are negligible, we obtain

$$\Delta \dot{p} \;=\; \Delta v \tag{1.5.3a}$$

$$M(t)\Delta \dot{v} \;=\; -K(t)\Delta p - D(t)\Delta v - G(t)\Delta \lambda + B(t)u(t) \tag{1.5.3b}$$

$$0 \;=\; G(t)\Delta p - z(t) \tag{1.5.3c}$$

with the

| | | |
|---|---|---|
| mass matrix | $M(t)$ | $:= M(p_N(t))$ |
| stiffness matrix | $K(t)$ | $:= (v_N^T \frac{\partial}{\partial p} M(p) -$ |
| | | $\frac{\partial}{\partial p} f_a(p, v_N(t), 0) + \frac{\partial}{\partial p} G(p)^T \lambda_N)_{p=p_N(t)}$ |
| damping matrix | $D(t)$ | $:= -(\frac{\partial}{\partial v} f_a(p_N(t), v, 0))_{v=v_N(t)}$ |
| constraint matrix | $G(t)$ | $:= (\frac{d}{dp} g(p))_{p=p_N(t)}$ |
| input matrix | $B(t)$ | $:= (\frac{\partial}{\partial u} f_a(p_N(t), v_N(t), u))_{u=0}$ |

While technically this linearization can be performed for any nominal solution as long as the functions are sufficiently smooth in its neighborhood, linearization makes only sense around *equilibrium solutions*. The linear system of differential equations describes the general system locally correct regarding its stability only in the neighborhood of an equilibrium solution [Arn78]. An equilibrium solution is defined by the requirements

$$p_N = p_N(t) = \text{const.}, \quad v_N = 0, \quad \dot{v}_N = 0, \quad \lambda_N = \lambda_N(t) = \text{const.}$$

Linearizing around an equilibrium results in a set of linear equations of motion with *constant coefficients*. A consequence of $p_N$ being an equilibrium is

$$f_a(p_N, 0, 0) - G(p_N)^T \lambda_N \;=\; 0 \tag{1.5.4a}$$

$$g(p_N) \;=\; 0. \tag{1.5.4b}$$

Chapter 3 will be devoted to the computation of equilibrium points.
For an equilibrium as a nominal solution we obtain after linearizing Eq. (1.5.2) a
system with constant coefficient matrices

$$\Delta \dot{p} \quad = \quad \Delta v \tag{1.5.5a}$$
$$M \Delta \dot{v} \quad = \quad -K \Delta p - D \Delta v - G^T \Delta \lambda + B u(t) \tag{1.5.5b}$$
$$0 \quad = \quad G \Delta p - z(t). \tag{1.5.5c}$$

Establishing the matrices $K, D$ and $B$ can be simplified by introducing a *force
distribution matrix*. This can be seen by relating the applied forces $f_a$ to the
individual interconnections in the multibody system

$$\bar{f}_a(f_e^0(p, v, u), p) = f_a(p, v, u) \tag{1.5.6}$$

with $f_e^0 = (f_{e_i}^0)_{i=1,\dots,n_e}$ being the elementary force vector assembling the components of the contributing forces of every interconnection between the parts of the
multibody system. For Example 1.3.1 we get

$$f_e^0 = (f_{10}^0, f_{12}^0, f_{30}^0, f_{23}^0, f_{24}^0, f_{42}^0, f_{25}^0, f_{2d}^0, f_{1d}^0)^T$$

with $n_e = 9$. (Note that the $f_{ij}^0$ were defined as scalars.)
Now applying the chain rule, we get

$$D = -\frac{\partial f_a}{\partial v} = -\sum_{i=1}^{n_e} \frac{\partial \bar{f}_a}{\partial f_{e_i}^0} \frac{\partial f_{e_i}^0}{\partial v} =: -T \frac{\partial f_e^0}{\partial v}$$

with all functions being evaluated at $p_N, v_N$ and $u(t_N)$ and the $n_p \times n_e$-force distribution matrix

$$T := \left( \frac{\partial \bar{f}_a}{\partial f_{e_i}^0} \right)_{i=1,n_e}.$$

Similarly we get

$$B = T \frac{\partial f_e^0}{\partial u} \quad \text{and} \quad K = -T \frac{\partial f_e^0}{\partial p} - \frac{\partial \bar{f}_a}{\partial p} + \dot{v}_N^T \frac{\partial}{\partial p} M + \frac{\partial}{\partial p} G^T \lambda_N$$

where, again, all functions are evaluated at $p_N, v_N$ and $u(t_N)$.

## 1.6   Nominal Interaction Forces

One of the basic tasks after setting up the equations of motion of a multibody system is the computation of the *nominal interaction forces*. These forces are related

to preloads or prestresses in the individual coupling elements of the system and they can be determined in such a way that the system is initially in an equilibrium position.

We assume that we are given a nominal position $p_N$, which satisfies the position constraint

$$g(p_N) = 0.$$

With this, the task is to determine the $n_e$ components of the nominal interaction forces $f_e^0$ and the nominal constraint forces $\lambda_N$ in such a way, that

$$0 = \bar{f}_a(f_e^0, p_N) - G(p_N)^T \lambda_N$$

holds.

We assume that $f_a$ depends linearly on the unknown parameters $f_e^0$, which is the common case. Then, we can write instead

$$0 = \bar{f}_a(0, p_N) + T(p_N) f_e^0 - G(p_N)^T \lambda_N.$$

This leads to a linear systems of $n_p$ equations for $n_e + n_\lambda$ unknowns to be solved

$$\underbrace{(T(p_N), -G(p_N)^T)}_{=:T} \begin{pmatrix} f_e^0 \\ \lambda_N \end{pmatrix} = - \underbrace{\bar{f}_a(0, p_N)}_{=:b} \qquad (1.6.1)$$

(see also [RS88]). We assume that the governing matrix $T$ of this system is nonsingular and exclude by this assumption redundancies in the construction or modeling of a mechanical system.

We have to distinguish three cases:

1. $n_e + n_\lambda = n_p$: In that case the nominal interaction forces can be uniquely determined. This can be done by applying a standard technique like first performing a decomposition of $T$ into upper and lower triangular matrices $T = LU$ and then solving the system by a forward and backward elimination.

2. $n_e + n_\lambda > n_p$: The linear system is underdetermined, i.e. the nominal interaction forces cannot be uniquely determined. We will see in a later section how underdetermined linear systems can be treated. But in that particular context the model is no longer meaningful and must be altered. The freedom in the choice of the nominal forces would otherwise result in differently behaving systems.

3. $n_e + n_\lambda < n_p$: The linear system is overdetermined and has in general no solution. We will later treat also these kind of linear systems, but again, in this particular context the model must be altered instead of trying to apply a numerical method.

**Example 1.6.1** *For the model of the constrained truck $n_e = 7, n_\lambda = 2$ and $n_p = 9$. Thus the system is square. We get for the nominal forces*

$$
\begin{aligned}
f_e^0 &= (f_{10}, f_{12}, f_{30}, f_{23}, f_{24}, f_{42}, f_{25})^T \\
&= (-2.33, -0.247, -1.15, -0.085, -0.128, -0.127, -0.855)^T \, 10^6
\end{aligned}
$$

*and $\lambda_N = (0.0, 6.81)^T \, 10^4$.*

# 2   Linear Systems

Many methods for linear system analysis are based on explicit linear ODEs. Constrained linear systems have therefore to be reduced first to an explicit linear ODE. In Sec. 1.4 such a reduction was obtained for tree structured systems by formulating the system in relative coordinates. For general systems this reduction has to be performed numerically. The reduction to this so-called *state space form* will be the topic of the first part of this chapter. Then, the exact solution of linear ODEs and DAEs is discussed.

For more details on linear system analysis in multibody dynamics we refer to [KL94].

## 2.1   State Space Form of Linear Constrained Systems

The transformation to state space form reduces the number of differential equations to its minimum by incorporating the constraint equations. In the resulting equations the generalized constraint force no longer appears and an additional equation must be provided if its value is required.

In this section the numerical generation of the state space form for linear time invariant systems will be discussed. We saw in Sec. 1.5 how the equations of motion (1.5.2) can be linearized to

$$\dot{p} = v \tag{2.1.1a}$$
$$M\dot{v} = -Kp - Dv - G^T\lambda \tag{2.1.1b}$$
$$Gp = z. \tag{2.1.1c}$$

Herein, $G$ is the $n_\lambda \times n_p$ constraint matrix, $K, D, M$ are the stiffness-, damping- and mass matrices, respectively, and $z$ are time dependent kinematical excitations. To keep formulas short we assume $u(t) = 0$. Furthermore, we want to exclude redundant constraints by assuming $\text{rank}(G) = n_\lambda$.

First, the constraint equation

$$Gp = z \tag{2.1.2}$$

is transformed into an explicit form. For $n_\lambda < n_p$, this equation defines an under-determined linear system with the general solution

$$p = p_h + p_p, \qquad (2.1.3)$$

where $p_h$ is a solution of the homogeneous system ($z \equiv 0$) and $p_p$ a particular solution of the inhomogeneous system (2.1.2). There is a linear space of homogeneous solutions of dimension $n_y = n_p - n_\lambda$.

We define an $n_p \times n_y$ matrix $V$ such that its columns span a basis of this space. The general form of all solutions of the homogeneous system is then obtained as

$$p_h = Vy \qquad (2.1.4)$$

with an arbitrary vector $y \in \mathbb{R}^{n_y}$.
The number $n_y$ of components of $y$ is the number of *degrees of freedom* of the system and the linear space spanned by $V$ is called the *space of free motion*.
By these definitions the matrix $V$ has the property

$$GV = 0. \qquad (2.1.5)$$

Due to the regularity of $G$ there is an $n_p \times n_\lambda$ matrix $G^{\text{part}}$ with the property

$$GG^{\text{part}} = I_{n_\lambda} \qquad (2.1.6)$$

with $I_{n_\lambda}$ being the $n_\lambda \times n_\lambda$ identity matrix.
A particular choice of such a matrix $G^{\text{part}}$ is

$$G^\dagger := M^{-1}G^T(GM^{-1}G^T)^{-1}. \qquad (2.1.7)$$

which can be viewed as the generalization of the *Moore–Penrose pseudo-inverse*

$$G^+ := G^T(GG^T)^{-1}$$

for an $M$ related inner product, $<x,y>_M := x^T My$. Note, that due to this choice

$$V^T M G^\dagger = 0. \qquad (2.1.8)$$

In the sequel we choose

$$G^{\text{part}} := G^\dagger.$$

A possible choice for the particular solution of Eq. (2.1.1c) is

$$p_p = G^\dagger z. \qquad (2.1.9)$$

Equations (2.1.4) and (2.1.9) substituted into (2.1.3) give

$$p = Vy + G^\dagger z \qquad (2.1.10)$$

and the corresponding time derivatives are

$$\dot{p} = V\dot{y} + G^\dagger \dot{z}$$

$$\ddot{p} = V\ddot{y} + G^\dagger \ddot{z}.$$

These equations are substituted into Eqs. (2.1.1a) - (2.1.1c). After a premultiplication of the result by $V^T$ we obtain by using Eq. (2.1.5) and Eq. (2.1.7)

$$V^T MV\ddot{y} = -V^T KVy - V^T KG^\dagger z - V^T DV\dot{y} - V^T DG^\dagger \dot{z}. \qquad (2.1.11)$$

If this equation is then written in first order form, the state space form of the linear mechanical system consisting of $2n_y$ first order differential equations is obtained

$$\begin{pmatrix} I & 0 \\ 0 & \widetilde{M} \end{pmatrix} \begin{pmatrix} \dot{y} \\ \ddot{y} \end{pmatrix} = \begin{pmatrix} 0 & I \\ -\widetilde{K} & -\widetilde{D} \end{pmatrix} \begin{pmatrix} y \\ \dot{y} \end{pmatrix} - \begin{pmatrix} 0 \\ V^T KG^\dagger z + V^T DG^\dagger \dot{z} \end{pmatrix} \qquad (2.1.12)$$

with $\widetilde{M} = V^T MV$, $\widetilde{D} = V^T DV$, and $\widetilde{K} = V^T KV$.
The same procedure but with premultiplication by $G^\dagger$ results in an expression for the generalized constraint forces

$$\lambda = -G^{\dagger^T}(MV\ddot{y} + DV\dot{y} + KVy + MG^\dagger \ddot{z} + DG^\dagger \dot{z} + KG^\dagger z)$$

which can be evaluated, once Eq. (2.1.11) has been solved.

## 2.2 Numerical Reduction to the State Space Form of Linear Time Invariant Systems

As $V$ is not uniquely defined, there are several alternative procedures for numerically determining the matrix $V$. The central criterion for comparing these procedures will be the condition of the matrix

$$\widetilde{M} = V^T MV, \qquad (2.2.1)$$

which is sometimes called *reduced mass matrix*. This matrix has to be regular in order to be able to solve for $\ddot{y}$ in Eq. (2.1.11). With the requirement, that the set of components of the new, reduced coordinates $y$ is chosen as subset of the components of $p$ in order to keep their technical interpretability, there are only two methods which can be taken into consideration. They are based on the $LU$ or $QR$ decomposition of $G$, [GL83].

As the $QR$ decomposition is more reliable, its use will be described here briefly. For details see [FW84].

The constraint matrix $G$ can be decomposed into

$$G = Q\,(R_1, S)\,P \tag{2.2.2}$$

with an $n_\lambda \times n_\lambda$ orthogonal matrix $Q$, an $n_\lambda \times n_\lambda$ upper triangular matrix $R_1$ with decreasing diagonal elements, an $n_\lambda \times (n_p - n_\lambda)$ matrix $S$, and a permutation matrix $P$. The number of nonzero diagonal elements in $R_1$ determines the rank of $G$ which was assumed to be $n_\lambda$.

As the components of $y$ are required to be a subset of those of $p$, the matrix $V$ has the general form

$$V = P \begin{pmatrix} V_{11} \\ I \end{pmatrix} \tag{2.2.3}$$

with $I$ being the $n_y$ identity matrix. With this form Eq. (2.1.5) becomes

$$QR_1V_{11} + QS = 0 \tag{2.2.4}$$

and consequently, using the orthogonality of $Q$,

$$V_{11} = -R_1^{-1}S$$

follows. Thus we have

$$V = P \begin{pmatrix} -R_1^{-1}S \\ I \end{pmatrix}.$$

As $R_1$ is triangular, $-R_1^{-1}S$ can be computed simply by a backward substitution [GL83].

Equations (2.1.6) and (2.1.8) give a condition for $G^\dagger$, once $V$ is computed

$$\begin{pmatrix} G \\ V^T M \end{pmatrix} G^\dagger = \begin{pmatrix} I_{n_\lambda \times n_\lambda} \\ 0_{n_y \times n_\lambda} \end{pmatrix}. \tag{2.2.5}$$

This linear equation can be solved by standard methods, if $G$ has maximal rank. Otherwise, the generalized constraint forces are no longer uniquely defined.

All numerical steps described in this section can be performed by subroutines of the LINPACK or LAPACK packages [DMBS79, And95]. They provide a condition estimator by which the condition with respect to inversion of $\widetilde{M} = V^T M V$ can be estimated. This may serve as a criterion for the quality of the computed basis $V$.

If the condition is not much higher than the condition of $M$, the described method was successful.

The condition estimator is usually used in conjunction with a solver for linear equations, which is used here to compute $\widetilde{M}^{-1}\widetilde{D}$, and $\widetilde{M}^{-1}\widetilde{K}$ in order to write Eq. (2.1.12) in the explicit form $\dot{x} = Ax$ with

$$x := \begin{pmatrix} y \\ \dot{y} \end{pmatrix} ; A := \begin{pmatrix} 0 & I \\ -\widetilde{M}^{-1}\widetilde{K} & -\widetilde{M}^{-1}\widetilde{D} \end{pmatrix}$$

**Example 2.2.1** *To summarize this procedure, we will apply it to the constrained truck example (Ex. 1.3.2). This results in the following MATLAB statements*

```
[Q,R,P]=qr(G);   % QR-Decomposition with Pivoting
R1=R(1:2,1:2);   % Splitting up R into a triangular part R1
S=R(1:2,3:9);    % and a rectangular S part
V=P*[-R1\S;eye(7)]; %
%
% transformation to state-space form
%
Mtilde=V'*M*V;  Ktilde=V'*K*V; Dtilde=V'*D*V;
A=[zeros(7),eye(7);-Mtilde\Ktilde,-Mtilde\Dtilde];
```

*As result for the constrained truck, linearized around its nominal position we obtain*

$$V = \begin{pmatrix}
1.0 & 0 & 0 & 0 & 0 & 0 & 0 \\
0 & 0 & 0 & 0 & 1.0 & 0 & 0 \\
0 & 0 & 0 & 0 & 0.32 & -0.32 & 0.52 \\
0 & 1.0 & 0 & 0 & 0 & 0 & 0 \\
0 & 0 & 1.0 & 0 & 0 & 0 & 0 \\
0 & 0 & 0 & 1.0 & 0 & 0 & 0 \\
0 & 0 & 0 & 0 & -0.05 & 0.05 & -0.83 \\
0 & 0 & 0 & 0 & 0 & 1.0 & 0 \\
0 & 0 & 0 & 0 & 0 & 0 & 1.0
\end{pmatrix}$$

*Thus, this procedure takes $p_1, p_2, p_4, p_5, p_6, p_8, p_9$ as state variables and $p_3, p_7$ as dependent ones. This can be verified by checking V or P.*

Alternative methods can be found in Sec. 2.3.2.

## 2.3   Constrained Least Squares Problems

We will frequently encounter in the sequel constrained least squares problems of the form

$$\|Fx - a\|_2^2 = \min_x \qquad (2.3.1a)$$
$$Gx - b = 0 \qquad (2.3.1b)$$

with an $n_x \times n_x$ regular matrix $F$ and a full rank $n_\mu \times n_x$ constraint matrix $G$. Typically, these systems occur when discretizing the linearized equations of motion of a constrained multibody system. In the nonlinear case, they occur inside an iteration process, cf. Sec. 5.3.

By introducing Lagrange multipliers $\mu$, the problem reads equivalently

$$F^T F x - F^T a + G^T \mu \;=\; 0 \qquad\qquad (2.3.2a)$$
$$G x - b \;=\; 0, \qquad\qquad (2.3.2b)$$

or in matrix notation

$$\begin{pmatrix} F^T F & G^T \\ G & 0 \end{pmatrix} \begin{pmatrix} x \\ \mu \end{pmatrix} = \begin{pmatrix} F^T a \\ b \end{pmatrix}. \qquad (2.3.3)$$

The solution of this equation can be expressed in two different ways (see [GMW81]):

**Range Space Formulation**
From the first equation we obtain

$$x = (F^T F)^{-1} \left( F^T a - G^T \mu \right).$$

Inserting this into the second one gives

$$G(F^T F)^{-1}(F^T a - G^T \mu) - b = 0.$$

By assuming $G(F^T F)^{-1} G^T$ to be regular we finally get

$$x \;=\; \left( I - (F^T F)^{-1} G^T \left( G(F^T F)^{-1} G^T \right)^{-1} G \right) (F^T F)^{-1} F^T a$$
$$\qquad + (F^T F)^{-1} G^T \left( G(F^T F)^{-1} G^T \right)^{-1} b \qquad\qquad (2.3.4a)$$
$$\mu \;=\; (G(F^T F)^{-1} G^T)^{-1}(G(F^T F)^{-1} F^T a - b). \qquad\qquad (2.3.4b)$$

**Null Space Formulation**
Defining $V$ as a matrix spanning the null space of $G$ and $x_p$ as a particular solution of the second equation of (2.3.3), we get

$$x = V y + x_p.$$

Inserting this into the first equation of (2.3.3) and premultiplying by $V^T$ results in

$$y = (V^T F^T F V)^{-1} V^T F^T (-F x_p + a),$$

and finally we obtain

$$x = V(V^T F^T FV)^{-1}V^T F^T a + (I - V(V^T F^T FV)^{-1}V^T F^T F)\, x_p. \qquad (2.3.5)$$

$x_p$ can be computed using the Moore–Penrose pseudo-inverse

$$x_p = G^+ b \text{ with } G^+ = G^T(GG^T)^{-1}.$$

We get

$$x = \left( \begin{array}{c} F \\ G \end{array} \right)^{\text{CLSQ}+} \left( \begin{array}{c} a \\ b \end{array} \right) \qquad (2.3.6)$$

with

$$\left( \begin{array}{c} F \\ G \end{array} \right)^{\text{CLSQ}+} := \left( \begin{array}{cc} V(FV)^+ & (I - V(FV)^+ F)\, G^+ \end{array} \right). \qquad (2.3.7)$$

Alternatively, a weighted generalized inverse $G^\dagger$ may be used instead of the Moore–Penrose pseudo-inverse:

$$x_p = G^\dagger b \text{ with } G^\dagger = M^{-1}G^T(GM^{-1}G^T)^{-1} \qquad (2.3.8)$$

with $M$ being an invertible $n_x \times n_x$ matrix.

The particular steps in a numerical computation of the solution of a constrained least squares problem will be discussed in Sec. 2.3.2.

Both approaches can be related to each other by first setting $b = 0$ and equating (2.3.4a) with (2.3.5):

$$\left(I - (F^T F)^{-1}G^T(G(F^T F)^{-1}G^T)^{-1}G\right)(F^T F)^{-1}F^T = V(V^T F^T FV)^{-1}V^T F^T. \qquad (2.3.9)$$

This relation will be used later.

On the other hand, by setting $a = 0$ we get

$$(I - V(V^T F^T FV)^{-1}V^T F^T F)G^\dagger = (F^T F)^{-1}G^T(G(F^T F)^{-1}G^T)^{-1}. \qquad (2.3.10)$$

## 2.3.1   Pseudo-Inverses

In the special case of an underdetermined linear system we may look for the minimum norm least squares solution of this system. With $F = I$, $a = 0$ we obtain from Eq. (2.3.4)

$$x = G^T(GG^T)^{-1}b.$$

For underdetermined linear systems with a full-rank matrix $G$ the coefficient matrix

$$G^+ := G^T(GG^T)^{-1} \tag{2.3.11}$$

is called *Moore–Penrose pseudo-inverse*. In the case of overdetermined linear equations it is defined as

$$G^+ = (G^T G)^{-1} G^T. \tag{2.3.12}$$

The Moore–Penrose pseudo-inverse fulfills all four properties of a pseudo-inverse:

**Definition 2.3.1 (Pseudo-Inverse)** *(cf. [CM79]) The $n \times m$ matrix $A^*$ is called $(i,j,k)$-inverse of the $m \times n$ matrix $A$, if the conditions $i, j, k$ are fulfilled*

1. $AA^*A = A$

2. $A^*AA^* = A^*$

3. $(AA^*)^T = AA^*$

4. $(A^*A)^T = A^*A.$

The $M$-related generalized inverse $G^\dagger$ is an $(1,2,3)$-inverse. The restricted pseudo-inverse [BIAG73] in Eq. (2.3.7) is a $(1,2,4)$ inverse.

## 2.3.2   Numerical Computation of the Solution of a Constrained Least Squares Problem*

We assume $F$ being nonsingular and set $F^T F =: M$ which then is positive definite.

**Range Space Method**
Using this method the matrix is factorized as

$$\begin{pmatrix} M & G^T \\ G & 0 \end{pmatrix} = \begin{pmatrix} L & 0 \\ B^T & -L_1 \end{pmatrix} \begin{pmatrix} L^T & B \\ 0 & L_1^T \end{pmatrix}. \tag{2.3.13}$$

The following MATLAB sequence illustrates the foregoing

```
function [L,B,L1]=rangespd(M,G)
L= chol(M)';   % 1. Cholesky decomposition of M
B= L \ (G');   % 2. Computation of B=L^{-1}G'
D=B'*B;        % 3. Computation of D=B' B=GM^{-1}G'
L1= chol(D);   % 4. Cholesky decomposition of D=L1 L1'
```

Steps 3 and 4 give the method its name: It is the matrix $GM^{-1}G^T$, which is factorized, i.e. $M^{-1}$ with respect to the range space $G$.
The *solution* is obtained by performing the following steps:

```
function [u1,u2]=rangesps(L,B,L1,a,b)%
x1= L\a;              % 1. Solve Lx1= a
x2 = L1\(-b+B'*x1);   % 2. Solve L1 x2=-b+B'x1
u2 = (L1')\x2;        % 3. Solve L1' u2=x2
u1 = (L')\(x1-B*u2);  % 4. Solve L' u1= x1-B u2
```

Steps 1 and 2 correspond to the system defined by the left matrix in Eq. (2.3.13), steps 3 and 4 correspond to the right matrix. The effort for the decomposition of the matrix is

$$\frac{1}{6}n_x^3 + \frac{1}{2}n_\mu n_x(n_\mu + n_x) + \frac{1}{6}n_\mu^3$$

additions and multiplications, the solution effort is

$$n_x^2 + 2n_x n_\mu + n_\mu^2$$

operations.

**Null Space Method**

The *decomposition* of the matrix using the null space method proceeds along the following steps:

1. *LU*-decomposition of the matrix $G$.[1] Thus the nullspace of $G$ is here computed using an $LU$ decomposition.

2. Elimination of $M$ (from the right and the left hand side).

3. Cholesky-decomposition of the resulting matrix.

The matrix is factorized in the form

$$
\begin{pmatrix} M_{11} & M_{12} & G_1^T \\ M_{21} & M_{22} & G_2^T \\ G_1 & G_2 & 0 \end{pmatrix} = \begin{pmatrix} I & 0 & M_{11}R_1^{-1} \\ R_2^T(R_1^{-1})^T & I & M_{21}R_1^{-1} \\ 0 & 0 & L \end{pmatrix} \cdot
$$
$$
\cdot \begin{pmatrix} 0 & \widetilde{M}_{12}\tilde{L}^{-T} & R_1^T \\ 0 & \tilde{L} & 0 \\ R_1 & R_2\tilde{L}^{-T} & 0 \end{pmatrix} \begin{pmatrix} I & 0 & 0 \\ 0 & \tilde{L}^T & 0 \\ 0 & 0 & L^T \end{pmatrix}
$$

(2.3.14)

The decomposition is carried out along the steps:

```
n1 = size(G); n2 = size(M)-n1; n = size(M); %1. set dimensions
M11 =M(1:n1, 1:n1); M12=M(1:n1, n1+1:n);    %2. split matrices
M21 =M(n1+1:n,1:n1);M22=M(n1+1:n,n1+1:n);
Gs=[G zeros(n2,n)];
```

---

[1] For ease of notation we assume the permutation matrix due to pivoting being the identity.

```
[Lh,R]=lu(Gs);                         %3. LU decomposition of G
R1=R(1:n1,1:n1); R2=R(1:n1,n1+1:n);
L=Lh(1:n1,1:n1);
S=R1\R2;                               %4. Computation of S=R^{-1}R_2
M22hat=M22-M21*S-S'*M12+S'*M11*S
Ltil=chol(M22hat); Ltil=Ltil';    %5. Cholesky decomposition of M22hat
M12til=M12-M11*S%
```

If the vectors $u_1 = \begin{pmatrix} u_{11} \\ u_{12} \end{pmatrix}$ and $a = \begin{pmatrix} a_1 \\ a_2 \end{pmatrix}$ are partitioned corresponding to the partitioning of $M$ the solution can be carried out using the following steps:

```
x3 =L\b;                    % (1)
z=R1\x3;                    % (2)
x1=a1-M11*z;                % (3)
x2=a2-S'*x1-M21*z;          % (4)
v2=Ltil\x2;                 % (5)
u12=(Ltil')\v2;             % (6)
u11=R1\(x3-R2*u12);         % (7)
v3=(R1')\x1-M12til*u12;     % (8)
u2=(L')\v3;                 % (9)
u1=[u11 u12];
```

Steps (1) to (4) correspond to the solution defined by the left matrix in Eq. (2.3.14), steps (5),(7),(8) to the matrix in the middle, steps (6) and (9) to the system defined by the right matrix. The effort for the decomposition of the system is

$$\frac{1}{6}n_x^3 + \frac{3}{2}n_\mu n_x(n_x - n_\mu) + \frac{1}{6}n_\mu^3$$

multiplications and additions, the solution effort is

$$n_x^2 - 2n_x n_\mu + 5n_\mu^2$$

operations.

Fig. 2.1 shows the decomposition effort dependent on the relative number $n_\mu/n_x$ of equality constraints and variables. Thus, for

- $n_\mu < \frac{n_x}{2}$ one should use the range space method,

- for $n_\mu > \frac{n_x}{2}$ one should use the null space method.

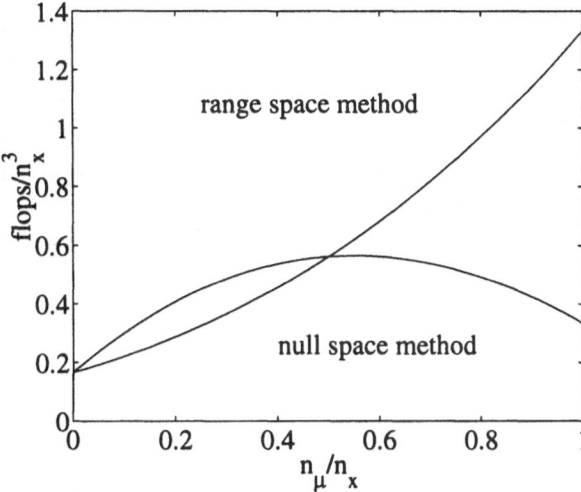

Figure 2.1: Decomposition effort for null- and range space method (flops=number of floating point operations)

### 2.3.3 Computing the Minimum Norm Solution of an Underdetermined Linear System*

In this section we summarize numerical methods for the minimum norm solution of underdetermined linear systems

$$Gx = b \qquad (2.3.15a)$$
$$\|x\|_2 = \min \qquad (2.3.15b)$$

where $G \in \mathbb{R}^{n_\mu \times n_x}$ is assumed to have full rank. An overview about numerical methods can be found in [CP76]. These methods are described in the sequel.

The task to solve an underdetermined linear system of equations occurs frequently in multibody dynamics, e.g. when using coordinate projection methods to stabilize the numerical integration of the index reduced system of equations of motion, see Sec. 5.3.1.

In the sequel let $R$ be an upper triangular matrix, $L$ a lower triangular matrix, $D$ a diagonal matrix, $Q$ an orthogonal matrix and $P$ a permutation matrix.

The effort counts the number of multiplications and divisions. In general, the number of additions has the same order of magnitude. Lower order terms are neglected.

### 2.3.3.1   Cholesky Decomposition of $GG^T$

The easiest way to solve Eq. (2.3.15a) is to use directly the representation of the Moore Penrose pseudo-inverse $G^+ = G^T(GG^T)^{-1}$. This corresponds to the solution of the normal equations.

1. Decomposition: Computation of a Cholesky decomposition of $GG^T$ into $LL^T$. The effort is $n_\mu^2(n_x/2 + n_\mu/6)$.

2. Solution:

   (a) Solve $LL^T w = b$.

   (b) $x = G^T w$.

   The effort is $n_\mu^2 + n_\mu n_x$ operations.

The disadvantage using this method is to use directly the matrix $GG^T$, the condition of which depends quadratically on the condition of $G$. For badly conditioned matrices accuracy losses have to be taken into account.

### 2.3.3.2   $LDR$ Decomposition of $G$ and Cholesky Decomposition of $RR^T$

1. Decomposition:
   $LDR$ decomposition of $G$:

   $$\hat{G} = P_1 G P_2 = LDR.$$

   The matrices $L, R$ are scaled such that their diagonal entries are ones. The operation count gives $\frac{1}{2}n_\mu^2(n_x - n_\mu/3)$ operations.

2. Solution:

   (a) Solve $Lw = P_1 b$ by forward substitution.
       The effort for this step is $n_\mu^2/2$ operations.

   (b) Compute the minimum norm solution $\tilde{z}$ of $DRz = w$.

   (c) $x = P_2 \tilde{z}$.

Step 2b has to be considered in more detail. One possibility is to compute first $RR^T$ (effort: $n_\mu^2(3/2n_x + 5/6n_\mu)$) and then to compute a Cholesky decomposition of this matrix (effort: $n_\mu^3/6$). Then the system

$$RR^T z = D^{-1}w$$

can be solved (effort: $n_\mu^2$). Then we have $\tilde{z} = R^T z$ (effort: $n_\mu n_x - n_\mu^2/2$).

The advantage of this foregoing related to the first is that only a well conditioned matrix, i.e. $R$ needs to be squared.
The total effort is

1. $n_\mu^2 (2n_x + 5/6n_\mu)$ operations for the decomposition

2. $n_\mu(n_\mu + n_x)$ operations for the solution.

### 2.3.3.3  $LDR$ Decomposition of $G$, Computation of $\tilde{z}$ Using Orthogonal Projection

This algorithm is similar to the previous one.

1. Decomposition:

   (a) $LDR$ decomposition of $G$ as in the previous algorithm

   (b) Determine a matrix $Q \in \mathbb{R}^{n_\mu \times n_z}$ such that $QQ^T = D_1$ and $RQ^T = R_1$ where $R_1$ is a quadratic upper triangular matrix with ones on the diagonal and $D_1$ is diagonal and nonsingular.

   The effort for the decomposition is $n_\mu^2 \left( \frac{3}{2}n_x - \frac{7}{6}n_\mu \right)$ when exploiting the special structure of $R$ [CP76].

2. Solution:

   (a) Solution of $Lw = P_1b$ using forward substitution.
       The effort is $n_\mu^2/2$ operations.

   (b) Solution of $R_1 z = D_1^{-1}w$. The effort for this step is $n_\mu/2$ operations.

   (c) $\tilde{z} = Q^T D_1^{-1}z$ (effort: $n_\mu(2n_x - n_\mu)$ operations).

   (d) $x = P_2 z$.

   The total solution effort is then $2n_x n_\mu$ operations.

### 2.3.3.4  $LDR$ Decomposition of $G$, Computation of an Orthogonal Basis of the Null Space of $G$ Using Gaussian Elimination

The idea of this algorithm can be found in [CP76] under "elimination and projection". It is based on projecting a particular solution of the system onto the image of $G^T$. Again, this foregoing is based on an $LDR$ decomposition, A matrix $S$ is determined, the columns of which form a basis of ker($G$). The matrix $I - S(S^T S)^{-1}S^T$ is then a projector onto im($G^T$). The computation of $S$ is done by using Gaussian elimination.

1. Decomposition:

   (a) $LDR$ decomposition of $G$: $\hat{G} = P_1 G P_2 = LDR$, $R = ( \ R_1 R_2 \ )$

(b) Because of $\ker(G) = \ker(RP_2^T)$ it is sufficient to determine an orthogonal basis of $R$. First we determine a vector $s^{(1)}$ belonging to the kernel of $G$. This can e.g. be done by solving

$$R_1^{(1)} s_1^{(1)} = -R_2^{(1)} s_2^{(1)}$$

with $R_i^{(1)} = R_i$, $s_i^{(1)} = s_i$.

Starting from $s^{(1)} = \begin{pmatrix} s_1 \\ s_2 \end{pmatrix}$ an orthogonal vector $s^{(2)}$ which belongs also to the null space of $G$ can be obtained as solution of

$$\begin{pmatrix} R_1^{(1)} & R_2^{(1)} \\ s_1^{(1)T} & s_2^{(1)T} \end{pmatrix} \begin{pmatrix} s_1^{(2)} \\ s_2^{(2)} \end{pmatrix} = 0.$$

It can be determined by using Gaussian elimination of the last row:

$$\left( R_1^{(2)}, R_2^{(2)} \right) \begin{pmatrix} s_1^{(2)} \\ s_2^{(2)} \end{pmatrix} = 0.$$

With this technique all basis vectors can be determined successively. This requires an effort of $\sum_{k=1}^{n_x - n_\mu} \sum_{i=1}^{n_\mu + k} (n_x - i) = \frac{1}{3}n_x^3 - \frac{1}{2}n_x n_\mu^2 + \frac{1}{6}n_\mu^3$ operations.

The total effort is $\frac{1}{3}n_x^3$.

2. Solution:

(a) Determine a particular solution $z$ with an effort of $n_\mu^2$ operations.

(b) $w = z - \sum_{i=1}^{n_x - n_\mu} \frac{s_i^T z}{\|s_i\|^2} s_i$.

The total solution effort is $n_\mu^2 - 2n_x n_\mu + 2n_x^2$ operations.

### 2.3.3.5  $LQ^T$ Decomposition of $G$

Let $G = LQ^T$ be a decomposition of $G$ into a lower triangular matrix $L$ and an orthogonal matrix $Q^T$ with

$$G = \begin{pmatrix} L_1 & 0 \end{pmatrix} \begin{pmatrix} Q_1^T \\ Q_2^T \end{pmatrix} = L_1 Q_1^T.$$

The Moore–Penrose pseudo-inverse then reads

$$G^+ = G^T (GG^T)^{-1} = Q_1 L_1^{-1}.$$

For the numerical computation we have to perform the following steps:

1. Decomposition:
   $QR$ decomposition of $G^T$, this corresponds to a $LQ^T$ decomposition of $G$.
   The effort for this step is $n_\mu^2 \left(n_x - \frac{n_\mu}{3}\right)$ operations.

2. Solution:

   (a) Solve $L_1 x_1 = b$

   (b) $\tilde{z} = Q_1 x_1$

   The solution effort is $n_\mu \left(2n_x - \frac{n_\mu}{2}\right)$ operations.

### 2.3.3.6  $LQ^T$ Decomposition of $G$, $GG^T = LL^T$

This foregoing corresponds to the first one, except that the Cholesky decomposition
of $GG^T$ is determined by applying a $LQ^T$ decomposition of $G$.

1. Decomposition: $LQ^T$ decomposition of $G$.
   This requires $n_\mu^2 \left(n_x - \frac{1}{3}n_\mu\right)$ operations.

2. Solution:

   (a) $LL^T w = b$

   (b) $z = G^T w$

   The effort is $n_\mu(n_\mu + n_x)$ operations.

### 2.3.3.7  $LU$ Decomposition and Cholesky Decomposition of $I + BB^T$, $B = -R_1^{-1} R_2$

In [DS80, Boc87] further variants are presented which are given here in slightly
modified way ($LU$-decomposition instead of $QR$-decomposition at the beginning).
Let $G = L \begin{pmatrix} R_1 & R_2 \end{pmatrix}$ be a $LU$-decomposition of the matrix $G$ then we have with
$\tilde{b} := R_1^{-1} L^{-1} b$ and $B := -R_1^{-1} R_2$

$$
\begin{aligned}
\|x\|_2 &= \min \\
Gx &= b
\end{aligned}
\quad \Longleftrightarrow \quad
\left\| \begin{pmatrix} \tilde{b} + Bx_2 \\ x_2 \end{pmatrix} \right\|_2 = \min \quad . \qquad (2.3.16)
$$
$$
x_1 = \tilde{b} + Bx_2
$$

Herein we dropped for simplicity permutations.
As solution we obtain

$$
x = \begin{pmatrix} B \\ I \end{pmatrix}^+ \begin{pmatrix} -\tilde{b} \\ 0 \end{pmatrix} = \begin{pmatrix} I - B(I + B^T B)^{-1} B^T \\ -(I + B^T B)^{-1} B^T \end{pmatrix} \tilde{b}.
$$

In [DS80] this solution is computed by using the normal equations and a Cholesky
decomposition of $I + B^T B$.

| | Method | effort | |
|---|---|---|---|
| | | decomposition | solution |
| 1. | Cholesky dec. of $GG^T$ | $n_\mu^2(\frac{n_x}{2} + \frac{n_\mu}{6})$ | $n_\mu(n_\mu + n_x)$ |
| 2. | $LDR$-dec. of $G$ and | $n_\mu^2(2n_x + \frac{5}{6}n_\mu)$ | $n_\mu(n_\mu + n_x)$ |
| | Cholesky-dec. of $RR^T$ | | |
| 3. | $LDR$-dec. of $G$ and | $n_\mu^2(\frac{3}{2}n_x - \frac{7}{6}n_\mu)$ | $2n_x n_\mu$ |
| | orthogonal dec. | | |
| 4. | $LDR$-dec. of $G$ and | $\frac{1}{3}n_x^3$ | $n_\mu^2 - 2n_x n_\mu + 2n_x^2$ |
| | elimination/projection | | |
| 5. | $LQ^T$-dec. of $G$, | $n_\mu^2(n_x - \frac{n_\mu}{3})$ | $n_\mu(2n_x - \frac{n_\mu}{2})$ |
| | $G^+ = Q^T L^{-1}$ | | |
| 6. | $LQ^T$-dec. of $G$ | $n_\mu^2(n_x - \frac{n_\mu}{3})$ | $n_\mu(n_\mu + n_x)$ |
| | $\rightarrow$ Cholesky-dec. of $GG^T$ | | |

Table 2.1: Effort for the $l_2$ solution of underdetermined linear systems

The effort is $\frac{1}{6}n_x^3 + \frac{1}{2}n_\mu^2 n_x - \frac{1}{3}n_\mu^3$ operations.
In [Boc87] the last step is solved by applying a $Q^T L$-decomposition taking the special structure of Eq. (2.3.16) into account.
Table 2.1 summarizes the effort. The decomposition effort is plotted in Fig. 2.2. Now, we have provided all linear algebra material necessary for the sequel and can start with the numerical solution of linear ODEs.

## 2.4  The Transition Matrix

We will assume that the reader is familiar with basics from the theory of linear, explicit, constant coefficient ODEs. For later reference we summarize here only some facts.
The solution of a linear constant coefficient ODE

$$\dot{x} = Ax + Bu, \text{ and } x(t_0) = x_0$$

has the form

$$x(t) = e^{A(t-t_0)}x_0 + \int_{t_0}^{t} e^{A(s-t)} Bu(s)ds. \tag{2.4.1}$$

The *matrix exponential function* is herein defined in terms of an infinite series as in the scalar case

$$e^{At} := I + \sum_{i=1}^{\infty} \frac{(At)^i}{i!}.$$

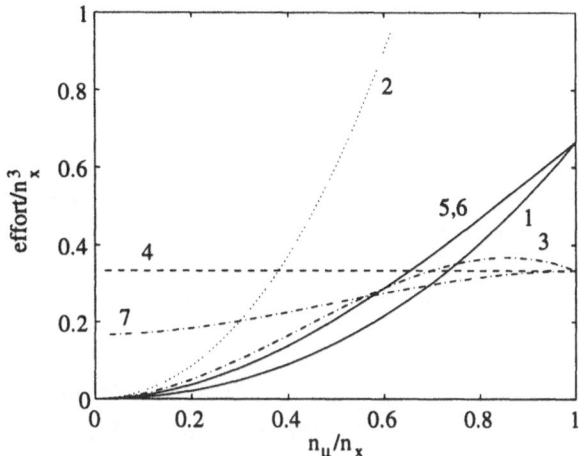

Figure 2.2: Effort to compute the minimum norm solution of underdetermined systems

The convergence of this series can be shown for any square matrix $A$. The function $e^{A(t-t_0)}$ is called the *transition matrix* or *transfer matrix* of the problem, because for the homogeneous case ($u \equiv 0$) it describes the transition from an initial value $x_0$ to the actual value $x(t)$.

It is rather rare that one has to numerically evaluate the transfer matrix. Computing the numerical solution of a linear system via Eq. (2.4.1) is replaced by modern integration methods which proved to be faster. In general an inhomogeneity $u$ would require to numerically approximate the integral expression in Eq. (2.4.1) anyway.

For a critical survey of methods for computing the transition matrix we refer the reader to [MvL78].

## 2.5   The Frequency Response

**Example 2.5.1** *In order to motivate the following we first consider two linear differential equations describing a simple spring/damper system excited by a sinusoidal function*

$$\begin{pmatrix} \dot{x}_1 \\ \dot{x}_2 \end{pmatrix} = \begin{pmatrix} 0 & 1 \\ -k & -d \end{pmatrix} \begin{pmatrix} x_1 \\ x_2 \end{pmatrix} + \begin{pmatrix} 0 \\ 1 \end{pmatrix} \sin \omega t \qquad (2.5.1)$$

*with positive values $d$ and $k$ satisfying $k - d^2/4 > 0$, so that the homogeneous system is weakly damped.*

*By standard techniques (e.g. [MK75]) the analytical solution of this system can be determined as*

$$x_1(t) = a \sin(\omega t + \varphi) + e^{-d/2\, t} \left(c_1 \cos \nu t + c_2 \sin \nu t\right) \qquad (2.5.2)$$

*with the amplitude $a = \left((k - \omega^2)^2 + d^2\omega^2\right)^{-1/2}$ and the phase $\sin \varphi = -d\omega/a$ and $\nu^2 = k - d^2/4$ and $x_2 = \dot{x}_1$. $c_1, c_2$ are given by the initial conditions.*
*This example shows that if $c_1 = c_2 = 0$ or for $t \to \infty$ the solution has the same frequency $\omega$ as the exciting function. Furthermore, there is a constant phase $\varphi$ in the solution with $-\pi \le \varphi \le 0$. It can be shown, that this is a general property of linear systems with constant coefficients.*
*In general, the amplitude $a$ and the phase $\varphi$ of the solution can be viewed as a function of the exciting frequency $\omega$. The amplitude $a$ becomes maximal for the resonance frequency $\omega_{res} = \sqrt{k - \frac{d^2}{2}}$. The functions $a(\omega)$ and $\varphi(\omega)$ are called the frequency response of the linear system (2.5.1).*

In the case of higher dimensional systems the analysis is not as easy as in this example, but the following theorem can be shown:

**Theorem 2.5.2** *Consider the linear differential equation*

$$\dot{x} = Ax + Bu \qquad (2.5.3)$$

*with the $n_x \times n_x$ matrix $A$ having eigenvalues $\lambda$ with $\text{Re}\lambda < 0$ and consider the scalar linear output function $y_m$ with*

$$y_m = Cx. \qquad (2.5.4)$$

*Then, there exists for all $\omega$ and all input functions $u = \sin \omega t$ a unique periodic output $y_m$ with*

$$y_m(t) = \text{Re}\,(G(i\omega)) \sin \omega t + \text{Im}\,(G(i\omega)) \cos \omega t \qquad (2.5.5)$$

*with $G(i\omega) = C(i\omega I - A)^{-1}B$.*

(For the proof see [Bro70]).
From Eq. (2.5.5) follows $y_m(t) = a \sin(\omega t + \varphi)$ with the amplitude

$$a = \sqrt{\text{Re}\,(G(i\omega))^2 + \text{Im}\,(G(i\omega))^2}$$

and the phase

$$\varphi = \arcsin\left(\frac{\text{Im}\,(G(i\omega))}{a}\right).$$

The complex valued $n_{y_m} \times n_u$ matrix function $G(s)$ is called the *frequency response* of the system. Here, we are only interested in its evaluations along the positive imaginary axis.

For the spring damper system of Ex. 2.5.1 we get with

$$
\begin{aligned}
G(i\omega) &= \begin{pmatrix} 1 & 0 \end{pmatrix} \begin{pmatrix} i\omega & 1 \\ -k & i\omega - d \end{pmatrix}^{-1} \begin{pmatrix} 0 \\ 1 \end{pmatrix} \\
&= \begin{pmatrix} 1 & 0 \end{pmatrix} \frac{1}{k - \omega^2 - i\omega d} \begin{pmatrix} i\omega - d & -1 \\ k & i\omega \end{pmatrix} \begin{pmatrix} 0 \\ 1 \end{pmatrix} \\
&= \frac{-(k - \omega^2) - i\omega d}{(k - \omega^2)^2 + \omega^2 d^2}.
\end{aligned}
$$

Consequently we get for $a$

$$
a^2 = (\mathrm{Re}(G(i\omega))^2 + \mathrm{Im}(G(i\omega))^2 = \frac{1}{(k - \omega^2)^2 + \omega^2 d^2}
$$

which is the same result as above.

**Example 2.5.3** *We consider the linearized unconstrained truck and assume that the spring between front wheel and ground is excited by $u_1(t) = \sin \omega t$ with $\omega \in [0, 40\pi]$. This corresponds to an input matrix*

$$
B = \begin{pmatrix} 0_9 & 0 & 0 & 0 & k_{30}/m_3 & 0 & 0 & 0 & 0 & 0 \end{pmatrix}^T.
$$

*We ask for the responding oscillation of the cabin rotation $p_6(t)$ which corresponds to an output matrix*

$$
C = \begin{pmatrix} 0 & 0 & 0 & 0 & 0 & 0 & 1 & 0 & 0 & 0 & 0_9 \end{pmatrix}.
$$

*The result of a frequency response analysis is shown in Fig. 2.3a.*

The situation becomes more realistic if also the front wheel's damper is taken into account. Then, also the first derivative of the excitation acts on the vehicle. We have therefore to distinguish between an excitation function $\xi(t)$ and an input function $u(t)$. For $\xi(t) = \sin \omega t$ the input to the front wheel is $u_1(t) = \xi(t)$ and $u_2(t) = \dot{\xi}(t)$. The input matrix then becomes

$$
B = \begin{pmatrix} 0_9 & 0 & 0 & 0 & k_{30}/m_3 & 0 & 0 & 0 & 0 & 0 \\ 0_9 & 0 & 0 & 0 & d_{30}/m_3 & 0 & 0 & 0 & 0 & 0 \end{pmatrix}^T.
$$

If we ask for the amplitude and phase ratio with respect to the excitation $\xi$ the frequency response becomes

$$
G(i\omega) = C(i\omega I - A)^{-1} B \tilde{B}(i\omega) \tag{2.5.6}
$$

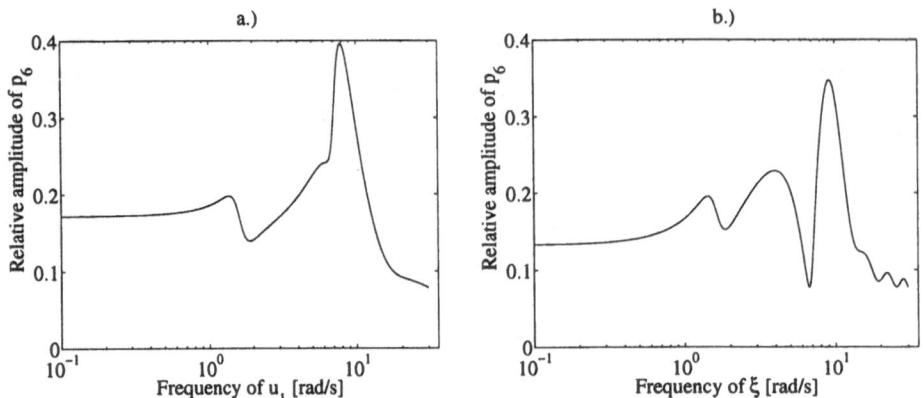

Figure 2.3: Frequency response $p_6$ versus frequency of an excitation $u_1$ of the front wheel spring only (a.) and versus a road irregularity $\xi$ acting with a time delay on the spring/damper system of the front and rear wheel (b.)

with $\tilde{B}(iw) = (b_1, \dots, b_{n_u})$ and $b_i = 1$ if $u_i = \xi$ and $b_i = iw$ if $u_i = \dot{\xi}$.

In vehicle dynamics $\xi$ can be the excitation due to road irregularities. Hence, the same excitation acts on the different wheels with a time delay. The time delay $T$ depends on the vehicle's geometry and the forward speed[2].

In this case the entries to $\tilde{B}(iw)$ become $b_i = 1$ if $u_i = \xi$, $b_i = iw$ if $u_i = \dot{\xi}$, $b_i = e^{-iwT}$ if $u_i = \xi(t - T)$ and $b_i = iwe^{-iwT}$ if $u_i = \dot{\xi}(t - T)$.

**Example 2.5.4** *For the truck with a forward speed $v_0 = 20$ m/s the corresponding frequency response $p_6$ versus $\xi$ is shown in Fig. 2.3b.*

## 2.5.1 Numerical Methods for Computing the Frequency Response

In vehicle dynamics the frequency range of interest is between 0.1 Hz and 20 Hz. These frequencies are important for discussing the systems basic behavior, the ride comfort and safety criteria. Within this range $G(iw)$ has to be evaluated for a dense grid of frequency values. Thus, several hundred matrix inversions might be necessary in order to get an accurate picture of this function. Therefore it pays off to discuss different possibilities to simplify this task as much as possible.

---

[2]In the truck example the forward motion is no degree of freedom. In this type of models it is usually assumed that the road moves with a constant speed under the truck.

**Direct Computation of $G(i\omega)$ in Real Arithmetic**

The problem can easily be transformed into the equivalent real one

$$\begin{pmatrix} -A & -\omega I \\ \omega I & -A \end{pmatrix} \begin{pmatrix} X \\ Y \end{pmatrix} = \begin{pmatrix} B \\ 0 \end{pmatrix} \tag{2.5.7a}$$

$$G = CX + iCY. \tag{2.5.7b}$$

The effort for performing this computation for $N$ different values of $\omega$ is

$$e_1 = N \left( \frac{8}{3} n_x^3 + 4 n_u n_x^2 + 2 n_{y_m} n_u n_x \right).$$

Alternatively, the problem can be transformed into

$$(\omega^2 I + A^2) Y = -\omega B \tag{2.5.8a}$$

$$X = \frac{1}{\omega} A Y \tag{2.5.8b}$$

$$G = CX + iCY. \tag{2.5.8c}$$

The effort for this foregoing, with $A$ being squared, is

$$e_2 = n_x^3 + N \left( \frac{1}{3} n_x^3 + 2 n_u n_x^2 + 2 n_u n_x + 2 n_{y_m} n_u n_x \right).$$

**Computation of $G(i\omega)$ After a Modal Transformation**

The minimal computational effort for computing $G(i\omega) = C(i\omega I - A)^{-1} B$ can be achieved by first transforming $A$ to a block diagonal form by means of similarity transformations. For this we have to assume that $A$ is diagonalizable

$$A = T \Lambda T^{-1}$$

with $\Lambda = \text{diag}(\ldots, \lambda_i, \ldots, \lambda_j, \ldots)$, where $\lambda_i$ is a real eigenvalue of $A$ and

$$\Lambda_j = \begin{pmatrix} \text{Re}\lambda_j & -\text{Im}\lambda_j \\ \text{Im}\lambda_j & \text{Re}\lambda_j \end{pmatrix}$$

for any conjugate complex eigenvalue pair $\text{Re}\lambda_j \pm \text{Im}\lambda_j$.

With this similarity transformation one obtains

$$G(i\omega) = C(i\omega I - A)^{-1} B = CT(i\omega I - \Lambda)^{-1} T^{-1} B.$$

Equation (2.5.7) takes now the form

$$\begin{pmatrix} -\Lambda & -\omega I \\ \omega I & -\Lambda \end{pmatrix} \begin{pmatrix} X \\ Y \end{pmatrix} = \begin{pmatrix} T^{-1}B \\ 0 \end{pmatrix} \tag{2.5.9a}$$

$$G = CTX + iCTY. \tag{2.5.9b}$$

The system (2.5.9a) can be transformed into block diagonal form with $4 \times 4$ blocks. This reduces significantly the computational effort for solving this system. The total effort is

$$e_3 = 15n_x^3 + (n_u + n_y)n_x^2 + N\left(\frac{32}{3}n_x + 8n_u n_x + 2n_y n_u n_x\right).$$

Again, there is a second alternative which is based on squaring the matrix $\Lambda$ (cf. Eq. (2.5.8))

$$(\omega^2 I + \Lambda^2)Y = -\omega T^{-1}B \tag{2.5.10a}$$

$$X = \frac{1}{\omega}\Lambda Y \tag{2.5.10b}$$

$$G = CTX + iCTY. \tag{2.5.10c}$$

The effort for this method is

$$e_4 = 15n_x^3 + (n_u + n_{y_m})n_x^2 + N\left(\frac{7}{3}n_x + 5n_u n_x + 2n_{y_m} n_u n_x\right),$$

see [Rug84].

## 2.6 Linear Constant Coefficient DAEs

Differential-algebraic equations (DAEs) differ in main aspects from explicit or regularly implicit ordinary differential equations. This concerns theory, e.g. solvability and representation of the solution, as well as numerical aspects, e.g. convergence behavior under discretization. Both aspects depend essentially on the *index* of the DAE. Thus, we first define the index.
We consider linear constant coefficient DAEs

$$E\dot{x}(t) = Ax(t) + f(t), \qquad x(t_0) = x_0, \tag{2.6.1}$$

where $A, E$ are $n_x \times n_x$ matrices which may be singular. In the case of mechanical systems with constraints on position level the matrices have the form

$$E := \begin{pmatrix} I & 0 & 0 \\ 0 & M & 0 \\ 0 & 0 & 0 \end{pmatrix}, \quad A := \begin{pmatrix} 0 & I & 0 \\ -K & -D & -G^T \\ G & 0 & 0 \end{pmatrix}. \tag{2.6.2}$$

The solution of these equations can be obtained explicitly by modal transformation, i.e. computing eigenvalues and eigenvectors. From the computation of the solution the main differences between explicit ODEs and DAEs are figured out.
We consider the following cases separately:

1. $E$ nonsingular;

2. $E$ singular, $A$ nonsingular;

3. $E$ and $A$ singular.

**Case 1: $E$ nonsingular**
If $E$ is a nonsingular matrix Eq. (2.6.1) is a regular implicit ordinary differential equation. We can compute the solution of Eq. (2.6.1) by inversion of $E$ and the transition matrix, see Sec. 2.4 with $\tilde{A} = E^{-1}A$.

**Case 2: $E$ singular, $A$ nonsingular**
For a nonsingular matrix $A$ Eq. (2.6.1) can be transformed into

$$A^{-1}E\dot{x}(t) = x(t) + A^{-1}f(t). \tag{2.6.3}$$

Let $J$ be the Jordan canonical form of the singular matrix $A^{-1}E$, i.e.

$$A^{-1}E = TJT^{-1}$$

with

$$J = \begin{pmatrix} R & 0 \\ 0 & N \end{pmatrix}, \tag{2.6.4}$$

where $R$ corresponds to the Jordan blocks with non-zero eigenvalues and $N$ to the zero eigenvalues. Then $R$ is nonsingular and $N$ is nilpotent.

**Definition 2.6.1** *The smallest number $k$ with $N^k = 0$ is called the* index *of nilpotency of $N$ and is identical to the size of the largest Jordan block of $N$.*

**Definition 2.6.2** *The* index *of the DAE (2.6.1) is defined as the index of nilpotency of the matrix $N$: $k = \mathrm{ind}(N)$.*

Substituting the definition of $J$ into Eq. (2.6.3), using the coordinate transformation $\bar{x} = T^{-1}x$ and premultiplying by $T^{-1}$ yields

$$\begin{pmatrix} R & 0 \\ 0 & N \end{pmatrix} \begin{pmatrix} \dot{\bar{x}}_1 \\ \dot{\bar{x}}_2 \end{pmatrix} = \begin{pmatrix} \bar{x}_1 \\ \bar{x}_2 \end{pmatrix} + \underbrace{T^{-1}A^{-1}f}_{\bar{f}}.$$

This system is decoupled

$$R\dot{\bar{x}}_1 = \bar{x}_1 + \bar{f}_1 \tag{2.6.5a}$$
$$N\dot{\bar{x}}_2 = \bar{x}_2 + \bar{f}_2. \tag{2.6.5b}$$

Because of the nonsingularity of $R$, Eq. (2.6.5a) is an explicit ODE, the *state space form* of the system.

From Eq. (2.6.5b) we get by $k - 1$-fold differentiation

$$
\begin{aligned}
\bar{x}_2 &= -\bar{f}_2 + N\dot{\bar{x}}_2 \\
&= -\bar{f}_2 + N(-\dot{\bar{f}}_2 + N\ddot{\bar{x}}_2) \\
&\;\;\vdots \\
&= -\sum_{i=0}^{k-1} N^i \bar{f}_2^{(i)}.
\end{aligned}
\tag{2.6.6}
$$

This is an explicit expression for the solution components $\bar{x}_2$. Differentiation leads together with Eq. (2.6.5a) to an ODE, the *underlying ordinary differential equation (UODE)*

$$
\dot{\bar{x}}_1 = R^{-1}(\bar{f}_1 + \bar{x}_1) \tag{2.6.7a}
$$

$$
\dot{\bar{x}}_2 = -\sum_{i=0}^{k-1} N^i \bar{f}_2^{(i+1)}. \tag{2.6.7b}
$$

From Eq. (2.6.6) we see that the index coincides with the number of differentiations necessary to obtain an ODE for all components.

The main differences between explicit ODEs and DAEs can be seen directly from the above computations:

1. The initial values for Eq. (2.6.1) cannot be chosen arbitrarily. Since Eq. (2.6.6) must be fulfilled for $t = t_0$, the initial values for $\bar{x}_2$ are determined by

$$
\bar{x}_2(t_0) = -\sum_{i=0}^{k-1} N^i \bar{f}_2^{(i)}(t_0).
$$

   Backtransformation yields the corresponding conditions for $x(t_0)$

$$
x(t_0) = T\bar{x}(t_0) = T \begin{pmatrix} (T^{-1}x(t_0))_1 \\ -\sum_{i=0}^{k-1} N^i \bar{f}_2^{(i)}(t_0). \end{pmatrix}.
$$

   Initial values satisfying these conditions are called *consistent*.

2. The solution depends on $k - 1$ derivatives of the functions $f, \bar{f}$.
   Note, that only components of $\bar{f}_2$ must be differentiable the corresponding number of times and that the differentiability requirements for each component may be different.

3. If parts of the right hand side of (2.6.6) are only continuous, the solution $\bar{x}_2$ is only continuous, too. In contrast, for explicit ODEs one gets one order of differentiability more for the solution than for the right hand side.

4. Solutions of the underlying ODE (2.6.7) need not be solutions of the DAE: The differentiation leads to a loss of information in form of integration constants.

5. Higher index DAEs have "hidden" algebraic constraints. This can be seen from the following example of an index-3 DAE.

**Example 2.6.3**

$$
\begin{aligned}
\dot{x}_1(t) &= f_1(t) - x_3(t) \\
\dot{x}_2(t) &= f_2(t) - x_1(t) \\
0 &= x_2(t) - f_3(t).
\end{aligned}
$$

*The solution is given as*

$$
\begin{aligned}
x_1 &= f_2 - \dot{f}_3 & \text{(2.6.8a)} \\
x_2 &= f_3 & \text{(2.6.8b)} \\
x_3 &= f_1 - \dot{f}_2 + \ddot{f}_3. & \text{(2.6.8c)}
\end{aligned}
$$

*From Eq. (2.6.8b) follows $\dot{x}_2 = \dot{f}_3 \Rightarrow f_2 - x_1 = \dot{f}_3$ and $\dot{x}_1 = f_2 - \ddot{f}_3 \Rightarrow f_1 - x_3 = f_2 - \ddot{f}_3$ which are constraints on $x_1$ and $x_3$. Thus the solution is fully determined and we have no freedom to chose any initial value.*

These hidden constraints are used in the computation of Eq. (2.6.6), when we inserted the derivatives of Eq. (2.6.5b).

**Case 3: $E$ singular, $A$ singular**
If $E$ and $A$ are singular, Eq. (2.6.1) need not to be uniquely solvable [Cam80]. The solvability is closely related to the regularity of the *matrix pencil* $\mu E - A$ with $\mu \in \mathbb{C}$.

**Definition 2.6.4** *The matrix pencil $\mu E - A$ is called* regular *if*

$$
d(\mu) := \det(\mu E - A)
$$

*is not identically zero.*

## 2.6.1 The Matrix Pencil*

**Lemma 2.6.5** *The matrix pencil $\mu E - A$ of the linear mechanical system (2.6.2) is regular if and only if the $n_\lambda \times n_p$ constraint matrix $G$ has full rank.*

*Proof:* Suppose the pencil is regular and $\text{rank}(G) < n_\lambda$. Then there exists a vector $\lambda \in \mathbb{R}^{n_\lambda}$ such that $G^T \lambda = 0$ and $\lambda \neq 0$. This implies

$$(\mu E - A) \begin{pmatrix} 0 \\ 0 \\ \lambda \end{pmatrix} = \begin{pmatrix} \mu I & -I & 0 \\ K & \mu M + D & G^T \\ -G & 0 & 0 \end{pmatrix} \begin{pmatrix} 0 \\ 0 \\ \lambda \end{pmatrix} = 0$$

for all $\mu \in \mathbb{C}$ and hence the pencil is singular in contradiction to the hypothesis. Conversely, suppose $\text{rank}(G) = n_\lambda$ and consider the system $(\mu E - A)x = 0$, $x^T = (p^T, v^T, \lambda^T)$, then

$$\begin{aligned} \mu p - v &= 0 \\ Kp + \mu M v + D v + G^T \lambda &= 0 \\ Gp &= 0. \end{aligned}$$

Substituting $v = \mu p$ yields

$$(K + \mu D + \mu^2 M)p = -G^T \lambda.$$

Since $M$ is regular, there is a $\mu_0$ such that $K + \mu D + \mu^2 M$ is regular for all $|\mu| > |\mu_0|$ and

$$\begin{aligned} (K + \mu D + \mu^2 M)^{-1} &= \frac{1}{\mu^2} M^{-1} \left( I + \frac{1}{\mu} M^{-1} (\frac{1}{\mu} K + D) \right)^{-1} \\ &= \frac{1}{\mu^2} \left( M^{-1} + \frac{1}{\mu} U \right) \end{aligned}$$

with

$$U = \sum_{i=1}^{\infty} \frac{1}{\mu^{i-1}} \left( -M^{-1} D - \frac{1}{\mu} M^{-1} K \right)^i M^{-1}.$$

Here we replaced the inverse matrix by its Neumann series. This can be done for $\mu$ being sufficiently large. Thus

$$0 = Gp = - \left( GM^{-1}G^T + \frac{1}{\mu} GUG^T \right) \lambda.$$

As $M^{-1}$ is symmetric positive definite, $GM^{-1}G^T$ is regular and $\lambda = 0$ is the only solution for $|\mu|$ large enough. Hence $p = 0$ and $v = 0$ and consequently $x = 0$ is the only solution of $(\mu E - A)x = 0$ for this choice of $\mu$. Thus $\mu E - A$ cannot be a singular matrix pencil.                                                                           □

If in Eq. (2.6.2) the constraints were formulated in terms of the velocity coordinates $v$ or in terms of the Lagrange multipliers $\lambda$, a result analogous to Lemma 2.6.5 can be proven in a similar way.

We will exclude the case of redundant constraints, so we can assume $\text{rank}(G) = n_\lambda$ and consequently all matrix pencils being regular.

## 2.6.2 The Matrix Pencil and the Solution of the DAE*

**Theorem 2.6.6** *Equation (2.6.1) is solvable iff the matrix pencil $\mu E - A$ is regular.*

*Proof:*As we already saw in Case 2, the Jordan canonical form plays an important role. It can be obtained by an appropriate similarity transformation applied to the system matrices. Here, a pair of matrices $(E, A)$ must be transformed simultaneously. By this similarity transformation the matrix pencil can be transformed into the so-called *Kronecker canonical form* $\mu \bar{E} - \bar{A}$, where both, $\bar{E}$ and $\bar{A}$, are block diagonal matrices consisting of Jordan blocks.

Choose $\mu$ such that $(\mu E - A)$ is nonsingular. Such a $\mu$ can be found, as the matrix pencil is assumed to be regular. Premultiplying the DAE by $(\mu E - A)^{-1}$ yields

$$\underbrace{(\mu E - A)^{-1} E}_{=:\bar{E}} \dot{x} = \underbrace{(\mu E - A)^{-1} A}_{=:\bar{A}} x + \underbrace{(\mu E - A)^{-1} f}_{=:\bar{f}}.$$

Note that

$$\bar{A} = -I + \mu \bar{E} \tag{2.6.9}$$

holds. Now we take $T_1$ such that $\bar{E}$ is transformed to its Jordan Form $J_{\bar{E}}$

$$T_1^{-1} \bar{E} T_1 := J_{\bar{E}} = \begin{pmatrix} R & 0 \\ 0 & N \end{pmatrix}$$

where $R$ is associated with the non-zero eigenvalues and $N$ is associated with the zero eigenvalues and therefore nilpotent. Due to Eq. (2.6.9) $J_{\bar{A}}$ has the structure of a Jordan canonical form of $\bar{A}$:

$$T_1^{-1} \bar{A} T_1 = J_{\bar{A}} = \begin{pmatrix} -I + \mu R & 0 \\ 0 & -I + \mu N \end{pmatrix}.$$

Clearly, this form still depends on $\mu$ as a free parameter.
We will add two additional transformations to give $(J_{\bar{A}}, J_{\bar{E}})$ a nicer structure:
By taking

$$T_2 := \begin{pmatrix} R & 0 \\ 0 & -I + \mu N \end{pmatrix}$$

$J_{\bar{E}}$ is transformed to

$$T_2^{-1} J_{\bar{E}} = \begin{pmatrix} I & 0 \\ 0 & (-I + \mu N)^{-1} N \end{pmatrix}$$

and $J_{\bar{A}}$ to

$$T_2^{-1} J_{\bar{A}} = \begin{pmatrix} -R^{-1} + \mu I & 0 \\ 0 & I \end{pmatrix}.$$

In the next and last transformation we use the fact that the blocks $(-I+\mu N)^{-1}N$ and $-R^{-1}+\mu I$ have their own Jordan canonical forms $T_N^{-1}(-I+\mu N)^{-1}NT_N = \tilde{N}$ with $\tilde{N}$ being nilpotent and $T_R^{-1}(-R^{-1}+\mu I)T_R = \tilde{R}$. Thus, a final transformation with

$$T_3 := \begin{pmatrix} T_R & 0 \\ 0 & T_N \end{pmatrix}$$

yields

$$J_{\tilde{E}} := T_3^{-1}T_2^{-1}T_1^{-1}\tilde{A}T_1T_3 = \begin{pmatrix} I & 0 \\ 0 & \tilde{N} \end{pmatrix}$$

$$J_{\tilde{A}} = T_3^{-1}T_2^{-1}T_1^{-1}\tilde{A}T_1T_3 = \begin{pmatrix} \tilde{R} & 0 \\ 0 & I \end{pmatrix}.$$

Applying the same transformations to the DAE system we obtain with $\bar{x} := T_3^{-1}T_1^{-1}x$ after premultiplication by $T^{-1}$ and $T := T_1T_2T_3$

$$\begin{pmatrix} I & 0 \\ 0 & \tilde{N} \end{pmatrix} \dot{\bar{x}} = \begin{pmatrix} \tilde{R} & 0 \\ 0 & I \end{pmatrix} \bar{x} + \underbrace{T^{-1}\bar{f}}_{=:\tilde{f}}. \tag{2.6.10}$$

We can partition $\bar{x}$ corresponding to the partitioning of the matrices into $\bar{x}_1$ and $\bar{x}_2$ and obtain an explicit ODE for $\bar{x}_1$

$$\dot{\bar{x}}_1 = \tilde{R}\bar{x}_1 + \tilde{f}_1$$

and an implicit one with a nilpotent leading matrix for $\bar{x}_2$

$$\tilde{N}\dot{\bar{x}}_2 = \bar{x}_2 + \tilde{f}_2.$$

Now, we restrict ourselves to the homogeneous case. Then we get $\bar{x}_2(t) \equiv 0$. Thus, we have as solution

$$\bar{x}_1(t) = e^{\tilde{R}t}\bar{x}_1(0) \quad \text{and} \quad \bar{x}_2(t) = 0$$

which can be written as

$$\bar{x}(t) = \begin{pmatrix} e^{\tilde{R}t} & 0 \\ 0 & 0 \end{pmatrix} \begin{pmatrix} I & 0 \\ 0 & 0 \end{pmatrix} \bar{x}(0).$$

Backtransformation leads to

$$
\begin{aligned}
x(t) &= T_1 T_3 \begin{pmatrix} e^{\tilde{R}t} & 0 \\ 0 & 0 \end{pmatrix} T_3^{-1} T_1^{-1} T_1 T_3 \begin{pmatrix} I & 0 \\ 0 & 0 \end{pmatrix} T_3^{-1} T_1^{-1} x(0) \\
&= T_1 \begin{pmatrix} e^{T_R \tilde{R} T_R^{-1} t} & 0 \\ 0 & 0 \end{pmatrix} T_1^{-1} T_1 \begin{pmatrix} I & 0 \\ 0 & 0 \end{pmatrix} T_1^{-1} x(0) \\
&= T_1 \begin{pmatrix} e^{(-R^{-1} + \mu I)t} & 0 \\ 0 & 0 \end{pmatrix} T_1^{-1} T_1 \begin{pmatrix} I & 0 \\ 0 & 0 \end{pmatrix} T_1^{-1} x(0) \\
&= \exp\left( T_1 \begin{pmatrix} R^{-1} & 0 \\ 0 & 0 \end{pmatrix} T_1^{-1} T_1 \begin{pmatrix} -I + \mu R & 0 \\ 0 & -I + \mu N \end{pmatrix} T_1^{-1} t \right) \cdot \\
&\quad \cdot T_1 \begin{pmatrix} I & 0 \\ 0 & 0 \end{pmatrix} T_1^{-1} x(0) \\
&= e^{\bar{E}^D \bar{A} t} \, \bar{E}^D \bar{E} x_0.
\end{aligned}
$$

with the matrix

$$
\bar{E}^D := T_1 \begin{pmatrix} R^{-1} & 0 \\ 0 & 0 \end{pmatrix} T_1^{-1}
$$

defining the *Drazin inverse* for

$$
\bar{E} = T_1 \begin{pmatrix} R & 0 \\ 0 & N \end{pmatrix} T_1^{-1}.
$$

Thus we have proven

**Lemma 2.6.7** *If the matrix pencil $\mu E - A$ is regular, the general solution of the linear homogeneous DAE can be written as*

$$
x(t) = e^{\bar{E}^D \bar{A} t} \bar{E}^D \bar{E} x_0. \tag{2.6.11}
$$

*The initial vector is consistent, iff $\bar{E}^D \bar{E} x_0 = x_0$ holds.*

Thus, for DAEs the transition matrix is given by $e^{\bar{E}^D \bar{A} t}$.
Formally, a Drazin inverse can be defined by the following properties:

**Definition 2.6.8** *A matrix $A^D$ is called a* Drazin inverse *of the matrix $A$, if there is a $k$ such that the following Drazin axioms are fulfilled:*

$$
\begin{array}{llll}
(D1) & A A^D &=& A^D A \\
(D2) & A^D A A^D &=& A^D, \\
(D3) & A^D A^{k+1} &=& A^k
\end{array}
$$

If $A$ is invertible, then $k = 0$ and $A^D = A^{-1}$.
The solution (2.6.11) of the linear homogeneous DAE

$$E\dot{x} = Ax \qquad \text{with} \qquad x(0) = x_0$$

solves also the so-called *Drazin ODE*:

$$\dot{x} = \bar{E}^D \bar{A} x \qquad \text{with} \qquad x(0) = \bar{E}^D \bar{E} x_0. \qquad (2.6.12)$$

Though the matrices $\bar{A}$ and $\bar{E}$ depend on the free parameter $\mu$, the system matrix $\bar{E}^D \bar{A}$ and the projector $\bar{E}^D \bar{E}$ are independent of $\mu$ and this ODE is well-defined. We will omit the proof for this statement and refer to [Cam80].
So, the linear homogeneous problem can be solved by first transforming to its Drazin ODE, the solution of which can be given in terms of the transition matrix

$$x(t) = e^{\bar{E}^D \bar{A} t} \bar{E}^D \bar{E} x_0.$$

In general this transformation step cannot be performed in a numerical robust way, as the process demands the transformation of matrices to its Jordan canonical form. This canonical form cannot be established by means of stable algorithms. In contrast, for mechanical systems the matrix $\hat{E}^D \hat{A}$ can be given explicitly [SFR93]. This will be shown in the next section.

## 2.6.3   Construction of the Drazin Inverse
##         for Mechanical Systems*

For this step we introduce the projection matrix

$$P := I - M^{-1} G^T (G M^{-1} G^T)^{-1} G = I - UG.$$

with $U := M^{-1} G^T (G M^{-1} G^T)^{-1}$.

**Remark 2.6.9** *P has the following properties*

   1. *P is a projector onto the nullspace of $G$.*

   2. *$I - P$ is an orthogonal projector with respect to the scalar product*

$$< x, y >_M := x^T M y$$

   3. *P is invariant with respect to coordinate changes, i.e. if $\phi : \mathcal{U}_{\bar{p}} \to \mathcal{U}_p$ is a diffeomorphism and $\bar{G}(\bar{p}) = \frac{\partial}{\partial \bar{p}} g(\phi(\bar{p}))$, then*

$$P = I - M^{-1} \bar{G}^T (\bar{G} M^{-1} \bar{G})^{-1} \bar{G}.$$

We will consider a constrained mechanical system in three different formulations

a.) formulation with constraints on acceleration level

$$E_1 = \begin{pmatrix} I & 0 & 0 \\ 0 & M & 0 \\ 0 & G & 0 \end{pmatrix} \text{ and } A_1 = \begin{pmatrix} 0 & I & 0 \\ -K & -D & -G^T \\ 0 & 0 & 0 \end{pmatrix} \qquad (2.6.13)$$

b.) formulation with constraints on the velocity level

$$E_2 = \begin{pmatrix} I & 0 & 0 \\ 0 & M & 0 \\ 0 & 0 & 0 \end{pmatrix} \text{ and } A_2 = \begin{pmatrix} 0 & I & 0 \\ -K & -D & -G^T \\ 0 & G & 0 \end{pmatrix} \qquad (2.6.14)$$

c.) formulation with constraints on the position level

$$E_3 = \begin{pmatrix} I & 0 & 0 \\ 0 & M & 0 \\ 0 & 0 & 0 \end{pmatrix} \text{ and } A_3 = \begin{pmatrix} 0 & I & 0 \\ -K & -D & G^T \\ G & 0 & 0 \end{pmatrix} \qquad (2.6.15)$$

## a) Constraints on Acceleration Level

We proceed now as follows: First we construct two matrices $S_1$ and $Q_1$, of which we believe, that these are the matrices $\bar{E}_1^D \bar{A}_1$ and $\bar{E}_1^D \bar{E}_1$ we are looking for. Then, we prove that this is true.

The equations of motion

$$\begin{aligned} \dot{p} &= v \\ M\dot{v} &= -Kp - Dv - G^T\lambda \\ 0 &= G\dot{v} \end{aligned}$$

can be solved for $\lambda$ and we get

$$\lambda = U^T(-Kp - Dv). \qquad (2.6.16)$$

Substituting this back gives

$$\begin{aligned} \dot{p} &= v \\ \dot{v} &= -PM^{-1}Kp - PM^{-1}Dv \end{aligned}$$

and by differentiating Eq. (2.6.16) with respect to time we obtain differential equations for $p, v$ and $\lambda$, which we claim is the Drazin ODE of the given problem

$$\begin{aligned} \dot{p} &= v & (2.6.17a) \\ \dot{v} &= -PM^{-1}Kp - PM^{-1}Dv & (2.6.17b) \\ \dot{\lambda} &= U^T(DPM^{-1}Kp - Kv + DPM^{-1}Dv). & (2.6.17c) \end{aligned}$$

**Theorem 2.6.10** *The linear problem $E_1 \dot{x} = A_1 x$ with $A_1$ and $E_1$ given by Eq. (2.6.13) has as corresponding Drazin ODE*

$$\dot{x} = S_1 x \qquad with \qquad x(0) = Q_1 x_0 \qquad\qquad (2.6.18)$$

*with*

$$S_1 := \begin{pmatrix} 0 & I & 0 \\ -PM^{-1}K & -PM^{-1}D & 0 \\ U^T DPM^{-1}K & U^T(-K+DPM^{-1}D) & 0 \end{pmatrix}$$

*and*

$$Q_1 := \begin{pmatrix} I & 0 & 0 \\ 0 & I & 0 \\ -U^T K & -U^T D & 0 \end{pmatrix}.$$

*Proof:* We have to show, that

$$x(t) := e^{S_1 t} Q_1 x_0 \qquad\qquad (2.6.19)$$

is a solution.

For this, we use the property $E_1 S_1 Q_1 = A_1 Q_1$ and $E_1 S_1^2 = A_1 S_1$ without showing this by performing the tedious block operations.

Exploring these properties after premultiplying $x(t)$ by $E_1$ and expanding the exponential function into Taylor series gives

$$
\begin{aligned}
E_1 \dot{x}(t) &= \left( E_1 S_1 Q_1 + \left( \sum_{i=1}^{\infty} E_1 S_1^{i+1} \right) Q_1 \right) x_0 \\
&= \left( A_1 Q_1 + \left( \sum_{i=1}^{\infty} A_1 S_1^{i} \right) Q_1 \right) x_0 \\
&= A_1 x(t)
\end{aligned}
$$

Thus, $x(t)$ defined by Eq. (2.6.19) is a solution of the problem and we have by the uniqueness of the solution

$$x(t) = e^{S_1 t} Q_1 x_0 = e^{\bar{E}_1^D \bar{A}_1 t} \bar{E}_1^D \bar{E}_1 x_0.$$

As this holds for arbitrary $x_0$ we can conclude by setting $t = 0$: $Q_1 = \bar{E}_1^D \bar{E}_1$. From the Taylor expansion of the matrix exponentials we get

$$S_1 Q_1 = \bar{E}_1^D \bar{A}_1 Q_1, \qquad\qquad (2.6.20)$$

so that it remains to show only, that $S_1 z = \bar{E}_1^D \bar{A}_1 z$ for all $z \in \mathrm{rg}(Q_1)^{\perp} = \ker(Q_1)$ holds. $\ker Q_1$ is spanned by all elements $x$ of the form $x^T = (0, 0, \lambda^T)^T$. Thus,

$S_1 z = 0$ due to the structure of $S_1$. In order to show that also the last block column of $\bar{E}_1^D \bar{A}_1$ is zero we note, that

$$Q_1 \begin{pmatrix} p_0 \\ v_0 \\ \lambda_0 \end{pmatrix} = \begin{pmatrix} a \\ b \\ 0 \end{pmatrix}$$

with $a$ and $b$ being linear combinations of $p_0$ and $v_0$ only. Therefore the transformation matrix $T$ in Eq. (2.6.10) has to be of the form

$$T = \begin{pmatrix} * & * & 0 \\ * & * & 0 \\ * & * & \alpha \end{pmatrix} \tag{2.6.21}$$

with $*$ denoting arbitrary blocks and $\alpha$ denoting a nonsingular block. Multiplication with $T$ keeps the $\lambda$ components decoupled from the $p$ and $v$ components. Thus, by the definition of $\bar{E}_1^D$ the matrix $\bar{E}_1^D \bar{A}_1$ has to have a zero last block column. This together with Eq. (2.6.20) gives $S_1 = \bar{E}_1^D \bar{A}_1$.  □.

**b) Constraints on Velocity Level**
In this paragraph we give expressions for $\bar{E}_2^D \bar{E}_2$ and $\bar{E}_2^D \bar{A}_2$, the system matrices for the Drazin ODE of a mechanical system with constraints formulated on velocity level.
For this end, let us consider first a formulation as a second order problem

$$\begin{align} M\ddot{p} &= -Kp - D\dot{p} - G^T \lambda \tag{2.6.22a} \\ 0 &= G\dot{p}. \tag{2.6.22b} \end{align}$$

Clearly, a solution of this problem also satisfies the acceleration constraint

$$0 = G\ddot{p}.$$

Thus, if this constraint is coupled to the equations of motion by introducing a new Lagrange multiplier, $\dot{\eta}$ say,

$$\begin{align} M\ddot{p} &= -Kp - D\dot{p} - G^T \lambda - G^T \dot{\eta} \tag{2.6.23a} \\ 0 &= G\dot{p}, \tag{2.6.23b} \end{align}$$

the additional constraint force $G^T \dot{\eta}$ must be identically zero. No additional force is necessary to enforce the acceleration to be satisfied. Nevertheless we will formally keep this term in the equations, when transforming now to a first order system

$$\begin{align} M\dot{p} &= Mv - G^T \eta \\ M\dot{v} &= -Kp - D\dot{p} - G^T \lambda \\ 0 &= G\ddot{p} \\ 0 &= G\dot{p}. \end{align}$$

Due to $\dot{\eta} = 0$ this system is equivalent to

$$M\dot{p} = Mv - G^T\eta \tag{2.6.24a}$$
$$M\dot{v} = -Kp - D\dot{p} - G^T\lambda \tag{2.6.24b}$$
$$0 = G\dot{v} \tag{2.6.24c}$$
$$0 = G\dot{p}. \tag{2.6.24d}$$

Note, that in this system $\dot{p}$ and $v$ differ by a constant vector, which is orthogonal to the direction of free motion.

Using the previously introduced projectors $U$ and $P$, we can compute from these equations the Lagrange multipliers

$$\lambda = U^T(-Kp - D\dot{p}) \text{ and } \eta = U^T Mv.$$

Due to $U^T G^T = I$, we indeed get $\dot{\eta} = 0$.

Introducing these expressions for the Lagrange multipliers into Eqs. (2.6.24a) and (2.6.24b) yields

$$\dot{p} = Pv \tag{2.6.25a}$$
$$\dot{v} = -PM^{-1}Kp - PM^{-1}DPv. \tag{2.6.25b}$$

Differentiating $\lambda$ and taking $\dot{p}$ from Eq. (2.6.25a) yields an differential equation for $\lambda$ also,

$$\dot{\lambda} = U^T(DPM^{-1}Kp - KPv + DPM^{-1}DPv). \tag{2.6.26}$$

We claim that Eqs. (2.6.25) together with Eq. (2.6.26) constitute the Drazin ODE for the velocity constrained mechanical problem

**Theorem 2.6.11** *The corresponding Drazin ODE of the DAE $E_2\dot{x} = A_2x$ with $A_2$ and $E_2$ defined in Eq. (2.6.14) is given by*

$$\dot{x} = S_2x \quad \text{with} \quad x(0) = Q_2x_0 \tag{2.6.27}$$

*with*

$$S_2 := \begin{pmatrix} 0 & P & 0 \\ -PM^{-1}K & -PM^{-1}DP & 0 \\ U^TDPM^{-1}K & U^T(-K+DPM^{-1}D)P & 0 \end{pmatrix}$$

*and*

$$Q_2 := \begin{pmatrix} I & 0 & 0 \\ 0 & P & 0 \\ -U^TK & -U^TDP & 0 \end{pmatrix}.$$

The proof is similar to the one of Theorem 2.6.10 and we refer to [SFR93].

## c) Constraints on Position Level

In the last paragraph we distinguished two different types of velocity variables $\dot{p}$ and $v$. The velocity constraint was based on $\dot{p}$, whereas the Drazin ODE was formulated in terms of $v$. This scheme will be repeated in this paragraph, when constructing the Drazin ODE of the position constrained problem. There, we will introduce two types of position variables $p$ and $q$. The position constraint will be formulated in terms of $p$, whereas the Drazin ODE is based on $q$-position variables. Let us start again from the second order formulation of the problem:

$$M\ddot{p} = -Kp - D\dot{p} - G^T\lambda$$
$$0 = Gp.$$

As the solution of this problem also satisfies the velocity and acceleration constraint, these constraints can be coupled to the equations by formally introducing new Lagrange multipliers $\dot{\eta}$ and $\ddot{\zeta}$

$$M\ddot{p} = -Kp - D\dot{p} - G^T\lambda - G^T\dot{\eta} - G^T\ddot{\zeta}$$
$$0 = G\ddot{p}$$
$$0 = G\dot{p}$$
$$0 = Gp.$$

Again, we note that $\dot{\eta} = \ddot{\zeta} = 0$. Using this fact, transforming this system to first order form gives

$$M\dot{q} = Mv - G^T\eta + DM^{-1}G^T\zeta \qquad (2.6.28a)$$
$$M\dot{v} = -Kp - D\dot{q} - G^T\lambda \qquad (2.6.28b)$$
$$Mp = Mq - G^T\zeta \qquad (2.6.28c)$$
$$0 = G\dot{v} \qquad (2.6.28d)$$
$$0 = G\dot{q} \qquad (2.6.28e)$$
$$0 = Gp. \qquad (2.6.28f)$$

From Eqs. (2.6.28f) and (2.6.28a) we get

$$\zeta = U^T Mq \qquad (2.6.29)$$

and similarly

$$\eta = U^T Mv + U^T DM^{-1}G^T\zeta. \qquad (2.6.30)$$

$\lambda$ can be obtained in the usual way by taking the acceleration constraint together with Eq. (2.6.28b)

$$\lambda = U^T(-Kp - D\dot{q}). \qquad (2.6.31)$$

Inserting these expressions for the Lagrange multipliers back into (2.6.28) gives

$$p = Pq \tag{2.6.32}$$

and

$$
\begin{aligned}
\dot{q} &= PM^{-1}DUGq + Pv \\
\dot{v} &= -PM^{-1}(KP + DPM^{-1}DUG)q - PM^{-1}DPv.
\end{aligned}
$$

This together with

$$\dot{\lambda} = \frac{d}{dt}U^T(-KPq - D\dot{q}) \tag{2.6.33}$$

constitutes the Drazin ODE of the position constrained system.
This is summarized in the following

**Theorem 2.6.12** *For the DAE $E_3\dot{x} = A_3x$ with $E_3, A_3$ given by Eq. (2.6.15) the corresponding Drazin ODE is*

$$\dot{x} = S_3x \qquad with \qquad x(0) = Q_3x_0, \tag{2.6.34}$$

*where the system matrices have the following form*

$$
S_3 = \begin{pmatrix}
PM^{-1}DUG & P & 0 \\
-PM^{-1}(KP + DPM^{-1}DUG) & -PM^{-1}DP & 0 \\
U^T(-K + DPM^{-1}D)PM^{-1}DUG & U^T(-K + DPM^{-1}D)P & 0 \\
+U^T DPM^{-1}KP & &
\end{pmatrix}
$$

$$
Q_3 = \begin{pmatrix}
P & 0 & 0 \\
PM^{-1}DUG & P & 0 \\
-U^T(KP + DPM^{-1}DUG) & -U^T DP & 0
\end{pmatrix}.
$$

For the proof we refer to [SFR93].

**Example 2.6.13** *For the example of the truck described in Ch. 1, Fig. 2.4 shows the solution of the Drazin ODE. We took the model linearized around the nominal position $p_N = (0.5, 2, 0, 0.5, 2.9, 0, -1.46, 2.9, 0)^T$ and an initial displacement of $p(0) = p_N + 0.1e_4$.*

**Remark 2.6.14** *By comparing the matrices $Q_1, Q_2$ and $Q_3$ one observes, that $\mathrm{rg}(Q_3) \subset \mathrm{rg}(Q_2) \subset \mathrm{rg}(Q_1)$, i.e. all initial conditions consistent for the position constrained problem are also consistent for the velocity or acceleration constrained system.*

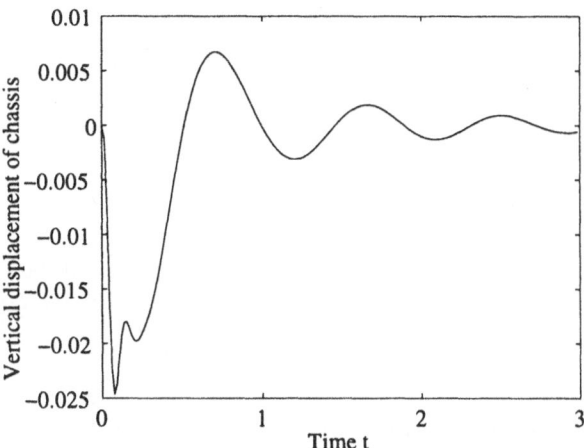

Figure 2.4: Solution of Drazin ODE, chassis

## 2.7 DAEs and the Generalized Eigenvalue Problem

The *generalized eigenvalues* play the same important role for the constrained problem as the eigenvalues for the unconstrained linear problem. After stating some fundamental properties of the generalized eigenvalue problem, we will see how the generalized eigenvalues of a constrained problem are related to the eigenvalues of its corresponding state space form. In this section we will follow the lines of [SFR93]. We consider again the following homogeneous linear problem

$$E\dot{x} = Ax \qquad \text{with} \qquad x(0) = x_0 \qquad (2.7.1)$$

and formally associate to Eq. (2.7.1) the eigenvalue[3] problem

$$\det(\mu E - A) = 0.$$

As the nonzero solutions $\bar{\mu}$ of

$$\det(E - \bar{\mu}A) = 0$$

have the property $\mu = \bar{\mu}^{-1}$, one sets the generalized eigenvalue $\mu := \infty$ if $\bar{\mu} = 0$. The possible appearance of infinite eigenvalues is the most important property of the generalized eigenvalue problem. The number of these infinite eigenvalues is strongly related to the structure of the DAE.
First we will consider three typical examples from [GL83]:

---

[3]In order to avoid a notational conflict with the Lagrange multipliers $\lambda$ we will denote the eigenvalues by $\mu$.

**Example 2.7.1**

$$A := \begin{pmatrix} 1 & 2 \\ 0 & 3 \end{pmatrix}, \quad E := \begin{pmatrix} 1 & 0 \\ 0 & 0 \end{pmatrix};$$

*with the corresponding DAE*

$$\begin{aligned} \dot{x}_1 &= x_1 + 2x_2 \\ 0 &= 3x_2. \end{aligned}$$

*The corresponding eigenvalue problem has only one finite solution $\mu = 1$, whereas the dimension of this problem is 2. It can easily be seen that the DAE is equivalent to the scalar ODE*

$$\dot{x}_1 = x_1$$

*having the same eigenvalue. The second example is given by*

$$A := \begin{pmatrix} 1 & 2 \\ 0 & 3 \end{pmatrix}, \quad E := \begin{pmatrix} 0 & 1 \\ 0 & 0 \end{pmatrix},$$

*which corresponds to*

$$\begin{aligned} \dot{x}_2 &= x_1 + 2x_2 \\ 0 &= 3x_2. \end{aligned}$$

*Here, the eigenvalue problem has no solution at all and the DAE turns out to consist simply of two algebraic equations (no dynamics!)*

$$3x_2 = 0 \qquad \text{and} \qquad x_1 = 0.$$

*The last of these three examples is*

$$A := \begin{pmatrix} 1 & 2 \\ 0 & 0 \end{pmatrix}, \quad E := \begin{pmatrix} 1 & 0 \\ 0 & 0 \end{pmatrix},$$

*corresponding to a DAE which evidently has an infinite number of solutions*

$$\begin{aligned} \dot{x}_1 &= x_1 + 2x_2 \\ 0 &= 0. \end{aligned}$$

*The eigenvalue problem can also take any complex number as solution.*

The properties of the generalized eigenvalues coincide with the "classical" case only if $E$ is a regular matrix:

**Lemma 2.7.2** *The generalized eigenvalue problem $\det(\mu E - A)$, with $E$ and $A$ being $n \times n$ matrices, has $n$ finite solutions if and only if $\operatorname{rank}(E) = n$.*

We rule out the case, where the eigenvalue problem has infinitely many solutions. This is the case if the matrix pencil is singular, which corresponds to the case that the corresponding DAE is not uniquely solvable.

## 2.7.1 State Space Form and the Drazin ODE*

In the last chapters we transformed linear DAEs to ODEs: the linear state space form and the Drazin ODE. Both are related to the equations of motion in DAE form in the sense, that they lead to the same solution, when initialized with corresponding initial conditions. One might wonder, if these ODEs share also other properties like stability properties: How are the eigenvalues of the various forms related?

For this we will compare the eigenvalues of the state space form

$$\begin{pmatrix} \dot{y}_p \\ \dot{y}_v \end{pmatrix} = \underbrace{\begin{pmatrix} 0 & I \\ -(V^T MV)^{-1} V^T KV & -(V^T MV)^{-1} V^T DV \end{pmatrix}}_{A_{\text{ssf}}} \begin{pmatrix} y_p \\ y_v \end{pmatrix}$$

(cf. Eq. (2.1.12)), to those of the different DAE formulations.

Note that, though the state space form is not unique due to the freedom in selecting $V$, two different state space forms always have similar system matrices $A_{\text{ssf}}$, so that their eigenvalues are independent of the particular choice of $V$.

We make the following observation [SFR93]:

**Theorem 2.7.3** Let $\Lambda = \{\mu_1, \ldots, \mu_{n_y}\} \subset \mathbb{C}$ be the set of eigenvalues of the state space form of the linear mechanical system $E_3 \dot{x} = A_3 x$ with $n_y := n_p - n_\lambda$, then the following statements about the generalized eigenvalues of the different formulations hold:

1. The set of generalized eigenvalues of $E_3 \dot{x} = A_3 x$ is $\Lambda \cup \{\infty\}$. The generalized eigenvalue $\infty$ has the multiplicity $3n_\lambda$ and its eigenspace the dimension $n_\lambda$.

2. The set of generalized eigenvalues of $E_2 \dot{x} = A_2 x$ is $\Lambda \cup \{\infty\} \cup \{0\}$. The generalized eigenvalue $\infty$ has the multiplicity $2n_\lambda$, its eigenspace the dimension $n_\lambda$, and the zero eigenvalue has the multiplicity $n_\lambda$ with an eigenspace having the dimension $n_\lambda$.

3. The set of generalized eigenvalues of $E_1 \dot{x} = A_1 x$ is $\Lambda \cup \{\infty\} \cup \{0\}$. The generalized eigenvalue $\infty$ has the multiplicity $n_\lambda$, its eigenspace the dimension $n_\lambda$, and the zero eigenvalue has the multiplicity $2n_\lambda$ with a defective eigenspace having the dimension $n_\lambda$.

4. The system matrix $S_3$ of the Drazin ODE (2.6.34) has the eigenvalues $\Lambda \cup \{0\}$ with the zero eigenvalue having the multiplicity $3n_\lambda$ and its eigenspace being of dimension $3n_\lambda$.

Proof: We prove statements (1) and (4). The others can be shown by applying similar techniques.

(i) Let $x_\mu = (p^T, v^T, \lambda^T)^T$ be a generalized eigenvector of $(E_3, A_3)$ corresponding to an eigenvalue $\mu$, i.e.

$$\begin{aligned}
\mu p &= v \\
\mu M v &= -Kp - Dv - G^T \lambda \\
0 &= Gp.
\end{aligned}$$

From the last equation we conclude $Gv = 0$ and consequently there exist vectors $y_p$ and $y_v = \mu y_p$ with

$$p = V y_p \text{ and } v = V y_v$$

where again, $V$ is a matrix spanning the null space of $G$. From this we get $\mu V^T M V y_v = -V^T K V y_p - V^T D V y_v$, which together with $\mu y_p = y_v$ means that $(y_p^T, y_v^T)^T$ is an eigenvector of $A_{\text{ssf}}$ to the eigenvalue $\mu$.

Conversely we show, that if $(y_p^T, y_v^T)^T$ is an eigenvector of $A_{\text{ssf}}$ to the eigenvalue $\mu$, the vector $x = (p^T, v^T, \lambda^T)^T$ with

$$p := V y_p, \quad v := V y_v, \quad \lambda := (GG^T)^{-1} G(\mu M v - Kp - Dv)$$

is a generalized eigenvector of $(E_3, A_3)$. First we note, $\mu p = \mu V y_p = V y_v = v$, and $0 = V^T(\mu M v + Kp + Dv) = (I - G^T(GG^T)^{-1})(\mu M v + Kp + Dv)$. This implies $\mu M v = -Kp - Dv - G^T \lambda$.

(ii) The relation between the generalized eigenvalues of $(E_3, A_3)$ and the eigenvalues of the Drazin ODE can be seen by considering the construction of the Drazin ODE again. Let $\mu$ be, like in Sec. 2.7, a complex number distinct from a generalized eigenvalue, then

$$\sigma E_3 x = A_3 \iff (\sigma \bar{E}_3 - A_3)x = 0 \iff$$

$$T^{-1}\left( \sigma \begin{pmatrix} R & 0 \\ 0 & N \end{pmatrix} - \begin{pmatrix} -I+\mu R & 0 \\ 0 & -I+\mu N \end{pmatrix} \right) Tx = 0 \iff$$

$$\begin{aligned} (\sigma - \mu)z_1 &= -R^{-1}z_1 \\ 0 &= z_2 \end{aligned} \quad \text{with} \quad \begin{pmatrix} z_1 \\ z_2 \end{pmatrix} = Tx.$$

On the other hand

$$\sigma x = \bar{E}_3^D \bar{A}_3 \iff$$

$$T^{-1}\left( \sigma I - \begin{pmatrix} R^{-1} & 0 \\ 0 & 0 \end{pmatrix} \begin{pmatrix} -I+\mu R & 0 \\ 0 & -I+\mu N \end{pmatrix} \right) Tx = 0 \iff$$

$$\begin{aligned} (\sigma - \mu)z_1 &= -R^{-1}z_1 \\ 0 &= \sigma z_2 \end{aligned} \quad \text{with} \quad \begin{pmatrix} z_1 \\ z_2 \end{pmatrix} = Tx.$$

This proves, that the generalized eigenvalues of $(E_3, A_3)$ are a subset of the eigenvalues of $\bar{E}_3^D A_3$. In addition, $\sigma = 0$ is always an eigenvalue of the Drazin ODE corresponding to the eigenvector

$$x = T^{-1} \begin{pmatrix} 0 \\ z_2 \end{pmatrix}$$

for any arbitrary $z_2$.

Finally, we note that there are $2n_p + n_\lambda - 2(n_p - n_\lambda) = 3n_\lambda$ linear independent vectors $z_2$. □

**Remark 2.7.4** *The second part (ii) of the proof makes no use of the particular structure of $E_3$ and $A_3$ and can therefore be applied to any nonsingular matrix pencil. From the last part of the proof it can be seen, that the eigenvectors corresponding to the additional zero eigenvalues are not consistent with the DAE. Thus, these eigenvalues cannot be excited by consistent initial values.*

**Example 2.7.5** *To summarize, the eigenvalues of the different linear formulations of the linear truck example are given in the following table. It can clearly be seen how the index reduction process from position level constraints to acceleration level constraints transforms infinity eigenvalues to zero eigenvalues.*

| Index 3 | Index 2 | Index 1 | Drazin ODE | St.Sp.Form |
|---|---|---|---|---|
| | | | 0 | |
| | | | 0 | |
| | | 0 | 0 | |
| | | 0 | 0 | |
| | 0 | 0 | 0 | |
| | 0 | 0 | 0 | |
| -2.47 ± 7.10 i | -2.47 ± 7.10 i | -2.47 ± 7.10 i | -2.47 ± 7.10 i | -2.47 ± 7.10 i |
| -0.67 ± 7.77 i | -0.67 ± 7.77 i | -0.67 ± 7.77 i | -0.67 ± 7.77 i | -0.67 ± 7.77 i |
| -9.20 ± 11.0 i | -9.20 ± 11.0 i | -9.20 ± 11.0 i | -9.20 ± 11.0 i | -9.20 ± 11.0 i |
| -14.9 | -14.9 | -14.9 | -14.9 | -14.9 |
| -8.62 ± 14.0 i | -8.62 ± 14.0 i | -8.62 ± 14.0 i | -8.62 ± 14.0 i | -8.62 ± 14.0 i |
| -21.4 ± 48.0 i | -21.4 ± 48.0 i | -21.4 ± 48.0 i | -21.4 ± 48.0 i | -21.4 ± 48.0 i |
| -7.95 ± 59.4 i | -7.95 ± 59.4 i | -7.95 ± 59.4 i | -7.95 ± 59.4 i | -7.95 ± 59.4 i |
| -75.7 | -75.7 | -75.7 | -75.7 | -75.7 |
| ∞ | ∞ | ∞ | | |
| ∞ | ∞ | ∞ | | |
| ∞ | ∞ | | | |
| ∞ | ∞ | | | |
| ∞ | | | | |
| ∞ | | | | |

Table 2.2: Eigenvalues of the different formulations of the constrained truck

# 3 Solution of Nonlinear Equations

## 3.1 Static Equilibrium Position

In this chapter we are seeking an *equilibrium position* of a multibody system. This is a position where the system does not move. The velocity and acceleration are zero.

There are mainly two different ways to transfer a given multibody system into an equilibrium state:

- to adjust the nominal forces in such a way that a *given position* becomes an equilibrium position;

- to modify the position for *given nominal forces*, so that with the modified position the system is in an equilibrium state.

We saw in Sec. 1.6 that the first task normally requires the solution of a system of linear equations. The second task yields a system of nonlinear equations of the form

$$F(x) := \begin{pmatrix} f_a(p,0) - G^T(p)\lambda \\ g(p) \end{pmatrix} = 0 \qquad (3.1.1)$$

with $x^T = (p^T, \lambda^T)$.

The determination of such an equilibrium position is important

- because it contains interesting information about the system's mechanical behavior;

- in order to obtain a point for linearization to perform a linear system analysis;

- in order to obtain consistent initial values for the integration of the equations of motion of the system.

Solving nonlinear systems is not only interesting in this context, but it is also a subtask of many other problems, e.g. simulation, optimization.

**Example 3.1.1** *If for the unconstrained truck example the individual bodies are assembled without respecting the influence of gravitational forces $m_i g_{gr}$, i.e. we assume $g_{gr} = 0$, the configuration shown in Fig. 3.1 is obtained. This configuration is called an assembling configuration, as it is the configuration which is described in a CAD environment by just assembling the individual construction elements without considering masses and forces. Adding gravitational forces moves the system into its equilibrium position represented by the dotted lines in Fig. 3.1.*

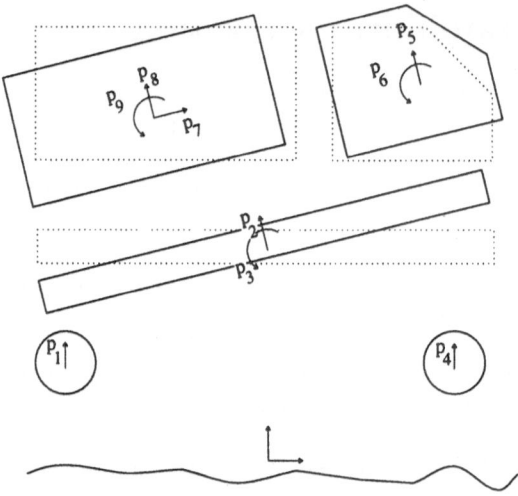

Figure 3.1: Assembling position vs. equilibrium position of the unconstrained truck

Often, there is a free parameter $s$ in the description of the problem and (3.1.1) takes the form

$$F(x, s) = \left( \begin{array}{c} f_a(p, 0, s) - G^T(p, s)\lambda \\ g(p, s) \end{array} \right) = 0. \tag{3.1.2}$$

Then, one is interested either in the curve $x(s)$ or in the solution for a particular parameter value $s_1$. In the example above this value is $s_1 = g_{gr}$.

**Example 3.1.2** *We construct a parameter dependent solution for the unconstrained truck when increasing $s$ from $0$ to $g_{gr} = 9.81 \, \text{m/s}^2$. This is shown in Figs. 3.2.*

Another typical example for a one-parametric problem is the determination of an equilibrium position for a mechanical system with a reference system which moves with constant speed along a curve with constant curvature. The curvature introduces centrifugal forces into the system, which affect the equilibrium position.

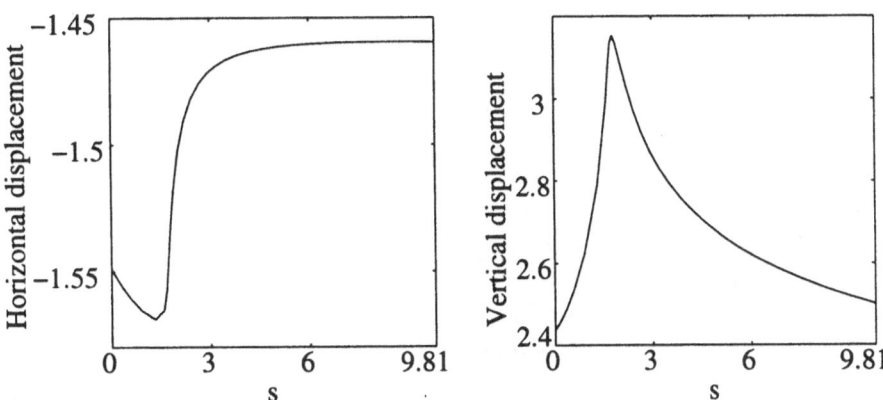

Figure 3.2: Parameter dependent solutions $p_7(s)$ and $p_8(s)$ for the unconstrained truck.

To distinguish this case from the case with a stationary reference system, the solution is sometimes called a *quasistatic equilibrium*.

Evaluating $F$ defined by (3.1.1) for an arbitrary point $x^{(0)}$ results in a *residual* $b^0 := F(x^{(0)})$. The first $n_p$ components of $b_0$ have the physical dimension of a force. This force must be added to the system to transform $x^{(0)}$ into an equilibrium. The goal of a numerical scheme is to iteratively reduce this force to zero by incrementally changing $x$.

In the sequel we will discuss the numerical solution of a set of nonlinear equations (3.1.1) by numerical methods such as Newton's method. Newton's method is the first method to be used if one has to solve nonlinear equations. There are many extensions of this method. We will focus on path following methods for one-parametric problems of type (3.1.2).

## 3.2 Solvability of Nonlinear Equations

The following theorem states conditions for the solvability of a system of nonlinear equations:

**Theorem 3.2.1 (Inverse Function Theorem)** *Let* $F : D \rightarrow \mathbb{R}^{n_x}$ *be continuously differentiable in an open set* $D \subseteq \mathbb{R}^{n_x}$ *and* $0 \in F(D)$. *If* $F'(x)$ *is regular for all* $x \in D$, *then*

1. *There is a locally unique* $x^* \in D$ *with* $F(x^*) = 0$.

2. *There is a neighborhood* $V(0)$ *of* $0$ *and a unique, continuously differentiable mapping* $G : V \rightarrow D$ *with* $F(G(y)) = y$ *for all* $y \in V$ *and especially* $G(0) = x^*$.

*3. For the derivative holds $G'(y) = F'(G(y))^{-1}$ in $\mathcal{V}(0)$.*

For the proof see e.g. [OR70]. A function $F$ with these properties is called a *local diffeomorphism*.

For the case of unconstrained multibody systems this theorem requires the regularity of the stiffness matrix

$$f_p(p,0) := \frac{\partial f}{\partial p}(p,0).$$

If the stiffness matrix is singular, then the mechanical system is statically undetermined and no (unique) equilibrium position can be found.

In the truck example this is e.g. the case if the springs between both wheels and the ground are removed.

Furthermore, the smoothness requirements can be a severe problem in practice, being often a reason for failure of numerical methods.

## 3.3   Fixed Point Iteration

Systems of nonlinear equations are solved by iterative methods. One of the simplest iterative methods is the *fixed point* or *functional iteration*, which is often the basis of more sophisticated approaches. A *fixed point* $x^*$ of a map $\varphi$ is defined by the condition $x^* = \varphi(x^*)$. The problem of finding a solution of $F(x) = 0$ is equivalent to finding a fixed point $x^*$ of the map

$$\varphi(x) = x - B(x)F(x) \qquad (3.3.1)$$

with a regular matrix $B(x)$. Starting from an initial guess $x^{(0)}$ the iterates of the fixed point iteration are given by

$$x^{(k+1)} = \varphi\big(x^{(k)}\big).$$

For stating convergence conditions for this sequence $\{x^{(k)}\}_{k \in \mathbb{N}}$ we need the following definition:

**Definition 3.3.1** *A function $\varphi : D \subseteq \mathbb{R}^{n_x} \to \mathbb{R}^{n_x}$ is called* Lipschitz continuous *on $D_0 \subseteq D$ if there is a constant $L(\varphi)$ such that*

$$\|\varphi(x) - \varphi(y)\| \leq L(\varphi)\|x - y\| \quad \forall x, y \in D_0. \qquad (3.3.2)$$

$L(\varphi)$ *is called a* Lipschitz constant *of $\varphi$. If $L(\varphi) < 1$ then $\varphi$ is called a* contraction.

Let $\rho(A)$ denote the *spectral radius* of $A$, i.e. $\rho(A) := \max\{|\lambda_i(A)|\}$, where $\lambda_i(A)$ is the $i^{\text{th}}$ eigenvalue of $A$. Then, for continuously differentiable functions $\varphi$ the condition

$$\rho(\varphi'(x^*)) < 1$$

is sufficient for $\varphi$ being a contraction in a convex neighborhood of a point $x^*$. For contractive functions the existence of a fixed point can be shown:

**Theorem 3.3.2 (Banach's Fixed Point Theorem)** *Let $\varphi : D \subseteq \mathbb{R}^{n_x} \to \mathbb{R}^{n_x}$ be contractive on a closed set $D_0 \subseteq D$ and suppose $\varphi(D_0) \subset D_0$. Then $\varphi$ has a unique fixed point $x^* \in D_0$. Moreover, for an arbitrary point $x^{(0)} \in D_0$ the iteration*

$$x^{(k+1)} = \varphi(x^{(k)})$$

*converges to the fixed point $x^*$. The distance to the solution is bounded by the a priori bound*

$$\|x^* - x^{(k)}\| \leq \frac{L(\varphi)^k}{1 - L(\varphi)} \|x^{(1)} - x^{(0)}\|.$$

*(For the proof, see e.g. [OR70]).*

We will apply this theorem now to construct Newton's method by an appropriate choice of $B$. Later, when dealing with discretized differential equations we will also apply the fixed point iteration directly to solve nonlinear systems.

## 3.4  Newton's Method

The contractivity of $\varphi$ in the form (3.3.1) can be enforced by setting

$$B(x) = F'(x)^{-1}.$$

In the solution point $x^*$ we then get $\varphi'(x^*) = 0$ , which is the best we can achieve. Inserting this into the iteration leads to *Newton's method*

$$x^{(k+1)} = \varphi(x^{(k)}) = x^{(k)} - F'(x^{(k)})^{-1} F(x^{(k)}),$$

or written in the form in which it is performed numerically

$$\begin{aligned} F'(x^{(k)})\Delta x^{(k)} &= -F(x^{(k)}) & \text{(3.4.1a)} \\ x^{(k+1)} &:= x^{(k)} + \Delta x^{(k)}. & \text{(3.4.1b)} \end{aligned}$$

In contrast to the fixed point iteration, this methods requires the evaluation of the Jacobian $F'(x)$ and the solution of linear systems in every step. Both tasks can be very time consuming. In multibody dynamics it is mainly the evaluation of the Jacobian, which is very costly, see Sec. 3.5 .

To save computation time one often refrains from recomputing $F'$ in every step. This leads to the *simplified Newton method*, which is characterized by

$$B(x) = F'(x^{(0)})^{-1}$$

Sometimes even only an approximation to the Jacobian is taken, i.e.

$$B(x) \approx F'(x^{(0)})^{-1}.$$

The main questions with such iterative methods are:

- Does the method converge to a zero of $F$,
  i.e. $\lim_{i \to \infty} x^{(i)} = x^*$ with $F(x^*) = 0$? (Convergence)

- How fast does the method converge? (Rate or speed of convergence)

- Is the method feasible, i.e. $x^{(0)} \in D_0 \overset{?}{\Longrightarrow} x^{(i)} \in D_0$ for all $i$? (Feasibility)

The main theorem for the convergence of Newton's method is the Newton Convergence Theorem, cf. [OR70], [Boc87], [DH79]. Since later on in this book we use other Newton type methods like *simplified Newton* and *Gauß-Newton methods* we present a sufficiently general form of the convergence theorem from [Boc89].

**Theorem 3.4.1** Let $D \subset \mathbb{R}^{n_x}$ be open and convex and $x^{(0)} \in D$.
Let $F \in C^1(D, \mathbb{R}^{n_F})$ and $B \in C^0(D, \mathbb{R}^{n_x \times n_F})$.
Assume that there exist constants $r, \omega, \kappa, \delta_0$ such that for all $x, y \in D$ the following properties hold:

1. *Curvature condition*

$$\|B(y)(F'(x + \tau(y - x)) - F'(x))(y - x)\| \le \tau\omega \|y - x\|^2 \qquad (3.4.2)$$

   with $\tau \in [0, 1]$.

2. *Compatibility condition*

$$\|B(y)R(x)\| \le \kappa\|y - x\| \qquad (3.4.3)$$

   with the compatibility residual $\quad R(x) := F(x) - F'(x)B(x)F(x)$ and $\kappa < 1$.

3. *Contraction condition*

$$\delta_0 := \kappa + \frac{\omega}{2}\|x^{(1)} - x^{(0)}\| < 1.$$

   with $x^{(1)} = x^{(0)} - B(x^{(0)})F(x^{(0)})$

4. *Condition on the initial guess*

$$D_0 := \{x| \|x - x^{(0)}\| \le r\} \subset D$$

   with $r := \dfrac{\|x^{(1)} - x^{(0)}\|}{1 - \delta_0}$.

*Then the iteration*

$$x^{(k+1)} := x^{(k)} - B(x^{(k)})F(x^{(k)})$$

*is well-defined with $x^{(k)} \in D_0$ and converges to a solution $x^* \in D_0$ of $B(x)F(x) = 0$. The speed of convergence can be estimated by*

$$\|x^{(k+j)} - x^*\| \leq \frac{\delta_k^j}{1 - \delta_k} \|\Delta x^{(k)}\| \tag{3.4.4}$$

*with $\delta_k := \kappa + \frac{\omega}{2} \|\Delta x^{(k)}\|$ and the increments decay conforming to*

$$\|\Delta x^{(k+1)}\| \leq \kappa \|\Delta x^{(k)}\| + \frac{\omega}{2} \|\Delta x^{(k)}\|^2. \tag{3.4.5}$$

**Remark 3.4.2** *By setting $k = 0$ the a priori error estimation formula*

$$\|x^{(j)} - x^*\| \leq \frac{\delta_0^j}{1 - \delta_0} \|x^{(1)} - x^{(0)}\| \tag{3.4.6}$$

*can be obtained.*

This theorem with its constants needs some interpretation:

1. *Curvature condition (1).* This is a weighted Lipschitz condition for $F'$, written in a way that it is invariant with respect to scaling (it is even affine invariant). It can be fulfilled with $\omega = L \sup \|B(y)\|$ if $F'$ is Lipschitz continuous with constant $L$ and $\|B\|$ is bounded on $D$. For linear problems $F'$ is constant and $\omega = 0$. Then, the method converges in one step. Hence $\omega$ is a measure for the nonlinearity of the problem. In the case of a large $\omega$, the radius of convergence $r$ may be very small because for a large $\omega$, $\|x^{(1)} - x^{(0)}\|$ has to be very small[1].

2. *Compatibility condition (2).* In the case of Newton's method, i.e. $B(x) = F'(x)^{-1}$ we have $R(x) = 0$ and thus $\kappa = 0$. Then the method converges quadratically.

   In the other cases this is a condition on the quality of the iteration matrix $B(x)$. It says how tolerant we can be, when replacing $F'(x^{(k)})^{-1}$ by an approximate $B(x)$. Due to $\kappa \neq 0$ we can expect only a linear convergence. The better the Newton iteration matrix is approximated the smaller $\kappa$. The compatibility condition can be transformed to

$$
\begin{aligned}
\|B(y)R(x)\| &= \|B(y)\left[F(x) - F'(x)B(x)F(x)\right]\| \\
&= \|B(y)\left(B(x)^{-1} - F'(x)\right) \underbrace{B(x)F(x)}_{=-(y-x)}\| \\
&\overset{!}{\leq} \kappa\|y - x\|, \quad \kappa < 1.
\end{aligned}
$$

---

[1] A formulation of this theorem with sharper bounds is given in [Bock87]

| $k$ | $\|\Delta x^{(k)}\|$ | $\dfrac{\|\Delta x^{(k)}\|}{\|\Delta x^{(k-1)}\|}$ | $\dfrac{\|\Delta x^{(k)}\|}{\|\Delta x^{(k-1)}\|^2}$ | $\dfrac{\|\Delta x^{(k)}\|}{\|\Delta x^{(k-1)}\|^3}$ |
|---|---|---|---|---|
| 1 | $8.4 \cdot 10^{-2}$ | | | |
| 2 | $1.2 \cdot 10^{-2}$ | $1.4 \cdot 10^{-1}$ | 1.7 | $2.0 \cdot 10^1$ |
| 3 | $9.9 \cdot 10^{-4}$ | $8.3 \cdot 10^{-2}$ | 7.0 | $5.8 \cdot 10^2$ |
| 4 | $5.0 \cdot 10^{-6}$ | $5.1 \cdot 10^{-3}$ | 5.1 | $5.1 \cdot 10^3$ |
| 5 | $1.9 \cdot 10^{-10}$ | $3.8 \cdot 10^{-5}$ | 7.6 | $1.5 \cdot 10^6$ |

Table 3.1: Newton increments and rate of convergence for the unconstrained truck

It can be fulfilled if $\| B(y) \left( B(x)^{-1} - F'(x) \right) \| \leq \kappa < 1$, i.e. if $B(x)^{-1} \approx F'(x)$ and $\| B(y) \|$ not too large.

3. *"δ"-conditions (3), (4)*. These conditions are restrictions on the initial guess $x^{(0)}$. Newton's method is only *locally convergent* and the size of the convergence region depends on $\omega$ and $\kappa$.

**Example 3.4.3** *The unconstrained truck is in the configuration*

$$p^{(0)} = (0.502, 2.06, 0.01, 0.502, 2.88, 0.01, -1.467, 2.85, 0.01)^T$$

*not in an equilibrium position. Starting with this configuration Newton's method results in the Newton increments given in Table 3.1. From the corresponding column in that table it can be seen that Newton's method converges quadratically. The solution is*

$$p^* = (0.5, 2., 0.0, 0.5, 2.9, 0.0, -1.46, 2.9, 0.0)^T .$$

## 3.5 Numerical Computation of Jacobians

Newton's method or its simplified versions require the knowledge of the Jacobian $F'(x)$ at some points. Unfortunately, today's multibody formalisms and programs don't provide this extra information, though at least by applying tools for automatic differentiation this would be possible in principle [Gri89]. Normally, the Jacobian is approximated by finite differences $\Delta^\eta_{e^j} F_i$ with $e^j$ being the $j^{\text{th}}$ unit vector

$$\frac{\partial F_i}{\partial x_j} \approx \Delta^\eta_{e^j} F_i = \frac{F_i(x + \eta e^j) - F_i(x)}{\eta}.$$

More generally, the derivative in the direction of $s$ can be obtained by using

$$F'(x)s \approx \Delta^\eta_s F(x) = \frac{F(x + \eta s) - F(x)}{\eta}, \qquad \|s\| = 1.$$

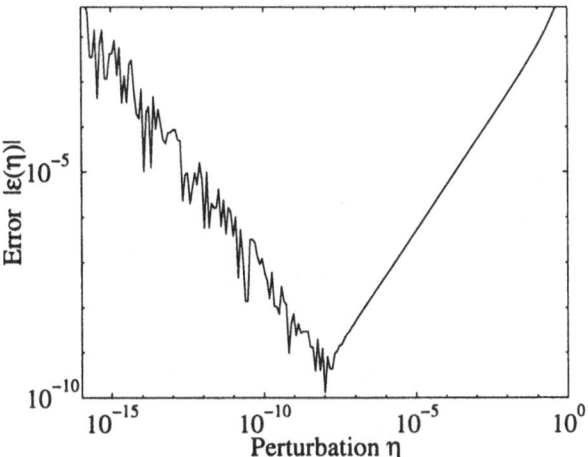

Figure 3.3: Error $|\varepsilon_{ij}(\eta)|$ in the numerical computation of $\sin'(1)$

The increment $\eta$ has to be chosen such that the influence of the approximation error $\varepsilon(\eta) = \Delta_s^\eta F(x) - \frac{\partial F}{\partial s}$ can be neglected. $\varepsilon(\eta)$ consists of *truncation errors* $(= \frac{1}{2}\|\frac{\partial^2 F}{\partial x^2}\|\eta + \mathcal{O}(\eta^2)))$ and *roundoff errors* in the evaluation of $F$. Let $\varepsilon_F$ be an upper bound for the error in the numerical computation of $F$, then

$$|\varepsilon_{ij}(\eta)| = \left| \Delta_{e_j}^\eta F_i - \frac{\partial F_i}{\partial x_j} \right| \leq \frac{2\varepsilon_F + \frac{1}{2}\|\frac{\partial^2 F_i}{\partial x_j^2}\|\eta^2 + \mathcal{O}(\eta^3)}{\eta}. \qquad (3.5.1)$$

In Fig. 3.3 the overall error for the example $\sin'(1)$ is given. In the left part of the figure the roundoff error dominates and in the right part the truncation error. The slopes in double logarithmic representation are -1 and +1 for the roundoff and approximation errors which can be expected from (3.5.1).

When neglecting $\eta^2$ and higher order terms this bound is minimized if $\eta$ is selected according the rule of thumb

$$\eta = 2\sqrt{\varepsilon_F \left| \frac{\partial^2 F_i}{\partial x_j^2} \right|^{-1}}$$

which in practice is often replaced by

$$\eta = 2\sqrt{\varepsilon_F}.$$

The effort for the computation of $F'(x)$ by numerical differentiation consists of additional $n_F$ additional evaluations of $F$. This high effort motivates the use of simplified Newton methods even when the convergence is no longer quadratic. Note that by numerically approximating $F'(x)$ this property is already lost ($\kappa \neq 0$).

## 3.6   Reevaluation of the Jacobian

In this section we want to discuss how a strategy for the reuse of the Jacobian works without loosing convergence.

The convergence rate $\delta_0$ can be estimated during the iteration. This estimate can be used to monitor the convergence of the method and to take the necessary actions (reevaluation of the Jacobian, reduction of the step length) if the convergence becomes poor.

An estimate $\tilde{\delta}_0$ for $\delta_0$ can be obtained after two Newton iterations from

$$\frac{\|\Delta x^{(1)}\|}{\|\Delta x^{(0)}\|} =: \tilde{\delta}_0 \leq \delta_0.$$

Requiring the error to be $\varepsilon$ after $it_{opt}$ iterations we obtain from the a priori estimate (3.4.6)

$$\frac{\varepsilon}{\|\Delta x^{(0)}\|} = \frac{\delta_0^{it_{opt}}}{1 - \delta_0}$$

From this a maximal value for $\tilde{\delta}_0$, $\bar{\delta}_0$, can be obtained. This suggests the following foregoing:

- If the estimated rate of convergence $\tilde{\delta}_0$ in an iteration step is larger than $\bar{\delta}_0$ the iteration is broken off as non convergent.

- Otherwise the iteration is continued until the error criterion is fulfilled.

If the iteration does not converge, i.e. $\tilde{\delta}_0 > \bar{\delta}_0$ this may have different reasons:

- The approximation $B(x)$ of the inverse Jacobian is too bad because the Jacobian changed too much ($\kappa$ is too large).

- The starting value is not in the domain of convergence ($\omega \|\Delta x^{(0)}\|$ is too large).

This leads to the following strategy if the iteration does not converge or $\tilde{\delta}_0$ is too large:

1. The Jacobian is computed at the actual point.

2. If this does not lead to convergence a better starting value $x^{(0)}$ must be achieved.

## 3.7 Limitations and Extensions

It is often a good idea to monitor the norm $\|F\|_2^2 = F^T F$ for controlling and *globalizing the convergence.*
With

$$g(x) = \frac{1}{2} F(x)^T F(x),$$

the Newton step is a descent direction of $g$ as

$$\nabla g \Delta x = (F^T F')(-F'^{-1} F) = -F^T F \leq 0.$$

Thus we may use the following strategy:

- Use the Newton step as long as it reduces $g$.

- If $g$ is not reduced by the Newton step a smaller step in the same direction is considered:

$$x^{(i+1)} := x^{(i)} + \lambda \Delta x, \quad \lambda \in (0, 1].$$

There are several ways to choose $\lambda$:

- The simplest strategy is to half $\lambda$ successively until $g$ is reduced sufficiently.
- A more sophisticated strategy tries to approximate $\tilde{g}(\lambda) = g\left(x^{(i)} + \lambda \Delta x^{(i)}\right)$ by a third order polynomial using the values of $\tilde{g}, \tilde{g}'$ at $\lambda = 0$ and $\lambda = 1$. Then, $\lambda$ is determined as the minimum of this approximation in $(0, 1]$ [PTVF92, SB93].

However, the problem with using $g$ to determine the step length is that the process can stop in a local minimum of $g$ which is not a zero of $F$. Thus, it is often better to use path following methods to achieve global convergence. These methods will be described in the next section.
If *multiple zeros* are present one has to start Newton's method with different starting values in order to obtain all zeros.
*Discontinuities* in $F$ or its derivative may cause trouble in the iteration process:

- The Newton Convergence Theorem does not hold any longer and thus all conclusions and numerical algorithms based on it are dubious.

- The numerical computation of the Jacobian may be wrong if there is a discontinuity between two values used for the difference approximation.

- If the Jacobian is reused for several steps including a discontinuity, the Newton direction may become totally misleading.

- Using the line search for $g$ may fail, because, firstly, with an old Jacobian $\Delta x$ may no longer be a descent direction and, secondly, $g$ may increase due to a jump in $F$.

## 3.8 Continuation Methods in Equilibrium Computation

### 3.8.1 Globalizing the convergence of Newton's Method

One of the main problems with Newton's method is the choice of a starting value $x^{(0)}$. In the case of highly nonlinear problems or due to a poor approximation $B(x)$ of $F'(x)^{-1}$ the method might converge only for starting values $x^{(0)}$ in a very small neighborhood of the unknown solution $x^*$. So, special techniques for generating good starting values have to be applied. One way is to embed the given problem $F(x) = 0$ into a parameter depending family of problems $H(x, s) = 0$ with

$$H(x, s) := F(x) - (1 - s)F(x_0) = 0 \qquad (3.8.1)$$

and a given value $x_0 \in D$. This family contains two limiting problems. On one hand we have

$$H(x, 0) = F(x) - F(x_0) = 0$$

which is a system of nonlinear equations with $x_0$ as a known solution. On the other hand we have

$$H(x, 1) = F(x) = 0$$

which is the problem we want to solve.

The basic idea of *continuation methods*[2] is to choose a partition $0 = s_0 < s_1 < s_2 < \cdots < s_m = 1$ of $[0, 1]$ and to solve a sequence of problems

$$H(x, s_i) = 0, \quad i = 1, \ldots, m$$

by Newton's method where the solution $x_i$ of the $i^{\text{th}}$ problem is taken as starting value for the iteration in the next problem. The key point is that if $\Delta s_i = s_{i+1} - s_i$ is sufficiently small, then the iteration process will converge since the starting value $x_i$ for $x$ will be hopefully in the region of convergence of the next subproblem $H(x, s_{i+1}) = 0$.

From the physical point of view $-F(x_0)$ is a force which has to be added to $F(x)$ in order to keep the system in a non-equilibrium position. The goal of the homotopy is then to successively reduce this force to zero by incrementally changing $s$.

The embedding chosen in (3.8.1) is called a *global homotopy*. It is a special case of a more general class of embeddings, the so-called *convex homotopy*

$$H(x, s) := (1 - s)G(x) + sF(x), \quad s \in [0, 1] \qquad (3.8.2)$$

---

[2] Often also called *homotopy* or *path following method*.

where $G \in C^1(D, \mathbb{R}^{n_F})$ is a function with a known zero, $G(x_0) = 0$. By taking $G(x) := F(x) - F(x_0)$ the global homotopy is obtained again.
We have

$$H(x, 0) = G(x), \quad H(x, 1) = F(x),$$

i.e. the parameter $s$ leads from a problem with a known solution to a problem with an unknown solution. It describes a path $x(s)$ with $x(0) = x_0$ and $x(1) = x^*$.
In the introduction to this chapter we saw already a third type of homotopy, which is based on a physical parameter in the physical problem. There, in Example 3.1.2 this parameter $s$ was related to the gravitational constant $g_{gr}$ in such a way that for $s = 0$ the truck was given in its known assembly configuration while for $s = 1$ the truck is in its unknown equilibrium position. $s$ reflects the amount of loading $s m_i g_{gr}$ considered in the system. A continuation method gradually increases this loading.
So, in general a homotopy function for a continuation method is defined as

$$H : \mathbb{R}^{n_F} \times \mathbb{R} \rightarrow \mathbb{R}^{n_F}$$

with $H(x_0, 0) = 0$ and $H(x, 1) = F(x)$, where $x_0$ is a given point $x_0 \in D$.

**Example 3.8.1** *We consider the unconstrained truck example, where the spring/ damper elements between the wheels and the chassis are replaced by nonlinear pneumatic springs, cf. Sec. A.2 and Fig. A.1. The assembling position of this system is given by*

$$x_0 = (.53129, 2.2426, .23904, .51769, 3.8280, .23614, -1.6435, 2.8912, .23926).$$

*With the following discretization of the loading parameter interval*

$$s := (0, 0.05, 0.1, 0.15, 0.16, 0.17, 0.18, 0.2, 0.25, 0.4, 0.5, 0.6, 0.65, 0.75, 0.85, 1.0)$$

*the equilibrium position can be obtained.*
*The following MATLAB sequence demonstrates the basic method*

```
x=x0;
for i=1:length(x)
  x=newton('equilib',x,s(i));
  xs(:,i)=x;
end;
```

*The resulting curve is shown in Fig. 3.2*

## 3.8.2   Theoretical Background

The method sketched so far is based on the assumption that there exists a smooth solution path $x(s)$ without bifurcation and turning points. Before describing continuation methods in more algorithmic details we look for criteria for an existence of such a solution path.

For this end we differentiate $H(x(s), s) = 0$ with respect to the parameter $s$ and obtain

$$H_x(x, s)x'(s) = -H_s(x, s)$$

with $H_x := \frac{d}{dx}H$, $H_s := \frac{d}{ds}H$.

If $H_x(x, s)$ is regular we get to the so-called *Davidenko differential equation* [Dav53]:

$$x'(s) = -H_x(x, s)^{-1}H_s(x, s), \quad x(0) = x_0 \tag{3.8.3}$$

The existence of a solution path $x(s)$ at least in a neighborhood of $(x(0), 0)$ can be ensured by standard existence theorems for ordinary differential equations as long as $H_x$ has a bounded inverse in that neighborhood. For the global homotopy (3.8.1) this requirement is met if $F$ satisfies the conditions of the Inverse Function Theorem 3.2.1.

For the global homotopy the Davidenko differential equation reads

$$x'(s) = -F'(x)^{-1}F(x_0).$$

We summarize these observations in the following Lemma [AG90].

**Lemma 3.8.2** *Let $H : D \times I \subset \mathbb{R}^{n_F+1} \to \mathbb{R}^{n_F}$ be a sufficiently smooth map and let $x_0 \in \mathbb{R}^{n_F}$ be such that $H(x_0, 0) = 0$ and $H_x(x_0, 0)$ is regular.*
*Then there exists for some open interval $0 \in I_0 \subset I$ a smooth curve $I_0 \ni s \mapsto x(s) \in \mathbb{R}^{n_F}$ with*

- $x(0) = x_0$;

- $H(x(s), s) = 0$;

- $\mathrm{rank}(H_x(x(s), s)) = n_F$ ;

*for all $s \in I_0$.*

The direction $x'(s)$ given by the right hand side of the Davidenko differential equation is just the tangent at the solution curve $x$ in $s$.

Fig. 3.4 motivates the following definition

**Definition 3.8.3** *A point $(x(s_i), s_i)$ is called a turning point if $\mathrm{rank}(H'(x(s_i), s_i)) = n_F$ and $\mathrm{rank}(H_x(x(s_i), s_i)) < n_F$.*

In the neighborhood of turning points the curve cannot be parameterized by the parameter $s$ and locally another parameterization must be taken into account. We will come to this point briefly later. First we will assume that there are no such points in $[0, 1]$.

Furthermore, the path can bifurcate into several branches. In this book we will not consider bifurcation points.

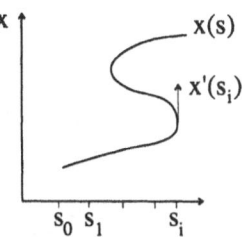

Having just seen that if no turning points are present, the path $x(s)$ is just the solution of an ordinary differential equation, the Davidenko differential equation, we could conclude this chapter by referring to numerical methods for ODEs. Unfortunately, this differential equation cannot be solved in a stable way. Small errors are not damped out. This differential equation belongs to the class of differential equations with invariants as it has $H(x(s), s) = 0$ as a first integral. This class will be studied later in Sec. 5.1.4 and we will see that drift-off effects are typical for this class of problems.

Figure 3.4: Curve with turning points

### 3.8.3 Basic Concepts of Continuation Methods

Most numerical path following methods are *predictor-corrector methods*. The predictor at step $i$ provides a starting value $x_i^{(0)}$ for the Newton iteration which attempts to "correct" this value to $x(s_i) := x_i^*$.

We have already seen the predictor of the *classical continuation method* which just takes

$$x_i^{(0)} := x_{i-1}^*.$$

There, the predictor is just a constant function in each step.

A more sophisticated predictor is used by the *tangential continuation method* where the predictor is defined as

$$x_i^{(0)} := x_{i-1}^* + (s_i - s_{i-1})x_{i-1}'(s_i).$$

This is just a step of the explicit Euler method for ODEs applied to the Davidenko differential equation, cf. Sec. 4.1.1.

The amount of corrector iteration steps depends on the quality of the prediction. First, we have to require that the predicted value is within the convergence region of the corrector problem. This region depends on the constants given by Theorem 3.4.1, mainly on the nonlinearity of the $F$. Even if the predicted value is within the domain of convergence, it should for reasons of efficiency be such that there are not too many corrector iteration steps needed. Both requirements demand an elaborated strategy for the step size control, i.e. for the location of the points $s_i$. We cite here the strategy of Seydel [Sey84] which is easy to implement. For other strategies we refer to [AG90].

Let $h_{i-1} := s_i - s_{i-1}$ be the step size in the $i - 1^{\text{st}}$ step. The step size $h_i$ in the next step is selected accordingly to the relation between the required number $n_{\text{it}}$

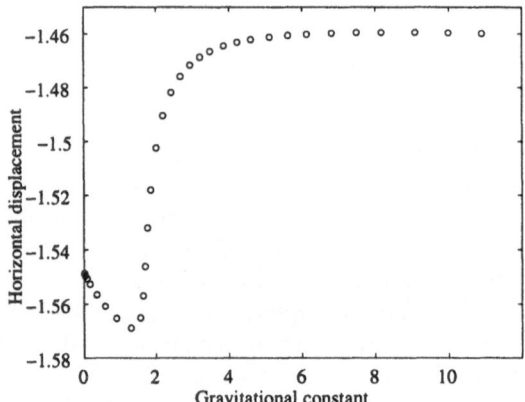

Figure 3.5: Homotopy steps for the unconstrained truck TOL=$10^{-3}$

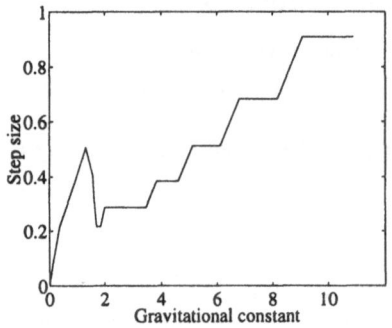

Figure 3.6: Homotopy step size for          Figure 3.7: Number of Newton iter-
TOL=$10^{-3}$                                ations for TOL=$10^{-3}$

of Newton iterations and the optimal number $n_{it_{opt}}$ of Newton iterations in the current step. $n_{it_{opt}}$ is a design parameter of the method. For multibody systems the choice $n_{it_{opt}} = 4$ has proven to be successful [Sim88]. The new step size is then determined by

$$h_i := \frac{n_{it_{opt}}}{n_{it_{i-1}}} h_{i-1}.$$

**Example 3.8.4** *We consider again the example of the constrained truck in the form described in Ex. 3.8.1 and increase the gravitational constant from 0 to 9.81 $m/s^2$. The homotopy steps obtained with this step size strategy and $n_{it_{opt}} = 4$ are shown in Fig. 3.5 and the step sizes and number of iterations in Fig. 3.6, 3.7.*

In a neighborhood of turning points the solution path $y = (x, s)$ has to be described in other parameters than $s$. Different strategies have been developed to find local parameters. They are based on considering $H(y) = 0$ as an underdetermined nonlinear system, which may be iteratively solved by Newton like methods. There, the matrix $B(y)$ of Theorem 3.4.1 is a generalized inverse of $H'(y)$.

An interesting application of these methods to determine the quasi static equilibrium of a railway boogie can be found in [SFR91].

# 4 Explicit Ordinary Differential Equations

Most ordinary differential equations (ODEs) cannot be solved analytically. Only for a restricted class of ODEs computing the analytical solution can be taken into consideration. But even in these cases it is often more efficient to apply numerical techniques to approximate the solution.

Numerical integration is the backbone of all simulation tools. Understanding the basics of these methods can help a lot in interpreting the results. Bad or unexpected performance of such a method can be a hint to modeling errors and can motivate either the choice of another method or a modification of the model at hand.

In this chapter the basics of numerical integration methods for explicit ODEs will be presented. The goal is to discuss the fundamental ideas, eg. order, stability, step size selection etc.. We will focus also on the main components of modern implementations of ODE solvers and point out special modifications which can exploit the structure of equations of motion of mechanical systems.

Often, models are considered to be "simple", if it is an easy task to set up their equations of motion, or if there are few equations which perhaps are nearly linear. Being simple in that sense does not necessarily imply that the integration task is easy. We will see this when discussing stiff ODEs.

It is important to note that numerical integration is not just a tool to solve an ODE but an exchange of models, i.e. a continuous model is replaced by a discrete one with its own properties and dynamics. There is one important parameter, the step size $h$, by which the discrete model can be tuned to be a good approximation to the continuous one. Numerical integration requires a selection of an appropriate discrete model (method selection) and a good tuning of this free parameter. This can be seen best by discussing linear multistep methods, with which we will start this chapter. We then focus on one-step methods and discuss special classes of ODEs, i.e. stiff ODEs and highly oscillatory ODEs.

## 4.1   Linear Multistep Methods

### 4.1.1   Adams Methods

The idea leading to *Adams methods* is quite simple. It is based on transforming
the initial value problem

$$\dot{x} = f(t, x) \qquad \text{with} \qquad x(t_0) = x_0 \tag{4.1.1}$$

into its integral form

$$x(t) = x_0 + \int_{t_0}^t f(\tau, x(\tau)) \, d\tau \tag{4.1.2}$$

and then approximating the integrand by an adequate polynomial.
We will assume that the time interval under consideration is partitioned into

$$t_0 < t_1 < \cdots < t_i < t_{i+1} = t_i + h_i < \cdots < t_e$$

with the step size $h_i$ at step $i + 1$. Let us assume for the moment that $k$ solution
points at successive time points are given

$$x_{n+1-i} := x(t_{n+1-i}), \quad i = 1, \ldots, k.$$

Then, by evaluating the function to be integrated (right hand side function in
(4.1.1) or simply rhs-function) the corresponding derivatives

$$f(t_{n+1-i}, x(t_{n+1-i})), \quad i = 1, \ldots, k$$

are known and can be used to define an interpolation polynomial $\pi_k^P$ of degree $k - 1$
with the property

$$\pi_k^P(t_{n+1-i}) = f(t_{n+1-i}, x_{n+1-i}), \quad i = 1, \ldots, k.$$

By this requirement this polynomial is uniquely defined though there are many
different ways to represent it. For theoretical purposes the Lagrange formulation
is convenient.   There, $\pi_k^P$ is combined of basis polynomials, so-called *cardinal
functions*, $l_i^k(t)$

$$\pi_k^P(t) = \sum_{i=1}^k l_i^k(t) f(t_{n+1-i}, x_{n+1-i}) \tag{4.1.3}$$

with

$$l_i^k(t) := \prod_{\substack{j=1 \\ j \neq i}}^k \frac{t - t_{n+1-j}}{t_{n+1-i} - t_{n+1-j}}.$$

They fulfill

$$l_i^k(t_{n+1-j}) := \delta_{ij} \qquad \text{(Kronecker symbol)}.$$

By integrating (4.1.3) from $t_n$ to $t_{n+1}$, the *Adams–Bashforth* scheme in Lagrange formulation for approximating $x(t_{n+1})$ is obtained:

$$x_{n+1}^{P} = x_n + h_n \sum_{i=1}^{k} \beta_{k-i}^{P} f(t_{n+1-i}, x_{n+1-i}) \qquad (4.1.4)$$

$$\text{with} \qquad \beta_{k-i}^{P} = \frac{1}{h_n} \int_{t_n}^{t_{n+1}} l_i^k(t)\, dt.$$

The number of previous values needed to approximate $x(t_{n+1})$ is called the number of steps of the method and all previous values and their derivatives sometimes are called the *trail* of the method.

In the sequel we will denote by $x_n$ the numerical approximation to $x(t_n)$ and set $f_n := f(t_n, x_n)$.

**Example 4.1.1** *For equal (constant) step sizes the Adams–Bashforth methods are given by the following formulas*

$$k = 1: \quad x_{n+1} = x_n + hf_n \qquad\qquad \text{explicit Euler method}$$

$$k = 2: \quad x_{n+1} = x_n + h\left(\tfrac{3}{2}f_n - \tfrac{1}{2}f_{n-1}\right)$$

$$k = 3: \quad x_{n+1} = x_n + h\left(\tfrac{23}{12}f_n - \tfrac{16}{12}f_{n-1} + \tfrac{5}{12}f_{n-2}\right).$$

As a consequence of the construction of the basic polynomials $l_i^k$ the coefficients $\beta$ depend on the spacings $h_n, \dots, h_{n-k}$. Methods with this property are called *variable coefficient methods* in contrast to *fixed coefficient methods*, where the coefficients do not depend on the step size history. In case of changing the step size $h$ to a new step size, $\bar{h}$ say, a fixed coefficient representation can be obtained by choosing a new set of ordinate values

$$\dot{\bar{x}}_{n+1-i} := \pi_k^{P}(t_{n+1} - i\bar{h})$$

and by defining an interpolation polynomial based on these values.

In contrast, when using the variable coefficient alternative, the ordinate values are kept unaffected, but the coefficients must be altered. This is the most significant difference between the various implementations of multistep methods. It influences efficiency and stability of the method, [JSD80].

To improve the accuracy of the method an *implicit multistep scheme* is taken into consideration:

Let us assume for the moment that $x_{n+1}$ and previous values are known. Then a polynomial $\pi^c_{k+1}$ of degree $k$ can be constructed by requiring

$$\pi^c_{k+1}(t_{n+1-i}) = f(t_{n+1-i}, x_{n+1-i}), \quad i = 0, 1, \dots, k.$$

Similar to eq. (4.1.4) this leads to the so-called *Adams–Moulton method* :

$$x_{n+1} = x_n + h_n \sum_{i=0}^{k} \beta^c_{k-i} f(t_{n+1-i}, x_{n+1-i}) \tag{4.1.5}$$

with

$$\beta^c_{k-i} \quad := \quad \frac{1}{h_n} \int_{t_n}^{t_{n+1}} \bar{l}^{k+1}_i(t)\, dt$$

$$\bar{l}^{k+1}_i(t) \quad := \quad \prod_{\substack{j=0 \\ j \neq i}}^{k} \frac{t - t_{n+1-j}}{t_{n+1-i} - t_{n+1-j}}.$$

**Example 4.1.2** *For equal (constant) step sizes the Adams–Moulton methods are given by the following formulas*

$$k = 0: \quad x_{n+1} = x_n + h f_{n+1} \qquad\qquad \text{\textit{implicit Euler method}}$$

$$k = 1: \quad x_{n+1} = x_n + h \left( \tfrac{1}{2} f_{n+1} + \tfrac{1}{2} f_n \right) \qquad\qquad \text{\textit{Trapezoidal rule}}$$

$$k = 2: \quad x_{n+1} = x_n + h \left( \tfrac{5}{12} f_{n+1} + \tfrac{8}{12} f_n - \tfrac{1}{12} f_{n-1} \right)$$

$$k = 3: \quad x_{n+1} = x_n + h \left( \tfrac{9}{24} f_{n+1} + \tfrac{19}{24} f_n - \tfrac{5}{24} f_{n-1} + \tfrac{1}{24} f_{n-2} \right).$$

In contrast to eq. (4.1.4) Adams–Moulton methods are defined by implicit equations, which must be solved iteratively for $x_{n+1}$.

The iteration process is started with the "predicted" value $x^P_{n+1}$ from the Adams–Bashforth scheme (4.1.4).

Both methods can be combined to one corrector equation by forming the difference of equations (4.1.4) and (4.1.5):

$$x_{n+1} = x^P_{n+1} + h_n \left( \bar{\beta}^c_k f(t_{n+1}, x_{n+1}) + \sum_{i=1}^{k} \bar{\beta}^c_{k-i} f(t_{n+1-i}, x_{n+1-i}) \right) \tag{4.1.6}$$

with $\bar{\beta}^c_i := \beta^c_i - \beta^P_i$.

This results in the Adams Predictor-Corrector scheme:

Predict (P)
$$x^P_{n+1} = x_n + h_n \sum_{i=1}^{k} \beta^P_{k-i} f(t_{n+1-i}, x_{n+1-i})$$
Evaluate (E)
$$\dot{x}_{n+1} = f(t_{n+1}, x^P_{n+1})$$
Correct (C)                                                                       (4.1.7)
$$x_{n+1} = x^P_{n+1} + h_n \left( \beta^c_k \dot{x}_{n+1} + \sum_{i=1}^{k} \bar{\beta}^c_{k-i} f(t_{n+1-i}, x_{n+1-i}) \right)$$
Evaluate (E)
$$\dot{x}_{n+1} = f(t_{n+1}, x_{n+1}).$$

This scheme is symbolized by the abbreviation $PECE$.
Frequently, the corrector is iterated in the following way:

$$x^{(i+1)}_{n+1} = x^P_{n+1} + h_n \left( \beta^c_k \dot{x}^{(i)}_{n+1} + \sum_{i=1}^{k} \bar{\beta}^c_{k-i} f(t_{n+1-i}, x_{n+1-i}) \right) \quad i = 0, \dots, m-1$$

(4.1.8)

with $\dot{x}^{(i)}_{n+1} := f(t_{n+1}, x^{(i)}_{n+1})$ and $x^{(0)}_{n+1} := x^P_{n+1}$. This implementation is symbolized by $P(EC)^m E$. It consists of $m$ steps of a fixed point iteration. The integration step is completed by assigning $x_{n+1} := x^{(m)}_{n+1}$.
Though the scheme (4.1.5) is an implicit one, the overall method is explicit if there is only a fixed number of corrector iteration steps taken, e.g. $m = 1$.
Alternatively, this iteration can be carried out "until convergence", i.e. the iteration is controlled and $m$ is kept variable. This approach will be discussed in more detail in Sec. 4.1.6.
If more than one iteration step is taken, some implementations refrain from performing the last evaluation and predict by

$$x^P_{n+1} = x_n + h_n \sum_{i=1}^{k} \beta^P_{k-i} f(t_{n+1-i}, x^{(m-1)}_{n+1-i}).$$

This is symbolized by $P(EC)^m$.
These versions differ with respect to their stability properties. In the sequel we will assume always corrector values obtained from iterating until convergence unless otherwise stated. For the properties of the other implementations we refer to [Lam91].

## 4.1.2   Backward Differentiation Formulas (BDF)

There is another important class of multistep methods, which is based on interpolating the solution points $x_{n+1-i}$ rather than the derivatives

$$\dot{x}_{n+1-i} := f(t_{n+1-i}, x_{n+1-i}).$$

Let $\pi_k^P$ be a polynomial of degree $(k-1)$, which interpolates the $k$ points

$$x_{n+1-i}, \quad i = 1, \ldots, k.$$

Again, using the Lagrange formulation it can be expressed by

$$\pi_k^P(t) = \sum_{i=1}^k l_i^k(t) x_{n+1-i}.$$

By just extrapolating this polynomial, a new solution point can be predicted as

$$x_{n+1}^P = \pi_k^P(t_{n+1}) = \sum_{i=1}^k l_i^k(t_{n+1}) x_{n+1-i}.$$

Introducing the coefficients $\alpha_{k-i}^P := -l_i^k(t_{n+1})$ the predictor equation

$$x_{n+1}^P = -\sum_{i=1}^k \alpha_{k-i}^P x_{n+1-i}$$

is obtained.

In this formula no information about the function $f$ is incorporated. It is only useful as a predictor in a predictor–corrector scheme.

The BDF corrector formula is obtained by considering the $k^{\text{th}}$ degree polynomial $\pi_{k+1}^C$ which satisfies the conditions:

$$\pi_{k+1}^C(t_{n+1-i}) = x_{n+1-i}, \quad i = 0, \ldots, k \qquad (4.1.9a)$$
$$\dot{\pi}_{k+1}^C(t_{n+1}) = f(t_{n+1}, x_{n+1}). \qquad (4.1.9b)$$

The first conditions are interpolation conditions using the unknown value $x_{n+1}$, which is defined implicitly by considering (4.1.9b).

With the coefficients

$$\alpha_{k-i}^C := \frac{\dot{l}_i^{k+1}(t_{n+1})}{\dot{l}_0^{k+1}(t_{n+1})}, \qquad \beta_k^C := \frac{1}{h_n \dot{l}_0^{k+1}(t_{n+1})}$$

equation (4.1.9b) can be expressed by

$$x_{n+1} = -\sum_{i=1}^k \alpha_{k-i}^C x_{n+1-i} + h_n \beta_k^C f(t_{n+1}, x_{n+1}), \qquad (4.1.10)$$

where the $l_i^{k+1}$ now correspond to the interpolation points $x_{n+1-i}$, $i = 0, \ldots, k$. This is the corrector scheme of the *backward differentiation formula*. We will see later that this method is of particular interest for stiff problems (see Sec. 4.4).

The predictor-corrector scheme for a BDF method has the form:

Predict (P)
$$x_{n+1}^P = -\sum_{i=1}^k \alpha_{k-i}^P x_{n+1-i}$$
Evaluate (E)
$$\dot{x}_{n+1} = f(t_{n+1}, x_{n+1}^P)$$
Correct (C)                                                            (4.1.11)
$$x_{n+1} = x_{n+1}^P - \sum_{i=1}^k \bar{\alpha}_{k-i}^c x_{n+1-i} + h_n \beta_k^c \dot{x}_{n+1}$$
Evaluate (E)
$$\dot{x}_{n+1} = f(t_{n+1}, x_{n+1})$$

with $\bar{\alpha}^c := \alpha^c - \alpha^P$. Again, the implicit formula can be solved iteratively by applying one of the schemes $P(EC)^m E$, $P(EC)^m$ with $m \geq 1$, though, in practice, BDF methods are mainly implemented together with Newton's method. We will see the reason for this later, when discussing stiff ODEs.

### 4.1.3 General Form of Multistep Methods

The general form of a linear multistep method reads

$$\sum_{i=0}^k \alpha_{k-i} x_{n+1-i} - h_n \sum_{i=0}^k \beta_{k-i} f(t_{n+1-i}, x_{n+1-i}) = 0. \qquad (4.1.12)$$

We associate with a $k$-step method a pair $(\rho, \sigma)$ of *generating polynomials* of order $k$ by setting

$$\rho(x) = \sum_{i=0}^k \alpha_{k-i} x^{k-i} \text{ and } \sigma(x) = \sum_{i=0}^k \beta_{k-i} x^{k-i}.$$

We will discuss later the stability of a multistep method in terms of roots of these polynomials.
We will also use $\rho$ and $\sigma$ as a compact notation for the sums in (4.1.12)

$$\rho x_n = \sum_{i=0}^k \alpha_{k-i} x_{n-i} \text{ and } \sigma x_n = \sum_{i=0}^k \beta_{k-i} x_{n-i}. \qquad (4.1.13)$$

Then, the multistep method takes the form

$$\rho x_n - h_n \sigma f(t_n, x_n) = 0.$$

Though this notation originates from the theory of difference operators, we refrain from formally introducing difference operators in this book.
It will be clear from the context if $\rho$ and $\sigma$ denote polynomials or are used in the sense of convention (4.1.13).

### 4.1.4  Accuracy of a Multistep Method

For starting a multistep method $k$ starting values $x_0, \dots, x_{k-1}$ are required. We will assume for the moment that these are values of one solution trajectory of the ODE, i.e.

$$x_i := x(t_i) \qquad \text{with} \qquad i = 0, \dots, k-1.$$

Values generated by a multistep method starting from these exact values can no longer be expected to be on this solution trajectory. They are erroneous due to the approximating nature of the integration scheme, due to round-off errors, and finally due to an *error transportation property* of the difference scheme which might differ significantly from that of the original differential equation. With $h$ and $k$ as tuning parameters one might attempt to reduce the approximation error and possibly influence the error propagation properties of the scheme. The way a small change of $h$ influences the approximation error defines the *order of the method* while the way the error is propagated defines its *stability*.

The quantity of interest is the *global error* of the method at a given time point $t_n$

$$e_n := x(t_n) - x_n,$$

with $n = t_n/h$.

If for exact starting values $e_n = \mathcal{O}(h)$, then the method is said to be *convergent*. More precisely, a method is *convergent of order* $p$, if $e_n = \mathcal{O}(h^p)$.

We first introduce the so-called *local residual* of the method:

**Definition 4.1.3** *Let $x$ be a differentiable function, then the quantity*

$$l(x, t_n, h) := \sum_{i=0}^{k} \alpha_{k-i} x(t_n - (i-1)h) - h \sum_{i=0}^{k} \beta_{k-i} \dot{x}(t_n - (i-1)h) \qquad (4.1.14)$$

*is called the* local residual *of the method.*

Sometimes one considers the *local residual per unit step* $l(x, t_n, h)/h$.

By forming the difference between (4.1.14) and (4.1.12) and by expanding $f$ we obtain a recursion formula for the global error:

$$\rho(e_{n+1}) = h\sigma(f_{x,n+1} e_{n+1}) + l(x, t_n, h)$$

with $f_{x,i} := \left( \frac{\partial f_j}{\partial x} (t_i, \xi_j) \right)_{j=1, \dots, n_x}$ and $\xi_j$ being appropriate values on the line between $x(t_i)$ and $x_i$.

By introducing the vector

$$E_n := \begin{pmatrix} e_n \\ e_{n-1} \\ \vdots \\ e_{n-k+1} \end{pmatrix} \in \mathbb{R}^{kn_x}$$

this recursion formula can be written in one-step form as

$$E_{n+1} = \Phi_n(h)E_n + M_n \qquad (4.1.15)$$

with

$$\Phi_n(h) := \begin{pmatrix} -A_k^{-1}A_{k-1} & -A_k^{-1}A_{k-2} & \cdots & -A_k^{-1}A_1 & -A_k^{-1}A_0 \\ I & 0 & \cdots & 0 & 0 \\ 0 & I & \cdots & 0 & 0 \\ \vdots & & \ddots & & \vdots \\ 0 & 0 & \cdots & I & 0 \end{pmatrix}, \cdot$$

$A_i := (\alpha_i I - h\beta_i f_{x,n+1-k+i})$ and

$$M_n := \begin{pmatrix} -A_k^{-1}l(x,t_n,h) \\ 0 \\ \vdots \\ 0 \end{pmatrix}.$$

From this formula we see how the global error of a multistep method is built up. There is in every step a (local) contribution $M_n$, which is of the size of the *local residual*. Therefore, a main task is to control the integration in such a way that this contribution is kept small. The effect of these local residuals on the global error is influenced by $\Phi_n(h)$. The local effects can be damped or amplified depending on the properties of the propagation matrix $\Phi_n(h)$. This leads to the discussion of the *stability properties* of the method and its relation to the stability of the problem.

**The Local Residual and the Global Error Increment**
First we will show how the local residual is connected to the step size.

**Definition 4.1.4** *A multistep method is called* consistent of order $p$, *if for all functions* $x \in \mathbb{C}^{p+1}[t_0, t_e]$

$$l(x,t,h) = \mathcal{O}(h^{p+1}) \quad \forall t \in [t_0, t_e].$$

Thus, by halving the step size the local residual is reduced by a factor $2^{p+1}$ if the method has the order of consistency $p$. Consequently, one is interested in methods of high order, as in these cases one expects a gain of (local) accuracy when decreasing step sizes.
The consistency of a method can easily be checked by expanding $l(x,t,h)$ in a Taylor series about $t$.

**Example 4.1.5** *The local residual of the two-step implicit Adams method is defined by*

$$l(x,t,h) = x(t+h) - x(t) - h\left(\frac{5}{12}\dot{x}(t+h) + \frac{8}{12}\dot{x}(t) - \frac{1}{12}\dot{x}(t-h)\right).$$

*Taylor expansion leads to*

$$
\begin{aligned}
l(x,t,h) \;=\;& h\dot{x}(t) + \frac{1}{2}h^2\ddot{x}(t) + \frac{1}{6}h^3 x^{(3)}(t) + \frac{1}{24}h^4 x^{(4)}(t) + \dots \\
& -\frac{5}{12}h\dot{x}(t) - \frac{5}{12}h^2\ddot{x}(t) - \frac{5}{24}h^3 x^{(3)}(t) - \frac{5}{72}h^4 x^{(4)}(t) + \dots \\
& -\frac{8}{12}h\dot{x}(t) \\
& +\frac{1}{12}h\dot{x}(t) - \frac{1}{12}h^2\ddot{x}(t) + \frac{1}{24}h^3 x^{(3)}(t) - \frac{1}{72}h^4 x^{(4)}(t) + \dots \\
=\;& -\frac{1}{24}h^4 x^{(4)}(t) + \dots
\end{aligned}
$$

*Thus the implicit two step Adams method has the order of consistency 3.*

This example shows that for a method to be consistent of order at least one, the coefficients $\alpha_i$ must sum up to zero, i.e. $\rho(1) = 0$.
Conditions for higher order of consistency are given by the following theorem:

**Theorem 4.1.6** *A linear multistep method has the order of consistency p if the following $p + 1$ conditions on its coefficients are met:*

$$
\sum_{i=0}^{k} \alpha_i \;=\; 0 \tag{4.1.16a}
$$

$$
\sum_{i=0}^{k} i\alpha_i - \beta_i \;=\; 0 \tag{4.1.16b}
$$

$$
\sum_{i=0}^{k} \frac{1}{j!} i^j \alpha_i - \frac{1}{(j-1)!} i^{j-1}\beta_i \;=\; 0 \text{ with } j = 2,\dots,p. \tag{4.1.16c}
$$

*The local residual of a method with order of consistency p takes the form*

$$
l(x,t,h) = c_{p+1} h^{p+1} x^{(p+1)}(t) + \mathcal{O}(h^{p+2}). \tag{4.1.17}
$$

(The proof is based on Taylor expansion, analogously to the above example.)

**Corollary 4.1.7** *Adams–Bashforth methods have order of consistency k, Adams–Moulton methods have order of consistency $k + 1$, and BDF methods have order of consistency k with k defined by (4.1.12).*

Two methods of the same order differ by their *error constant*, which is defined to be the constant $c_{p+1}$ of the first non vanishing term in the Taylor expansion of $l(x(t),h)/\sigma(1)$. (The normalization by $\sum \beta_i = \sigma(1)$ is performed to get a constant which is independent of the scaling of the method.) For the implicit two-step Adams method, the error constant is $-\frac{1}{24}$ (see the example above) and for the three-step BDF method, which has the same order of consistency, the error constant is $-\frac{1}{4}$. Thus, for obtaining the same local accuracy a three-step BDF method would require a step size which is smaller by a factor of $1.57 = 6^{1/4}$.

The error constants of the different methods are displayed in Fig. 4.1.

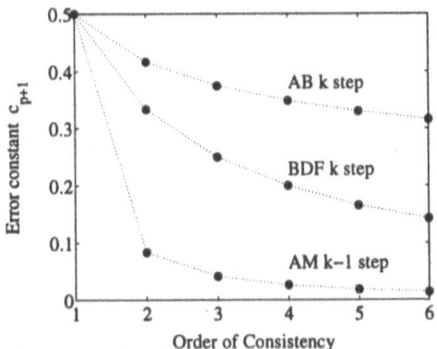

Figure 4.1: Absolute value of error constants for different multistep methods. (For $k = 1$ the Adams–Moulton method is the implicit Euler method.)

**Definition 4.1.8** *Let $\bar{x}_n$ be the numerical solution after one step initialized with exact values, i.e.*

$$
\begin{aligned}
0 = {} & \alpha_k \bar{x}_n - h\beta_k f(t_n, \bar{x}_n) + \\
& \sum_{i=1}^{k} \alpha_{k-i} x(t_{n-i}) - h \sum_{i=1}^{k} \beta_{k-i} f(t_{n-i}, x(t_{n-i})),
\end{aligned}
$$

*then*

$$
\varepsilon_n := x(t_n) - \bar{x}_n
$$

*is called the* global error increment *of the multistep method, [Söd92].*

The name of this quantity is motivated by viewing the definition of $M_n$ in (4.1.15) and

$$
\varepsilon_n = A_k^{-1} l(x, t_n, h) + \quad \text{higher order terms.} \tag{4.1.18}
$$

Both, the global error increment and the local residual, have the same asymptotic behavior for $h$ tending to zero but with a different magnitude for finite $h$. We will see later that for so-called stiff ODEs the difference in this constant can be of importance.

## Stability

Let us assume for the moment that the driving term $M_n$ in (4.1.15) is zero. Then we would like that the sequence $E_n$ started with $E_0$ remains bounded, when $n$ tends to infinity: an initial error should not blow up. This leads to the *stability requirement*

$$\left\| \prod_{i=0}^{n} \Phi_i(h) \right\| < C$$

with a constant $C$, which is independent of $n := (t_e - t_0)/h$.

It depends on the actual problem $f$ and the given step size $h$ if this condition is fulfilled.

Therefore, we will first study this stability requirement for a special class of problems, the *linear test equation*,

$$\dot{x} = \lambda x, \qquad (4.1.19)$$

where $\lambda$ may admit complex values, reflecting the fact that the linear test equation might be viewed as one of the resulting equations after a modal transformation of an arbitrary system of linear differential equations. In this case $\Phi_n(h)$ does not depend on time and the individual linearization points, i.e. $\Phi_i(h) = \Phi(h)$ and $A_i = (\alpha_i - h\beta_i\lambda)$.

The stability requirement then simplifies to

$$\|\Phi(h)^n\| < C$$

with $C$ being independent of $n$.

A sufficient condition for this is that the eigenvalues of $\Phi(h)$ are located in the unit circle of the complex plane, with those on its boundary being non defective. The eigenvalues of $\Phi(h)$ are just the roots of the polynomial

$$\rho(x) - h\lambda\sigma(x) = 0, \qquad (4.1.20)$$

as $\Phi(h)$ is the *companion matrix* of that polynomial.

If for a given value $z = h\lambda$ the roots of this polynomial are inside the unit circle and those on the unit circle are simple, then the method is stable for this particular step size $h$ and the given problem $\lambda$.

In Fig. 4.2 for different multistep methods regions of the complex plane are displayed in which $z = h\lambda$ leads to a stable discretization. Given a method and $\lambda$, then the step size $h$ has to be restricted in such a way, that $z = h\lambda$ lies in the corresponding *stability region*.

It is of particular interest to study these restrictions of the step size for stable problems, i.e. $\mathrm{Re}\lambda < 0$. One sees from Fig. 4.2 that the trapezoidal rule (AM-1) gives for all linear, stable problems a stable discretization independent of the

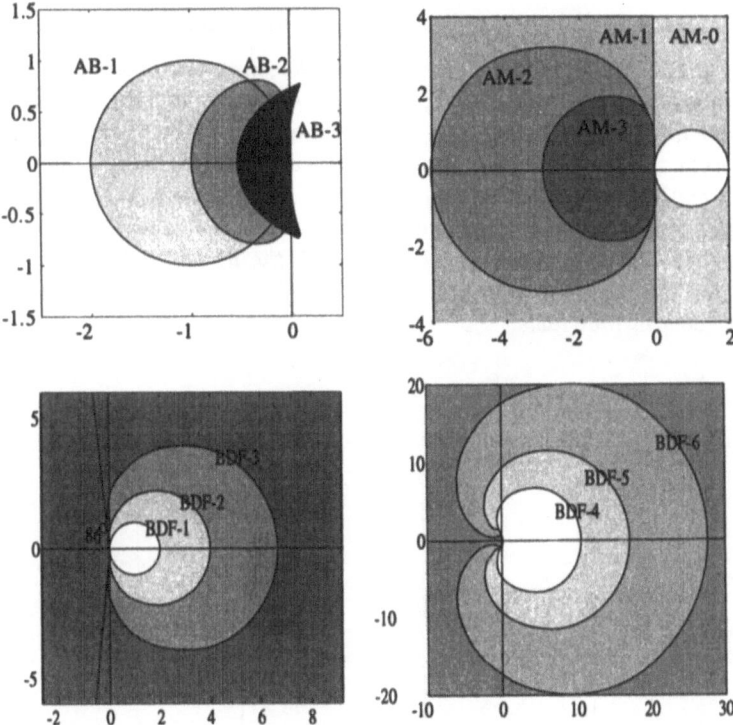

Figure 4.2: Stability regions for some multistep methods . The methods are stable outside the white or light gray areas, resp. I.e. BDF-1 is stable in the entire plane except in the white circle, BDF-5 is unstable in the light gray and white circle etc.

selected step size. This property is called *A-stability*. The implicit Euler method (BDF-1, AM-0) is also A-stable, but in contrast to the trapezoidal rule it might lead to stable discretizations also for unstable problems, if the step size is large enough.

BDF methods are stable for all $\lambda$ with $\mathrm{Re}\lambda < 0$ and $|\arg(-\lambda)| < \alpha$ with $\alpha$ given in Tab. 4.1.

| BDF-$k$ | 1 | 2 | 3 | 4 | 5 | 6 |
|---|---|---|---|---|---|---|
| $\alpha$ in degrees | 90 | 90 | 86 | 73 | 51 | 17 |

Table 4.1: Angle of A($\alpha$)-stability of BDF methods

This property often is called A($\alpha$)-stability. The effect of a discretization with a too large step size is demonstrated by the following example.

**Example 4.1.9** *We investigate the model of the unconstrained truck in its stable equilibrium position perturbed by $10^{-8}$ in $p_1$. Applying the explicit Euler method to the problem shows for step sizes larger than $h_0 = 0.0045$ clearly an unstable behavior while for smaller step sizes the numerical solution is stable, Fig. 4.3. This effect can also be confirmed by considering the eigenvalues $\lambda_i$ of the linearized truck model and noting that the stability region of the explicit Euler method is $S(h) := \{z| \ |1 + z| < 1, z = h\lambda\}$. In Tab. 4.2 the largest values of $|1 + h\lambda|$ are displayed for the different step sizes. The observations from the simulation are clearly reflected in that table.*

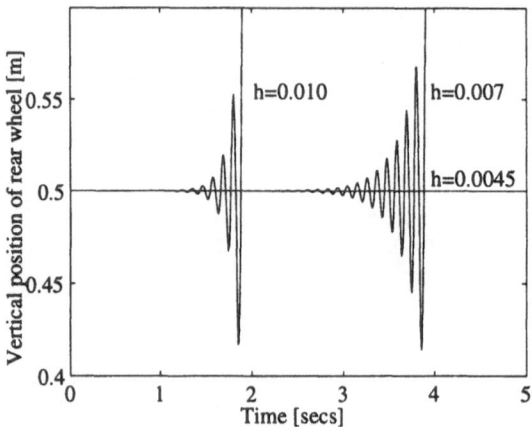

Figure 4.3: Stability behavior of the explicit Euler method when applied to the unconstrained truck with different step sizes

| $\lambda$ | h=0.01 | h=0.007 | h=0.0045 |
|---|---|---|---|
| -7.7040 ± 57.9308 i | 1.0897 | 1.0293 | 0.9999 |

Table 4.2: $|1 + h\lambda|$ for the linearized truck model in Ex. 4.1.9

An important property of the stability regions is that the origin is located at the stability boundary. This guarantees that an unstable discretization becomes more and more stable by decreasing the step size.

**Definition 4.1.10** *If there is a constant $C_0 > 0$ such that for all $n$*

$$\|\Phi(0)^n\| < C_0 ,$$                                          (4.1.21)

*then the multistep method is called* zero stable.

By setting in (4.1.20) $\lambda = 0$, one sees that the roots of the $\rho$-polynomial determine zero stability:
If the roots of $\rho(x) = 0$ lie inside the complex unit circle and those on its boundary are simple, then the method is stable.
As for Adams methods $\rho(x) = x^k - x^{k-1}$ all Adams methods are zero stable.
Evidently, a method not sharing this property cannot be used even for solving the differential equation $\dot{x} = 0$. An example of a non zero stable discretization is BDF-7. Indeed, no BDF method with an order larger than six is zero stable [HNW87].

**Example 4.1.11** *Finally, we demonstrate zero stability of the two step Adams–Bashforth method (AB-2) by discretizing the unconstrained truck. In Fig. 4.4 the seven largest eigenvalues of $\Phi(h)$ are traced as a function of the step size.*

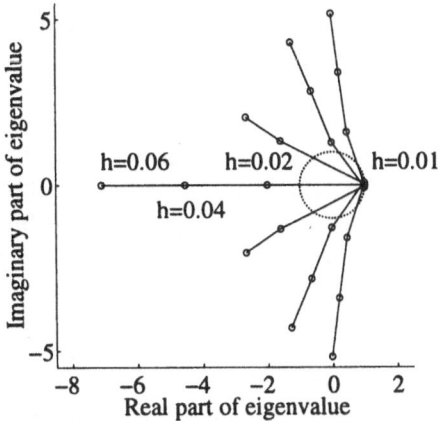

Figure 4.4: The seven largest eigenvalues of $\Phi(h)$ in dependency of $h$ for the two step Adams–Bashforth method applied to the unconstrained truck in nominal configuration.

**Convergence**
A consistent method guarantees that by decreasing the step size the local contribution to the global error, i.e. the global error increment, decreases also. If this process would lead to an unstable discretization, then the global error might grow due to instability regardless how small the local contributions are. Thus, in order to achieve a convergent method, at least consistency and zero stability have to be required.
The following theorem says that this requirement is also sufficient:

**Theorem 4.1.12** *A linear multistep method (4.1.12) with constant step size $h$, which is consistent of order $p$ and zero stable, is convergent of order $p$.*

## 4.1.5   Order and Step Size Selection

We saw that the error in the numerical solution depends on the problem $f$ and on the step size $h$. The step size influences both the dynamic behavior of the error equation and the size of the global error increment (4.1.15).

In order to control the global error a step size is selected in such a way that the global error increment is approximately of the size of a given tolerance $TOL$. Meeting such a tolerance bound without overshooting it or being over accurate is the aim of a step size control mechanism.

This requires in control theoretic terms

- a good model of the functional relationship between step size (control input) and global error increment (system response),

- a good estimator for the global error increment (observer) and

- a step size control strategy (control law).

An illustrative example for the tasks of error control has been given in [Hau89]:

*The process of changing the step size and order may be related to the example of driving a vehicle backward, looking at the road previously traversed. This analogy is meaningful, since numerical integration ahead in time in fact uses past information. Along a straight stretch of road the driver may move with large velocity, traversing long patches of road in unit time. This is analogous to using a large numerical step size $h$. At the same time, the driver is relatively relaxed and is viewing a long patch of the road over which he has come. This is analogous to using many previously passed data points. If the road changes direction abruptly the prudent driver will take two actions. First, the driver will slow down, which is analogous to reducing the step size in the integration process. Second, the driver will concentrate his or her attention on road curvature near the vehicle.*

This example also shows the problems, when running a simulation with a fixed step size code. (Considering an obstacle on the road this example also demonstrates the problems when integrating a non smooth problem, see Ch. 6).

### Estimating the Global Error Increment

For estimating the global error increment $\varepsilon_n$ at step $n$, this step is made by two different methods, started with the same data. If we assume that this data is taken to be points of a solution trajectory, a comparison of these two numerical results can give an estimate of the global error increment.

First, we will consider two methods of order $p$, distinguished in notation by superscripts (1) and (2). By Definition 4.1.8 and Eqs. (4.1.18), (4.1.17) we get

$$\varepsilon_n^{(1)} - \varepsilon_n^{(2)} = (A_k^{(1)}{}^{-1} c_{p+1}^{(1)} - A_k^{(2)}{}^{-1} c_{p+1}^{(2)}) h^{p+1} x^{(p+1)}(t_n) + \mathcal{O}(h^{p+2}). \qquad (4.1.22)$$

Consequently,

$$h^{p+1}x^{(p+1)}(t_n) = \left(A_k^{(1)-1}c_{p+1}^{(1)} - A_k^{(2)-1}c_{p+1}^{(2)}\right)^{-1}\left(\bar{x}_n^{(2)} - \bar{x}_n^{(1)}\right) + \mathcal{O}(h^{p+2}).$$

Inserting this in (4.1.17) and (4.1.18) results in

$$\varepsilon_n^{(2)} = A_k^{(2)-1}c_{p+1}^{(2)}\left(A_k^{(1)-1}c_{p+1}^{(1)} - A_k^{(2)-1}c_{p+1}^{(2)}\right)^{-1}\left(\bar{x}_n^{(2)} - \bar{x}_n^{(1)}\right) + \mathcal{O}(h^{p+2}).$$
(4.1.23)

Let the method denoted by the superscript (1) be the explicit predictor method, then $A_k^{(1)} = I$. We obtain by using the matrix relation $A^{-1}B^{-1} = (BA)^{-1}$

$$\varepsilon_n^{(2)} = \left(\frac{c_{p+1}^{(1)}}{c_{p+1}^{(2)}}A_k^{(2)} - I\right)^{-1}\left(\bar{x}_n^{(2)} - \bar{x}_n^{(1)}\right) + \mathcal{O}(h^{p+2})$$

$$= \left(\frac{c_{p+1}^{(1)}}{c_{p+1}^{(2)}}\left(I - h\beta_k^{(2)}J\right) - I\right)^{-1}\left(\bar{x}_n^{(2)} - \bar{x}_n^{(1)}\right) + \mathcal{O}(h^{p+2})$$

$$= \frac{c_{p+1}^{(2)}}{(c_{p+1}^{(1)} - c_{p+1}^{(2)})}A_k^{(1,2)-1}\left(\bar{x}_n^{(2)} - \bar{x}_n^{(1)}\right) + \mathcal{O}(h^{p+2}) \qquad (4.1.24)$$

with $J := f_x(t_n, \xi)$, cf. (4.1.18) and

$$A_k^{(2)} := \left(I - h\beta_k^{(2)}\frac{c_{p+1}^{(1)} - c_{p+1}^{(2)}}{c_{p+1}^{(1)}}J\right).$$

For $h$ small compared to the largest eigenvalue of the Jacobian $J$ we can also set approximately $A_k^{(2)} \approx I$ and get a simpler error estimator:

$$\varepsilon_n^{(2)} = \frac{c_{p+1}^{(2)}}{(c_{p+1}^{(1)} - c_{p+1}^{(2)})}\left(\bar{x}_n^{(2)} - \bar{x}_n^{(1)}\right) + \mathcal{O}(h^{p+2}). \qquad (4.1.25)$$

If method (2) is of order $p+1$ while method (1) is of order $p$, the above analysis must be modified. One obtains instead of (4.1.22)

$$\varepsilon_n^{(1)} - \varepsilon_n^{(2)} = \bar{x}_n^{(2)} - \bar{x}_n^{(1)} = A_k^{(1)-1}c_{p+1}^{(1)}h^{p+1}x^{(p+1)}(t_n) + \mathcal{O}(h^{p+2}).$$

and consequently if method (1) is the explicit predictor method

$$\varepsilon_n^{(1)} = \bar{x}_n^{(2)} - \bar{x}_n^{(1)} + \mathcal{O}(h^{p+2}). \qquad (4.1.26)$$

In this case only the error of the lower order method can be estimated. Nevertheless, this quantity is frequently used to control the step size of the higher order method (2). This technique is called *local extrapolation.*

In practice the values $\bar{x}_n$ cannot be computed. When estimating the error they are replaced by $x(t_n)$, so that the error estimation is based on the *predictor - corrector difference.*

To check the success of a step the estimated error is checked against a given tolerance

$$|\varepsilon_{ni}| \leq TOL_i \quad i = 1, \ldots, n_x. \tag{4.1.27}$$

The tolerance is usually a combination of a relative tolerance bound $RTOL$ and an absolute tolerance bound $ATOL$ by setting:

$$TOL_i := RTOL|x_i(t)| + ATOL.$$

If the components of the solution $x$ differ very much in magnitude $ATOL$ is given as a vector.

If the error criterion (4.1.27) is not met, the step is rejected and will be recomputed with a smaller step size. If on the other hand this criterion is fulfilled, it will be checked if it would have been fulfilled even with a larger step size. If so, and if the gain is significant, the larger step size is taken for the next step.

The step size $\bar{h}_n$ corresponding to $TOL$ is computed from the actual step size $h_n$ and the error estimate by

$$\bar{h}_n = \left(\frac{TOL}{\|\varepsilon_n\|}\right)^{\frac{1}{p+1}} h_n.$$

The result is often multiplied with an additional safety factor.

**Example 4.1.13** *In order to illustrate the estimates of the global error increment, the unconstrained truck is simulated running over a 12 cm high road hump described by the input function $u(t) = \frac{1}{100}(t - t_0)^6 e^{-(t-t_0)}$ and its derivative (see Sec. 1.3.1). The simulation is performed by using the explicit Euler method as predictor and the implicit Euler method as corrector (with fixed point iteration iterated "until convergence", see Sec. 4.1.1). In Fig. 4.5 the results for the rotation of the chassis obtained with $h = 10^{-2}$ are plotted. The thickness of the band around this solution curve corresponds to the estimated global error increment magnified by $5 \cdot 10^3$. This figure reflects clearly the influence of the second derivative of the solution on the size of the error.*

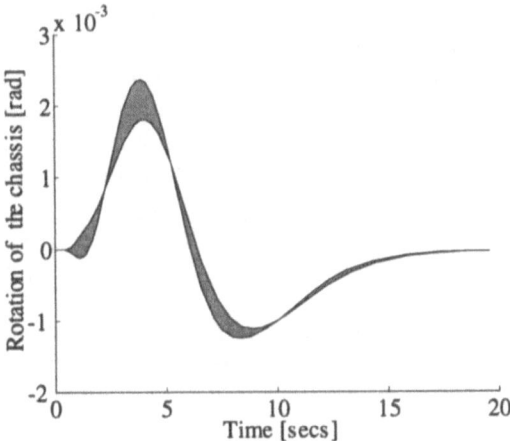

Figure 4.5: Simulation result of the truck running over an obstacle together with the estimated global error increment (magnified by $5 \cdot 10^3$).

**Stiffness**

The aim of step size control is to select a step size in such a way, that the global error increment is bounded by some given tolerance. This limits the step size. It might happen, that for a given problem the step size is limited additionally by a stability bound like the bound demonstrated in Example 4.1.9. If this restricts the step size more than the tolerance bound, the problem is called *stiff*. In connection with mechanical systems the term stiffness might be misleading as not only the stiffness of the mechanical system but also its damping might cause these numerical effects. Using a method which has for a given problem a step size restriction due to stability might result in a very inefficient computation. This will be discussed in a Sec. 4.4.

**Order Selection and Starting a Multistep Method**

The local residual of a $p^{\text{th}}$ order method depends on the $p + 1^{\text{st}}$ derivative of the solution in the actual time interval $[t_n, t_{n+1}]$. If

$$\|c_p h^p x^{(p)}\| < \|c_{p+1} h^{p+1} x^{(p+1)}\|$$

it might be advantageous to take a method of order $p - 1$ instead. Similarly, one might consider to raise the order if the higher derivative is smaller in that sense. Adams and BDF methods allow to vary the order in an easy way. They are defined in terms of interpolation polynomials based on a certain number of points. Raising or lowering the number of interpolation points raises or lowers the order of the interpolation polynomial and the order of the method. After every successful step it is considered how many points of the past should be taken to perform the next step.

As mentioned earlier in this chapter, interpolation polynomials can be defined in different ways. Especially the definition based on finite differences, like in Newton's interpolation method, permits in an efficient way to vary the order of an interpolation polynomial by adding or taking away additional interpolation points. The Adams multistep code DE/STEP by Shampine and Gordon [SG75] and the BDF code DASSL by Petzold [BCP89] are based on a modification of Newton interpolation polynomials.

In order to get an idea of the size of $x^{(p+1)}$, the $p^{\text{th}}$ resp. $p + 1^{\text{st}}$ derivative of the $k^{\text{th}}$ order interpolation polynomial is taken. This quantity is accessible in a cheap way in these codes.

The automatic order variation is also used for starting multistep methods:

The starting points are successively obtained by starting with a one step method, then proceeding with a two step method and so on until an order is reached which is appropriate for the given problem.

Starting a multistep method in this way can be costly. If frequent restarts are necessary, e.g. when treating problems with discontinuities (see Ch. 6), a starting procedure based on a higher order one step method might be advantageous [SB95].

## 4.1.6　Solving the Corrector Equations

In this section we discuss the corrector iteration. The corrector iteration described so far has the general form

$$x_{n+1}^{(i+1)} = \xi + h_n \beta_k^c f(t_{n+1}, x_{n+1}^{(i)}) \tag{4.1.28}$$

with $\xi$ being the contributions based on "old" data, which is

$$\xi = x_{n+1}^p + h_n \sum_{i=1}^{k} \bar{\beta}_{k-i}^c f(t_{n+1-i}, x_{n+1-i})$$

in the case of Adams–Moulton methods, see Eq. (4.1.8), and

$$\xi = \sum_{i=1}^{k} \bar{\alpha}_{k-i}^c x_{n+1-i}$$

in the case of implicit BDF methods, cf. (4.1.11). Eq. (4.1.28) describes a fixed point iteration, which was introduced in Ch. 3.

By Theorem 3.3.2 a necessary condition for the convergence of this iteration is that the corresponding mapping

$$\varphi(x) = \xi + h_n \beta_k^c f(t, x)$$

is a contraction, cf. Def. 3.3.1. As the Lipschitz constants of $\varphi$ and $f$ are related by

$$L(\varphi) = h_n \beta_k^c L(f),$$

it is obvious that $\varphi$ is contractive and the iteration convergent if the step size is sufficiently small. In many cases $L(f)$ is of moderate size such that the step size for the required local accuracy is small enough to ensure fast convergence of the fixed point iteration.

On the other hand, if the Jacobian $\frac{d}{dx}f(t,x)$ has eigenvalues being large in modulus, the step size might be restricted much more by the demand for contractivity than by the required tolerance and stability. This situation is to be expected when dealing with stiff systems. In this case it is appropriate to switch from fixed point iteration to Newton iteration for solving the implicit corrector equation.

When applying Newton's method, the nonlinear equation

$$F(x_{n+1}) = x_{n+1} - (\xi + h_n\beta_k^c f(t_{n+1}, x_{n+1})) = 0 \qquad (4.1.29)$$

is considered. Newton's method then defines the iteration

$$J(t_{n+1}, x_{n+1}^{(i)})\,\Delta x^{(i)} = -\left(x_{n+1}^{(i)} - (\xi + h_n\beta_k^c f(t_{n+1}, x_{n+1}^{(i)}))\right) \qquad (4.1.30)$$

with $x_{n+1}^{(i+1)} := x_{n+1}^{(i)} + \Delta x^{(i)}$ and

$$J(t, x) := I - h\beta_k^c \frac{d}{dx}f(t, x).$$

Like in the case of fixed point iteration the predictor solution is taken as starting value: $x_{n+1}^{(0)} := x_{n+1}^p$. The method demands an high computational effort, which is mostly spent when computing the Jacobian $J$ and solving the linear system.

**Controlling the Update of the Jacobian**

For the solution of Eq. (4.1.30) the matrix $J(t_{n+1}, x_{n+1}^{(i)})$ has to be decomposed. This is in general done using an $LU$ decomposition. There are cases where the effort required for this task can be reduced by exploiting the structure of the system before decomposing it. For many applications the costs spent in computing the Jacobian and solving the linear system dominates the integration effort. Since normally the Jacobian changes very little with respect to time, computing time can be saved by modifying the iteration so that the Jacobian is updated only after several integration steps:

$$J(t_m, x_m^p)\Delta x^{(i)} = -\left(x_{n+1}^{(i)} - (\xi + \gamma f(t_{n+1}, x_{n+1}^{(i)}))\right)$$

with $t_m$ being the time when $J$ was updated last and $\gamma := h_n\beta_k^c$. As long as $J(t_m, x_m^p)$ is close enough to $J(t_{n+1}, x_{n+1}^p)$, the iteration is expected to converge if the starting value, i.e. the predictor, is sufficiently good.

Different integration programs use different strategies to decide when to update the Jacobian. Though of central importance for the performance of an integration code, these strategies are mainly based on heuristics. When presenting the basics of a control strategy, we only give examples for these heuristics, in order to give an idea of the concept.

To monitor the convergence behavior of this simplified Newton method the rate of convergence can be estimated by

$$\frac{\|\Delta x^{(i+1)}\|}{\|\Delta x^{(i)}\|} =: \tilde{\delta}_i \leq \delta_i$$

with $\delta_i$ defined by Theorem 3.4.1. It is this estimation which is the basis for the actions to be taken next:

- If the convergence rate in one step is larger than an upper limit $\bar{\delta}_0$, e.g. $\bar{\delta}_0 = 0.7$ it is assumed that the iteration fails to converge. The choice of this constant depends on what is cheaper, to evaluate and decompose the Jacobian or to perform some additional iterations. However, for $\tilde{\delta}_i$ near or even greater than one the iteration has to be stopped.

- Otherwise the iteration is continued until the error $\|x^{(i)} - x^*\|$ estimated by the error estimation formulae (3.4.4), (3.4.6) is below a given bound, e.g. TOL/10.

Divergence of the iteration may occur due to too bad predictor values or due to a too old Jacobian $J$.

A predictor value, which is too bad, corresponds to a high nonlinearity in the problem. $\omega$ in Theorem 3.4.1 is too large.

Errors in the Jacobian $J(t, x, h) := I - h\beta_k^c \frac{d}{dx} f(t, x)$ may have several reasons:

- The method coefficients $h\beta_k^c$ changed significantly.

- $\frac{d}{dx} f(t, x)$ changed due to changes in $x$ or $t$.

In the case of non convergence, the following steps can be carried out:

1. The Jacobian is re-decomposed for the actual $h\beta_k^c$.

2. If this action does not lead to convergence, $\frac{d}{dx} f(t, x)$ is recomputed and the entire matrix is decomposed.

3. If again this does not lead to convergence, the step size is decreased in order to adapt the step size to the small radius of convergence.

As for multibody systems normally no information about the Jacobian is available, the Jacobian must be computed numerically by finite differences, see below and Sec. 3.5.

**Exploiting the Structure of Multibody Equations**

Frequently, the equations of motion occur in such a form that the effort for solving the corrector equation (4.1.29) can be reduced significantly. This concerns the solution of the linear systems (4.1.30) and the computation the Jacobian by finite differences. This can be done for all ODEs in first order form originating from higher order ODEs.

To explain the idea, we consider system (1.3.11) with $T(p) = I$. As we discuss implicit ODEs later (see Ch. 5), we will assume for the moment a system without constraints and with the mass matrix being the identity or inverted to the right hand side of the differential equation. For details also concerning implementation of the approach we refer to [Eic91].

The system under consideration has the form

$$\dot{p} = v \tag{4.1.31a}$$
$$\dot{v} = f_a(t, p, v). \tag{4.1.31b}$$

For this system Eq. (4.1.30) becomes

$$\begin{pmatrix} I & -\gamma I \\ -\gamma K^{(i)} & I - \gamma D^{(i)} \end{pmatrix} \begin{pmatrix} \Delta p_{n+1}^{(i+1)} \\ \Delta v_{n+1}^{(i+1)} \end{pmatrix} = - \begin{pmatrix} p_{n+1}^{(i)} - (\xi_p + \gamma v_{n+1}^{(i)}) \\ v_{n+1}^{(i)} - (\xi_v + \gamma f_a(p_{n+1}^{(i)}, v_{n+1}^{(i)})) \end{pmatrix}$$

$$p_{n+1}^{(i+1)} = p_{n+1}^{(i)} + \Delta p_{n+1}^{(i+1)}$$
$$v_{n+1}^{(i+1)} = v_{n+1}^{(i)} + \Delta v_{n+1}^{(i+1)},$$

with $\xi$ partitioned in $\xi = (\xi_p^T, \xi_v^T)^T$ and $K^{(i)} := \frac{\partial}{\partial p} f_a(p_{n+1}^{(i)}, v_{n+1}^{(i)})$ and $D^{(i)} := \frac{\partial}{\partial v} f_a(p_{n+1}^{(i)}, v_{n+1}^{(i)})$ being matrices corresponding to stiffness and damping matrices in the linear case.

The linear system can be reduced by solving for $\Delta p_{n+1}^{(i+1)}$ first. This leads to the system

$$(I - \gamma D^{(i)} - \gamma^2 K^{(i)}) \Delta v_{n+1}^{(i+1)} = -v_{n+1}^{(i)} + \xi_v + \gamma(f_a(p_{n+1}^{(i)}, v_{n+1}^{(i)}) +$$
$$+ \gamma K^{(i)} r_{n+1_p}^{(i)} \tag{4.1.32a}$$
$$v_{n+1}^{(i+1)} = v_{n+1}^{(i)} + \Delta v_{n+1}^{(i+1)} \tag{4.1.32b}$$
$$p_{n+1}^{(i+1)} = \gamma v_{n+1}^{(i+1)} + \xi_p, \tag{4.1.32c}$$

Note that for the residual of the position part

$$r_{n+1_p}^{(i)} := - \left( p_{n+1}^{(i)} - (\xi_p + \gamma v_{n+1}^{(i)}) \right)$$

we have $r_{n+1_p}^{(i)} = 0$ for $i \geq 1$. If we modify the predictor such that $p_{n+1}^{(0)} := \xi_p + \gamma v_{n+1}^{(0)}$, then $r_{n+1_p}^{(i)} = 0$ for all $i$.

Consequently, the linear system (4.1.32a) simplifies to

$$(I - \gamma D^{(i)} - \gamma^2 K^{(i)})\Delta v_{n+1}^{(i+1)} = - \left( v_{n+1}^{(i)} - (\xi_v + \gamma f_a(p_{n+1}^{(i)}, v_{n+1}^{(i)})) \right). \qquad (4.1.33)$$

This system has half the dimension of the original system, so the effort for decomposing is decreased by a factor eight.

Furthermore, the number of function evaluations for computing the Jacobian $f'$ by finite differences can be halved, if the structure is exploited:

The classical way to approximate $f'$ makes use of (see Sec. 3.5)

$$f'(x) = \left( \frac{f(x + e_i\eta) - f(x)}{\eta} \right)_{i=1,\dots,n_x} + \mathcal{O}(\eta). \qquad (4.1.34)$$

Herein $e_i$ is the $i^{\text{th}}$ unit vector and $\eta$ is a scalar increment, which may be different for different $i$. In the context discussed here, this procedure requires $n_p + n_v = 2n_p$ evaluations of $f$.

Using Eq. (4.1.33) permits the use of

$$(I - \gamma D^{(i)} - \gamma^2 K^{(i)}) = \left( \frac{f_a(p + \gamma e_i\eta, v + e_i\eta) - f_a(p, v)}{\eta} \right)_{i=1,\dots,n_v} + \mathcal{O}(\eta)$$
$$(4.1.35)$$

instead. This reduces the number of function evaluations. Note, the order of accuracy is the same as in the unmodified approach.

## 4.2  Explicit Runge–Kutta Methods

Runge–Kutta methods are *one-step methods*, i.e. they have the generic form

$$x_{n+1} := x_n + h\phi_h(t_n, x_n) \qquad (4.2.1)$$

with a method dependent *increment function* $\phi_h$. In contrast to multistep methods, the transition from one step to the next is based on data of the most recent step only.

The basic construction scheme is

$$X_1 = x_n \qquad (4.2.2a)$$

$$X_i = x_n + h\sum_{j=1}^{i-1} a_{ij} f(t_n + c_j h, X_j) \quad i = 2, \dots, s \qquad (4.2.2b)$$

$$x_{n+1} = x_n + h\sum_{i=1}^{s} b_i f(t_n + c_i h, X_i). \qquad (4.2.2c)$$

$s$ is called the number of stages.

**Example 4.2.1** *By taking* $s = 2, a_{21} = 1/2, b_1 = 0, b_2 = 1, c_1 = 0,$ *and* $c_2 = 1/2$
*the following scheme is obtained*

$$X_1 \quad = \quad x_n \tag{4.2.3a}$$

$$X_2 \quad = \quad x_n + \frac{h}{2}f(t_n, X_1) \tag{4.2.3b}$$

$$x_{n+1} \quad = \quad x_n + hf(t_n + \frac{h}{2}, X_2) \tag{4.2.3c}$$

*For this method the increment function reads*

$$\phi_h(t, x) := f\left(t + \frac{h}{2}, x + \frac{h}{2}f(t, x)\right).$$

Normally, Runge–Kutta methods are written in an equivalent form by substituting
$k_i := f(t_n + c_i h, X_i)$

$$k_1 \quad = \quad f(t_n, x_n)$$

$$k_i \quad = \quad f(t_n + c_i h, x_n + h\sum_{j=1}^{i-1} a_{ij} k_j) \qquad i = 2, \dots, s$$

$$x_{n+1} \quad = \quad x_n + h\sum_{i=1}^{s} b_i k_i.$$

The coefficients characterizing a Runge–Kutta method are written in a compact
form using a so-called *Butcher tableau*:

| $c_1$ | | | | |
|---|---|---|---|---|
| $c_2$ | $a_{21}$ | | | |
| $c_3$ | $a_{31}$ | $a_{32}$ | | |
| $\vdots$ | $\vdots$ | $\vdots$ | $\ddots$ | |
| $c_s$ | $a_{s1}$ | $a_{s2}$ | $\cdots$ | $a_{s,s-1}$ |
| | $b_1$ | $b_2$ | $\cdots$ | $b_{s-1}$ $b_s$ |

or

$$\frac{c \mid A}{\phantom{c} \mid b^T}$$

with $A = (a_{ij})$ and $a_{ij} = 0$ for $j \geq i$.
The classical 4-stage Runge–Kutta method reads in this notation

| 0 | | | | |
|---|---|---|---|---|
| $\frac{1}{2}$ | $\frac{1}{2}$ | | | |
| $\frac{1}{2}$ | 0 | $\frac{1}{2}$ | | |
| 1 | 0 | 0 | 1 | |
| | $\frac{1}{6}$ | $\frac{2}{6}$ | $\frac{2}{6}$ | $\frac{1}{6}$ |

$$\tag{4.2.4}$$

An $s$-stage Runge–Kutta method usually requires $s$ function evaluations. If

$$c_s = 1, a_{sj} = b_j \text{ and } b_s = 0 \tag{4.2.5}$$

the function evaluation for the last stage at $t_n$ can be used for the first stage at $t_{n+1}$.

The higher amount of function evaluations per step in Runge–Kutta methods compared to multistep methods is often compensated by the fact that Runge–Kutta methods may be able to use larger step sizes.

A *non autonomous differential equation* $\dot{x} = f(t, x)$ can be written in autonomous form where the right hand side of the differential equation is not explicitly depending on time, by augmenting the system by the trivial equation $\dot{t} = 1$:

$$\dot{y} = \begin{pmatrix} \dot{t} \\ \dot{x} \end{pmatrix} = \begin{pmatrix} 1 \\ f(t, x) \end{pmatrix} = F(y).$$

Applying a Runge–Kutta method to the original and to the reformulated system should lead to the same equations. This requirement relates the coefficients $c_i$ to the $a_{ij}$:

$$c_i = \sum_{j=1}^{s} a_{ij}. \tag{4.2.6}$$

We will consider only methods fulfilling this condition and assuming for the rest of this chapter autonomous differential equations for ease of notation.

## 4.2.1   The Order of a Runge–Kutta Method

The global error of a Runge–Kutta method at $t_n$ is defined in the same way as for multistep methods

$$e_n := x(t_n) - x_n.$$

with $n = t_n/h$. A Runge–Kutta method has order $p$ if $e_n = \mathcal{O}(h^p)$.
Using (4.2.1) we get

$$
\begin{aligned}
e_{n+1} &= x(t_{n+1}) - x_n - h\phi_h(t_n, x_n) \\
&= x(t_{n+1}) - x(t_n) + e_n - h(\phi_h(t_n, x(t_n)) - \phi_h(t_n, x(t_n)) + \phi_h(t_n, x_n)).
\end{aligned}
$$

Setting $\varepsilon(t, x, h) := x(t + h) - x(t) - h\phi_h(t, x(t))$ and applying the mean value theorem[1] to $\phi$ gives

$$e_{n+1} = \Phi_n(h)e_n + \varepsilon(t_n, x, h) + \mathcal{O}(e_n^2) \tag{4.2.7}$$

---

[1] For notational simplicity we restrict the presentation to the scalar case.

with $\Phi_n(h) := 1 + h\frac{d}{dx}\phi_h(t_n, x)\big|_{x(t_n)}$. Thus,

$$\Phi_n(h) = 1 + \mathcal{O}(h).$$

$\varepsilon$ is called the *local error* or, due to its role in (4.2.7), the *global error increment* of the Runge–Kutta method.

Viewing the error propagation formula (4.2.7) we have to require

$$\varepsilon(t_n, x, h) = \mathcal{O}(h^{p+1}) \tag{4.2.8}$$

to get a method of order $p$.

**Example 4.2.2** *For the Runge–Kutta method (4.2.3) we get by Taylor expansion*

$$\varepsilon(t_{n-1}, h) = \frac{h^3}{24}(f_{xx}f^2 + 4f_x^2 f) + \mathcal{O}(h^4). \tag{4.2.9}$$

*using the notation* $f_x := \frac{d}{dx}f(x)\big|_{x(t_{n-1})}$ *for the elementary differentials.*
*Thus (4.2.3) is a second order method.*

The goal when constructing a Runge–Kutta method is to choose the coefficients in such a way that all elementary differentials up to a certain order cancel in a Taylor series expansion of $\varepsilon$. This requires a special symbolic calculus with elementary differentials, which is based on relating these differentials to labeled trees. The order conditions are much more complicated than in the multistep case. They consist of an underdetermined system of nonlinear equations for the coefficients $A, b, c$, which is solved by considering additional *simplifying assumptions* [But87]. The example above shows that in the Runge–Kutta case a sum of elementary differentials with different weights, the *principle error term*, determines the accuracy of the method. While comparing the accuracy of two methods of the same order requires in the multistep case only the comparison of two numbers, $c_{p+1}$, we have here to compare the weights of the elementary differentials corresponding to the $p+1^{\text{st}}$ derivative of the solution.

## 4.2.2   Embedded Methods for Error Estimation

Variable step size codes adjust the step size in such a way, that the global error increment is kept below a certain tolerance threshold $TOL$. This requires a good estimation of this quantity. Like in the multistep case, the error can be estimated by comparing two different methods. Here, a Runge–Kutta method of order $p$ and another method of order $p+1$ is taken to perform a step from $t_n$ to $t_{n+1}$, say. The global error increment of the $p^{\text{th}}$ order method is

$$\varepsilon^{(p)}(t, x) = C^{(p)}(t, x)h^{p+1} + \mathcal{O}(h^{p+2})$$

while the global error increment of the $p + 1^{st}$ order method is

$$\varepsilon^{(p+1)}(t, x) = C^{(p+1)}(t, x)h^{p+2} + \mathcal{O}(h^{p+3}),$$

where the coefficients $C^{(i)}$ depend on error coefficients and elementary differentials, cf. (4.2.9).

The difference of both quantities is

$$x_n^{(p)} - x_n^{(p+1)} = C^{(p)}(t, x)h^{p+1} + \mathcal{O}(h^{p+2}) = \varepsilon^{(p)}(t_n, x(t_n)) + \mathcal{O}(h^{p+2})$$

where the superscripts indicate the respective method.

Directly evaluating this formula for estimating the error would require additional function evaluations according to the number of stages of the higher order method. This would be much too expensive. This extra work can be avoided by using *embedded methods*. These are pairs of Runge–Kutta methods using the same coefficients $A$ and $c$. Thus, the stage values $k_i$ of both methods coincide. The only difference is in the $b$ coefficients, which are determined in such a way, that one method has order $p + 1$ while the other has order $p$. The two methods are described by the tableau

$$
\begin{array}{c|ccccc}
c_1 & & & & & \\
c_2 & a_{21} & & & & \\
c_3 & a_{31} & a_{32} & & & \\
\vdots & \vdots & \vdots & \vdots & & \\
c_s & a_{s1} & a_{s2} & \cdots & a_{s,s-1} & \\
\hline
& b_1^p & b_2^p & \cdots & b_{s-1}^p & b_s^p \\
& b_1^{p+1} & b_2^{p+1} & \cdots & b_{s-1}^{p+1} & b_s^{p+1} \\
\end{array}
$$

where $p$ indicates the order of the method and where

$$x_{n+1}^{(p)} = x_n + h \sum_{i=1}^{s} b_i^p k_i$$

$$x_{n+1}^{(p+1)} = x_n + h \sum_{i=1}^{s} b_i^{p+1} k_i.$$

**Example 4.2.3** *One method of low order in that class is the RKF2(3) method*

$$
\begin{array}{c|ccc}
0 & & & \\
1 & 1 & & \\
\frac{1}{2} & \frac{1}{4} & \frac{1}{4} & \\
\hline
& \frac{1}{2} & \frac{1}{2} & 0 \\
& \frac{1}{6} & \frac{1}{6} & \frac{4}{6} \\
\end{array}
\qquad (4.2.10)
$$

*It uses 3 stages.*

Note that unlike the multistep case, for Runge–Kutta methods an error estimation always requires two methods of different order.

**Local Extrapolation**
Though it is always the error of the lower order method which is estimated, one often uses the higher order and more accurate method for continuing the integration process. This foregoing is called *local extrapolation*. This is also reflected in naming the method, i.e.

- in RK2(3) the integration process is carried out with a second order method and in

- in RK3(2) local extrapolation is used, thus a third order method is taken for the integration.

When designing a Runge–Kutta method one is interested in minimizing the weights in the principle error term. In the context of embedded methods the difference of both methods must be as large as possible to give a good estimate, so it might not be a good idea to seek for a lower order method with minimal weights in the principle error term, when applying local extrapolation. Coefficients optimal with respect to this and some other design criteria have been found by Dormand and Prince leading to one of the most effective explicit Runge–Kutta formulas [DP89]. We give here the coefficients for the 5,4-pair:

**Example 4.2.4**

$$
\begin{array}{c|ccccccc}
0 \\
\frac{1}{5} & \frac{1}{5} \\
\frac{3}{10} & \frac{3}{40} & \frac{9}{40} \\
\frac{4}{5} & \frac{44}{45} & -\frac{56}{15} & \frac{32}{9} \\
\frac{8}{9} & \frac{19372}{6561} & -\frac{25360}{2187} & \frac{64448}{6561} & -\frac{212}{729} \\
1 & \frac{9017}{3168} & -\frac{355}{33} & \frac{46732}{5247} & \frac{49}{176} & -\frac{5103}{18656} \\
1 & \frac{35}{384} & 0 & \frac{500}{1113} & \frac{125}{192} & -\frac{2187}{6784} & \frac{11}{84} \\
\hline
x_n^{(p+1)} & \frac{35}{384} & 0 & \frac{500}{1113} & \frac{125}{192} & -\frac{2187}{6784} & \frac{11}{84} & 0 \\
x_n^{(p)} & \frac{5179}{57600} & 0 & \frac{7571}{16695} & \frac{393}{640} & -\frac{92097}{339200} & \frac{187}{2100} & \frac{1}{40}
\end{array}
\qquad (4.2.11)
$$

*This method uses six stages for* 5th *order result and one more for obtaining the* 4th *order result. One clearly sees that the method saves one function evaluation by meeting the requirement (4.2.5) for the 5th order method which is used for local extrapolation.*

## 4.2.3   Stability of Runge–Kutta Methods

Similar to the discussion in the multistep case we investigate the stability of Runge–Kutta methods for finite step sizes $h$ by considering the linear test equation $\dot{x} = \lambda x$, cf. (4.1.19).

Applying the Runge–Kutta method (4.2.2) to that equation results in

$$X_1 = x_n$$

$$X_i = x_n + h \sum_{j=1}^{i-1} a_{ij} \lambda X_j \quad i = 2, \dots, s$$

$$x_{n+1} = x_n + h \sum_{i=1}^{s} b_i \lambda X_i.$$

By inserting the stage values into the final equation we get

$$x_{n+1} = \left(1 + h\lambda \sum_{i=1}^{s} b_i + (h\lambda)^2 \sum_{i=2}^{s} b_i a_{i1} + (h\lambda)^3 \sum_{i=3}^{s} b_i a_{i2} a_{i-1,1} + \dots \right.$$
$$\left. \dots + (h\lambda)^s b_s a_{s,s-1} \right) x_n$$
$$:= R(h\lambda) x_n. \tag{4.2.12}$$

The function $R : \mathbb{C} \to \mathbb{C}$ is called *stability function* of the Runge–Kutta method, cf. the definition of $\Phi$ for multistep methods Sec. 4.1.4.

A Runge–Kutta method is stable for a given step size $h$ and a given complex parameter $\lambda$ iff $|R(h\lambda)| \leq 1$. Then, an error in $x_0$ is not increased by applying (4.2.12).

In Fig. 4.6 the stability regions of the methods DOPRI4 and DOPRI5 are displayed. Again, one realizes that lower order methods tend to have the larger stability region. When applying embedded methods with local extrapolation this fact must be envisaged.

As $R(0) = 1$, all Runge–Kutta methods are *zero stable*. This fact is also reflected in Fig. 4.6, where the origin is located at the stability boundary, see Sec. 4.1.4.

## 4.3   Implicit Runge–Kutta Methods

In this section we introduce the concept of implicit Runge–Kutta methods of collocation type without aiming for completeness. The use of implicit Runge–Kutta methods is quite recent in multibody dynamics. There are higher order A-stable methods in this class for simulating highly oscillatory problems as well as methods specially designed for constrained multibody systems, see Sec. 5.4.2.

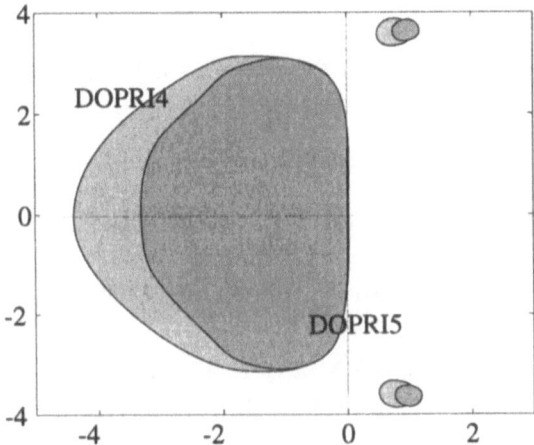

Figure 4.6: Stability regions for the Runge–Kutta pair DOPRI4 and DOPRI5.The methods are stable inside the gray areas.

## 4.3.1   Collocation Methods

**Definition 4.3.1** *The polynomial u of degree s defined by the conditions*

$$u(t_n) = x_n \tag{4.3.1a}$$
$$\dot{u}(t_n + c_i h) = f(t_n + c_i h, u(t_n + c_i h)) \tag{4.3.1b}$$

*and given distinct values $c_i \in [0, 1], i = 1, \ldots, s$ and a step size h is called a collocation polynomial of the differential equation $\dot{x} = f(t, x), x(t_n) = x_n$. The $c_i$ are called* collocation points.

The idea of collocation methods is to approximate $x(t_{n+1})$ by $x_{n+1} := u(t_n + h)$. $\dot{u}$ can be expressed using the Lagrange formulation (see Eq. (4.1.3)):

$$\dot{u}(t_n + \theta h) = \sum_{i=1}^{s} f(t_n + c_i h, u(t_n + c_i h)) \, l_i(\theta)$$

with

$$l_i(\theta) = \prod_{\substack{j=1 \\ j \neq i}}^{s} \frac{\theta - c_j}{c_i - c_j}.$$

We get by integration

$$u(t_n + c_i h) = x_n + h \int_0^{c_i} \dot{u}(t_n + \theta h) d\theta = x_n + h \sum_{j=1}^{s} a_{ij} f(t_n + c_j h, u(t_n + c_j h))$$

with

$$a_{ij} = \int_0^{c_i} l_j(\theta) d\theta. \tag{4.3.2}$$

By setting $X_i := u(t_n + c_i h)$ we can express the collocation method as

$$X_i = x_n + h \sum_{j=1}^{s} a_{ij} f(t_n + c_j h, X_j) \quad i = 1, \ldots, s \tag{4.3.3a}$$

$$x_{n+1} = x_n + h \sum_{i=1}^{s} b_i f(t_n + c_i h, X_i) \tag{4.3.3b}$$

with

$$b_i = \int_0^1 l_i(\theta) d\theta. \tag{4.3.4}$$

A method defined by (4.3.3) is called an *implicit Runge–Kutta method*. Unlike (4.2.2) the stages $X_i$ are defined here implicitly.
Again, the method can be represented by a *Butcher tableau*:

$$\begin{array}{c|c} c & A \\ \hline & b^T \end{array}$$

with $c = (c_1, \ldots, c_s)$, $b = (b_1, \ldots, b_s)$, and $A = (a_{ij})_{i,j=1,\ldots,s}$.
Collocation methods can be related to quadrature formulas by applying them to the special case: $\dot{x}(t) = f(t), x(t_n) = x_n$ or, equivalently, to

$$x(t_{n+1}) = x_n + \int_{t_n}^{t_{n+1}} f(\theta) d\theta.$$

We define $I_n^{n+1}(f) := x_n + \int_{t_n}^{t_{n+1}} f(\theta) d\theta$ and $\widehat{I}_n^{n+1}(f) := x_{n+1}$ with $x_{n+1}$ given by (4.3.3b). The collocation method is then constructed in such a way that for a $k$ as large as possible the following requirement is met:

$$I_n^{n+1}(\pi) = \widehat{I}_n^{n+1}(\pi) \qquad \text{for all polynomials } \pi \text{ up to degree } k.$$

This requirement defines the collocation points $c_i$ and by (4.3.2) and (4.3.4) also the other Runge–Kutta coefficients $a_{ij}$ and $b_i$. The maximal degree $k$ can be achieved by the so-called *Gauß points*, where $k = 2s - 1$. The $c_i$ are the roots of the *Legendre polynomials*. For $s = 3$ these coefficients are $c_1 = 1/2 - \sqrt{15}/10$, $c_2 = 1/2$, and $c_3 = 1/2 + \sqrt{15}/10$. This gives the 3-stage *Gauß integration method*:

$$
\begin{array}{c|ccc}
\frac{1}{2} - \frac{\sqrt{15}}{10} & \frac{5}{36} & \frac{2}{9} - \frac{\sqrt{15}}{15} & \frac{5}{36} - \frac{\sqrt{15}}{30} \\
\frac{1}{2} & \frac{5}{36} + \frac{\sqrt{15}}{24} & \frac{2}{9} & \frac{5}{36} - \frac{\sqrt{15}}{24} \\
\frac{1}{2} + \frac{\sqrt{15}}{10} & \frac{5}{36} + \frac{\sqrt{15}}{30} & \frac{2}{9} + \frac{\sqrt{15}}{15} & \frac{5}{36} \\
\hline
& \frac{5}{18} & \frac{4}{9} & \frac{5}{18}
\end{array}
$$

By selecting the collocation points as roots of the polynomial

$$
\frac{d^{s-1}}{dt^{s-1}} \left( t^{s-1}(t-1)^s \right)
$$

one obtains the class of Radau IIa methods. A widely used implementation of a three stage Radau method is the code RADAU5 by Hairer [HW96]. Its coefficients are

$$
\begin{array}{c|ccc}
\frac{4-\sqrt{6}}{10} & \frac{88-7\sqrt{6}}{360} & \frac{296-169\sqrt{6}}{1800} & \frac{-2+3\sqrt{6}}{225} \\
\frac{4+\sqrt{6}}{10} & \frac{296+169\sqrt{6}}{1800} & \frac{88+7\sqrt{6}}{360} & \frac{-2-3\sqrt{6}}{225} \\
1 & \frac{16-\sqrt{6}}{36} & \frac{16+\sqrt{6}}{36} & \frac{1}{9} \\
\hline
& \frac{16-\sqrt{6}}{36} & \frac{16+\sqrt{6}}{36} & \frac{1}{9}
\end{array}
\tag{4.3.5}
$$

## 4.3.2 Corrector Equations in Implicit Runge–Kutta Methods

In order to solve the set of $ns$ nonlinear equations (4.3.3a) Newton's method is applied. Again, like in the multistep case, Newton's method is implemented in a simplified manner, which saves evaluations (and decompositions) of the Jacobian: With $J$ being an approximation to the Jacobian

$$
J \approx \frac{\partial f}{\partial x}(t_m, x(t_m))
$$

at some time point $t_m \le t_n$ the $ns \times ns$ iteration matrix is

$$
\begin{pmatrix}
I - ha_{11}J & \cdots & -ha_{1s}J \\
\vdots & & \vdots \\
-ha_{s1}J & \cdots & I - ha_{ss}J
\end{pmatrix}.
$$

With the Kronecker product notation

$$
A \otimes B := (a_{ij}B)_{i,j=1,\dots,\dim(A)}
$$

the iteration matrix reads

$$(I_{ns} - hA \otimes J).$$

With this the iterates for the stages are defined by

$$(I_{ns} - hA \otimes J)\Delta X^{(k)} = -X^{(k)} + x_n + h(A \otimes I_n)F(X^{(k)}) \quad (4.3.6a)$$
$$X^{(k+1)} := X^{(k)} + \Delta X^{(k)} \quad (4.3.6b)$$

with the vector of the stage iterates

$$X := (X_1^T, \ldots, X_s^T)^T \in \mathbb{R}^{ns}$$

and

$$F(X) := (f(t_n + c_1 h, X_1)^T, \ldots, f(t_n + c_s h, X_s)^T)^T.$$

Using the matrix multiplication rule for Kronecker products

$$(A \otimes B)(C \otimes D) = AC \otimes BD$$

we can transform (4.3.6a) into

$$(h^{-1}A^{-1} \otimes I_n - I_s \otimes J)\Delta X^{(k)} = (h^{-1}A^{-1} \otimes I_n)(x_n - X^{(k)}) + F(X^{(k)}). \quad (4.3.7)$$

Now, the constant, method dependent matrix $A$ is separated from the state and problem dependent matrix $J$. $A^{-1}$ can be transformed by means of a similarity transformation into a simpler form, i.e. a block diagonal form $\Lambda$

$$T^{-1}A^{-1}T =: \Lambda.$$

Making a coordinate transformation $Z := (T^{-1} \otimes I_n)X$, $z_n := (T^{-1} \otimes I_n)x_n$ finally simplifies (4.3.7) to

$$(h^{-1}\Lambda \otimes I_n - I_s \otimes J)\Delta Z^{(k)} =$$
$$(h^{-1}\Lambda \otimes I_n)(z_n - Z^{(k)}) + (T^{-1} \otimes I_n)F((T \otimes I_n)Z^{(k)}). \quad (4.3.8)$$

The coefficient matrix $A^{-1}$ of the three stage Radau IIa method (4.3.5) has one real eigenvalue $\gamma = 3.6378$ and a conjugate pair of eigenvalues $\alpha \pm \beta = 2.6811 \pm 3.0504\,i$, thus $A^{-1}$ can be transformed into

$$\Lambda = \begin{pmatrix} \gamma & 0 & 0 \\ 0 & \alpha & -\beta \\ 0 & \beta & \alpha \end{pmatrix}.$$

The linear system (4.3.8) then takes the form

$$\begin{pmatrix} h^{-1}\gamma I_n - J & 0 & 0 \\ 0 & h^{-1}\alpha I_n - J & -h^{-1}\beta I_n \\ 0 & h^{-1}\beta I_n & h^{-1}\alpha I_n - J \end{pmatrix} \begin{pmatrix} \Delta Z_1 \\ \Delta Z_2 \\ \Delta Z_3 \end{pmatrix} = \begin{pmatrix} R_1 \\ R_2 \\ R_3 \end{pmatrix} \qquad (4.3.9)$$

with $R_i$ denoting the corresponding parts of the right hand side of the linear system. From this we get to solve two decoupled sets of linear equation, a real

$$\left(h^{-1}\gamma I_n - J\right)\Delta Z_1 = R_1$$

and a complex one

$$\left(h^{-1}(\alpha + i\beta)I_n - J\right)\Delta Z_2 + i\Delta Z_3 = R_2 + iR_3. \qquad (4.3.10)$$

The numerical effort for solving these systems consists of

1. $n^3/3 + \mathcal{O}(n^2)$ multiplications for the LU decomposition of the real matrix

2. $2n^3/3 + \mathcal{O}(n^2)$ multiplications for the LU decomposition of the complex matrix

3. $n^2 + \mathcal{O}(n)$ multiplications for the backward/forward substitution in the real case

4. $2n^2 + \mathcal{O}(n)$ multiplications for the backward/forward substitution in the complex case

5. $6n^2 = 2sn^2$ multiplications for forming $(T^{-1} \otimes I_n)X$ and $(T^{-1} \otimes I_n)F$, and

6. $n^2$ multiplications for the back transformation $(T \otimes I_n)Z$.

The first two steps are only necessary if $J$ or $h$ changed significantly, the other steps are part of every Newton iteration.
To start the Newton iteration the collocation polynomial used in the last step is extrapolated to obtain guesses for the stage values:

$$X_i^{(0)} := u_{n-1}(t_n + c_i h)$$

with $u_{n-1}$ being the collocation polynomial for the step $t_{n-1} \to t_n$, see also Fig. 4.7. Alternative choices of starting values are discussed in [OS98].

**Example 4.3.2** *We consider a simulation with RADAU5 of the unconstrained truck running over a 12 cm high road hump described by the input function* $u(t) = 6.653 \cdot 10^{10}(t-t_0)^6 e^{-200(t-t_0)}$ *and its derivative (see Sec. 1.3.1). In Fig. 4.7 two typical integration steps (circled area in the left plot) are zoomed out to show the stage values and the collocation polynomial. In order to visualize the difference between extrapolated values of the collocation polynomial and the numerical result in the next step (arrow in the right plot) a rather crude tolerance, of* $RTOL=ATOL=10^{-1}$, *was chosen.*

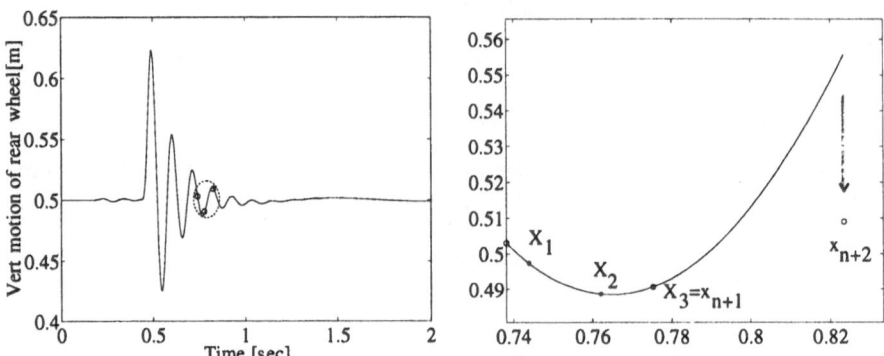

Figure 4.7: The collocation polynomial in a typical step of RADAU5 acting on the unconstrained truck

### 4.3.3   Accuracy of Implicit Runge–Kutta Methods

As we consider in this book only implicit Runge–Kutta methods which are based on collocation methods, accuracy results of these methods are very much related to the accuracy properties of interpolation polynomials.
We cite from [HNW87] the following central result

**Theorem 4.3.3** *Let $p(t)$ be a polynomial of degree $s$ with roots at $c_1, \ldots, c_s$. If it is orthogonal to all polynomials of degree $r - 1$, i.e.*

$$\int_0^1 p(\tau)\tau^q d\tau = 0 \quad q = 0, \ldots, r - 1, \tag{4.3.11}$$

*then the method has order $p = s + r$.*

For Radau IIA methods $p(t) = \frac{d^{s-1}}{dt^{s-1}} \left( t^{s-1}(t-1)^s \right)$ and it can be checked, that for a three stage method ($s = 3$) the orthogonality condition gives $r = 2$, thus the order is 5 . On the other hand, a three stage Gauß method has $r = 3$ and the order is 6. The requirement for Radau IIa methods that the end point of the integration interval is a collocation point had to be paid by a lower order. We will see when considering implicit ODEs (Sec. 5.4.1) that this order drop is compensated by other advantages.
Note, that the order is defined via the global error in the discretization points $t_n$. If we evaluate the collocation polynomial between these points and compare the result with the exact solution of the differential equation one usually gets a lower order (cf. Sec. 4.5). This phenomenon is called *superconvergence* of collocation methods.

### 4.3.4 Stability of Implicit Runge–Kutta Methods

The main reason for using implicit Runge–Kutta methods is due the excellent stability properties of some of the methods in this class. Again, we consider stability for the linear test equation and obtain by applying (4.3.3)

$$X_i = x_n + h \sum_{j=1}^{s} a_{ij}\lambda X_j \quad i = 1, \dots, s$$

$$x_{n+1} = x_n + h \sum_{i=1}^{s} b_i \lambda X_i.$$

Solving this linear system for the stage values $X_i$ and inserting these values into the last equation results in

$$x_{n+1} = x_n + h\lambda b^T (I_s - h\lambda A)^{-1}(x_n, \dots, x_n)^T,$$

where the last vector consists of $s$ repetitions of $x_n$. By introducing a *stability function* this can be rewritten as

$$x_{n+1} = R(h\lambda)x_n = (1 + h\lambda b^T (I_s - h\lambda A)^{-1}\mathbf{1})x_n$$

with $\mathbf{1} := (1, \dots, 1)^T \in \mathbb{R}^s$. The stability is now a rational function in $z := h\lambda$ in contrast to explicit methods where it is a polynomial, see (4.2.12).

In Fig. 4.8 the stability region of the three stage Radau IIa method is displayed. One realizes that the stability of the Radau method is much alike the stability of the implicit Euler method, though the three stage Radau method has order 5. Again we note the property $R(0) = 1$ which corresponds to zero stability in the multistep case.

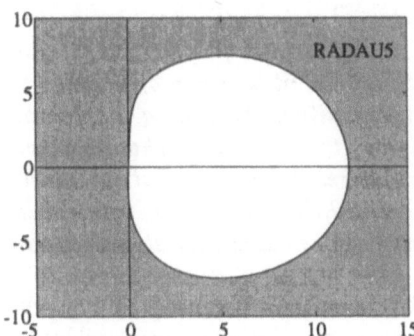

Figure 4.8: Stability region for the three stage Radau IIa (the shaded area).

## 4.4   Stiff Systems

We described earlier (cf. Sec. 4.1.5) stiffness as a property of a method, when applied to a certain differential equation with a given tolerance.

Often, the term *stiff differential equation* is used to indicate that special methods are used for numerically solving them. These methods are called *stiff integrators* and are characterized by $A$-stability or at least $A(\alpha)$-stability. They are always implicit and require a corrector iteration based on Newton's method. For example BDF methods or some implicit Runge–Kutta methods, like the Radau method are stiff integrators in that sense.

Though there is no rigorous definition of a stiff differential equation we will give here some characteristics and demonstrate some phenomena.

Stiff differential equations are characterized by a solution composed by *fast and slow modes*. If a fast mode is excited, this leads to a steep transient to the solution component given by the slow mode, see Fig. 4.9.

When using a numerical method with a bounded stability region and a too large step size, a fast mode causes an instability in the numerical solution, so the step size must be restricted in this case, often drastically.

We have to distinguish two cases when treating stiff differential equations:

- The transient phase is of interest and should be computed.

- The transient phase is caused only by negligible excitations of the system, e.g. numerical errors, and is of no practical interest.

**Example 4.4.1** *We consider the unconstrained truck and change its damping coefficients $d_{10}$ and $d_{30}$ successively by factors $fac = 1000, 10000$, cf. Tab. A.4. As initial conditions the steady state with a slightly perturbed initial velocity of the rear wheel is taken. Depending on the size of the damping coefficients there is a fast transient from this perturbed state to the steady state, see Fig. 4.9. When simulating this situation by a nonstiff method, like DOPRI45, the transition phase and the asymptotic phase lead to different behavior of the step size selection, when changing tolerances. During transition sharpening the tolerance requirement results in smaller step sizes, while in the asymptotic phase a change of the tolerance has no effect on the step size. There, stability is what determines the step size, Fig. 4.10. In contrast in a stiff method, like RADAU5, the tolerance affects the step size in the entire range. Furthermore much larger step sizes can be taken*[2].

A special case of stability problems is caused by systems with highly oscillatory modes. There, a slowly varying solution is overlayed by a highly oscillating solution. Again, one has to decide what the goal of the numerical computation is:

---

[2]For this example the FORTRAN codes given in [HW96] were used together with default control parameters.

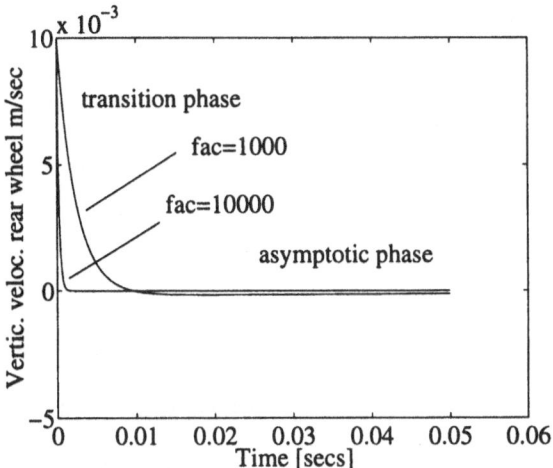

Figure 4.9: The transient and asymptotic phase of a solution

- resolving all oscillations

- computing an envelop

- neglecting the high frequencies as their amplitudes are negligible, i.e. less than the given tolerance.

In multibody dynamics one encounters often the last situation, while the first two alternatives occur in celestial mechanics or molecular dynamics, [PJY97].

High frequencies with negligible amplitudes occur due to the modeling of coupled rigid and elastic systems, [Sim96, Sac96]. Also, this situation occurs due to extremely stiff springs with poor damping. When modeling penetration only bad knowledge of the material's damping is known, whereas its stiffness properties are well known. A typical example is a metal roller bearing, where the ball may penetrate the inner or outer ring, causing high oscillations with negligible amplitudes, [Sjö96].

**Example 4.4.2** *This situation also can be modeled by an elastic pendulum, which we will consider here as an example, [Lub93].*

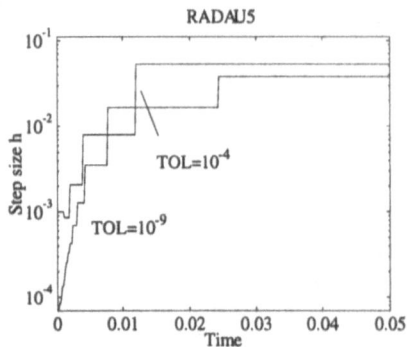 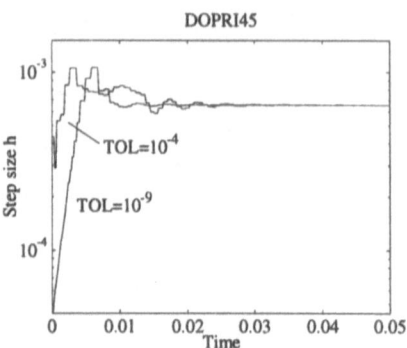

Figure 4.10: Step sizes of RADAU5 and DOPRI45 when applied to the stiff truck example.

$$\dot{p}_1 = v_1$$

$$\dot{p}_2 = v_2$$

$$\dot{v}_1 = -\frac{1}{\epsilon}\frac{p_1}{\sqrt{p_1^2 + p_2^2}}\left(\sqrt{p_1^2 + p_2^2} - 1\right)$$

$$\dot{v}_2 = -\frac{1}{\epsilon}\frac{p_2}{\sqrt{p_1^2 + p_2^2}}\left(\sqrt{p_1^2 + p_2^2} - 1\right) - g_{gr}$$

*Due to the spring this pendulum has a non constant length. In this numerical experiment we simulate the pendulum with different integration methods and tolerances in order to get an idea about its length when $\epsilon = 10^{-6}$ has been chosen.*

*First we consider an integration method with a bounded stability area, DOPRI45. The results for different tolerances are shown in Fig. 4.11. One clearly sees the influence of the tolerances on the results. Without running two simulations with different tolerances one would have assumed that the solution contains oscillations with an amplitude of $6 \cdot 10^{-4}$ as depicted in the left figure. But as the influence of the tolerance on the results indicates, the oscillations of this size are rather a "numerical effect": Due to its bounded stability region, there is an oscillatory instability in the numerical solution, which is detected by the error control when the amplitude exceeds the tolerance limits. When using a method with unbounded stability region and internal damping, like RADAU5, this effect disappears, see Fig. 4.12.*

*Table 4.3 shows the different computational effort of these two methods. It becomes evident which price must be paid to limit the oscillatory instability by controlling the error and restricting the step size.*

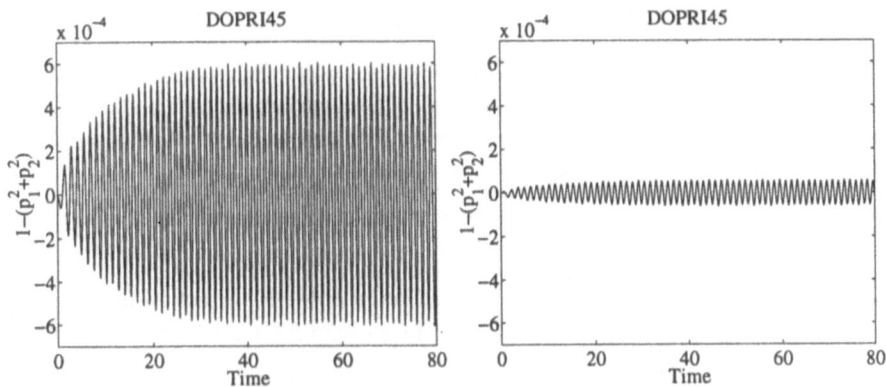

Figure 4.11: Length of the elastic pendulum, when simulated with DOPRI45 and ATOL=RTOL=$10^{-4}$ (left) and ATOL=RTOL=$10^{-5}$ (right)

| Method | TOL | NST | NRJ | NFE | CPU |
|--------|-----|-----|-----|-----|-----|
| DOPRI45 | $10^{-4}$ | 97911 | 16218 | 587468 | 2.73 |
| DOPRI45 | $10^{-5}$ | 98551 | 16472 | 591308 | 2.76 |
| RADAU5 | $10^{-4}$ | 5619 | 1 | 37209 | 1 |
| RADAU5 | $10^{-5}$ | 8576 | 1 | 57963 | 5.2 |

Table 4.3: Computational effort of DOPRI45 and RADAU5 when applied to the stiff pendulum example: Tolerance (TOL), number of steps (NST), of rejected steps (NRJ), of function evaluations (NFE) and normalized computing time (CPU)

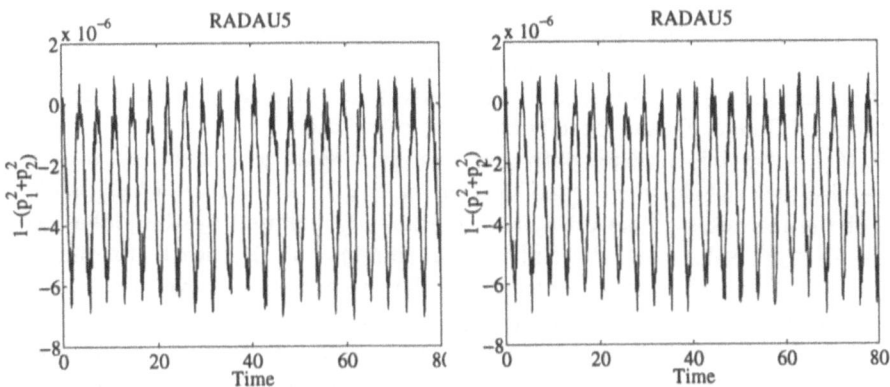

Figure 4.12: Length of the elastic pendulum, when simulated with RADAU5 and ATOL=RTOL=$10^{-4}$ (left) and ATOL=RTOL=$10^{-5}$ (right)

## 4.5    Continuous Representation of the Solution

Numerical integration methods discussed so far provide information about the solution of the ODE at certain discrete points. The location of these points is determined by the step size control. Often however, the numerical solution is needed at other points, so-called *output points*. This situation occurs due to

- physical requirement, e.g. the model might change its characteristics at certain points.

- post processing requirements, e.g. often graphical output is based on an equidistant grid, solution values must be provided for post processing etc..

In the first case the location of the output points is often not known explicitly and is determined iteratively by solving an implicit relation. This will be the topic of Ch. 6.

In all cases one is interested in obtaining a functional description of the numerical approximation to the solution rather than in its values only at discrete points.

**Definition 4.5.1** *A function* $X_n : [0, 1] \longrightarrow I\!\!R^{n_x}$ *with the properties*

$$X_n(0) = x_n, \quad X_n(1) = x_{n+1}$$

*and*

$$\|X_n(\theta) - \hat{x}(t_n + \theta h_n)\| = \mathcal{O}(h^{q+1})$$

*is called a* continuous representation of the numerical solution *of order* $q$, *where* $\hat{x}(t)$ *is the local exact solution, i.e. the solution of the differential equation passing through* $x_n$.

The goal is to construct and to evaluate a continuous representation without additional evaluations of the right hand side function $f$. The construction of a continuous representation is straight forward for integration methods based on a polynomial representation of the solution or its derivative, like Adams or BDF multistep methods or Runge–Kutta methods based on collocation.

If $X_n(\theta)$, $\theta \neq 1$ is used for output or post processing purposes only, it suffices to require that the order $q$ is such that the error $\|X_n(\theta) - \hat{x}(t_n + \theta h_n)\|$ is of the size of the method's global error, as $x_n$ is already polluted by the global error. Thus, for a $p^{\text{th}}$ order method it suffices to require $q = p - 1$. Clearly, for the point $x_{n+1} = X_n(1)$, which is used for continuing the integration, we have to require that the order of $X_n(1)$ is $q = p$ in that point.

However, for the use in switching algorithms, see Sec. 6, the interpolation error is propagated as it influences the determination of the switching point and the point of restart. Thus in this case, the interpolation error must be controlled. In [SG84] it is shown for Adams methods that the continuous representation is error controlled, for BDF methods the corresponding result can be found in [Eich91].

**Adams–Moulton Schemes**

From Eq. (4.1.5) a continuous representation for Adams–Moulton methods can easily be derived:

$$X_n(\theta) = x_n + \theta h_n \sum_{i=0}^{k} \beta^c_{k-i}(\theta) f(t_{n+1-i}, x_{n+1-i}) \tag{4.5.1}$$

with

$$\beta^c_{k-i}(\theta) \quad := \quad \frac{1}{\theta h_n} \int_{t_n}^{t_n+\theta h_n} \bar{l}_i^{k+1}(t)\, dt$$

$$\bar{l}_i^{k+1}(t) \quad := \quad \prod_{\substack{j=0 \\ j \neq i}}^{k} \frac{t - t_{n+1-j}}{t_{n+1-i} - t_{n+1-j}}.$$

By construction the order of the continuous representation is the order of the Adams–Moulton method, i.e. $q = k + 1$.

It should be noted that the way an Adams method is implemented influences the continuous representation of the solution. Special care has to be taken in $P(EC)^m$ implementations, see Sec. 4.1.1. In that case $X_n(1) \neq x_{n+1}$. For this type of implementation (4.5.1) defines no continuous representation of the solution. Using (4.5.1) in that case may lead to problems when passing from one interval to the next in connection with switching algorithms, which will be described in Sec. 6.

**BDF Schemes**

For BDF methods the polynomial defined by the corrector can be used as a continuous representation of the numerical solution. The values $X_n(\theta)$ are obtained by evaluating this polynomial at the point $t_n + h\theta$:

$$X_n(\theta) = \pi^c_{k+1}(t_n + \theta h_n)$$

with $\pi^c$ defined as in Sec. 4.1.2. The order is $q = k$.

**Explicit Runge–Kutta methods**

In order to have a continuous representation of the numerical solution without additional evaluations of the rhs-function $f$, we seek for Runge–Kutta methods of the form

$$k_1 \quad = \quad f(t_n, x_n)$$

$$k_i \quad = \quad f(t_n + c_i h, x_n + h \sum_{j=1}^{i-1} a_{ij} k_j) \quad i = 2, \ldots, s$$

$$X_n(\theta) \quad = \quad x_n + h \sum_{i=1}^{s} b_i(\theta) k_i$$

with $b_i(1) = b_i$ and $b_i(0) = 0$.

This kind of methods allows the use of the same stage values $k_i$ for all $X_n(\theta) \in$ $[x_n, x_{n+1}]$. If the order of the continuous representation is required to be the order of the global error of the method, then there is such a formulation only for very few Runge–Kutta methods. For the DOPRI45 method a continuous representation of order 4 exists:

$$b_1(\theta) = \theta\left(1 + \theta\left(-\frac{1337}{480} + \theta\left(-\frac{1163}{1152}\right)\right)\right)$$

$$b_2(\theta) = 0$$

$$b_3(\theta) = \frac{100}{3}\theta^2\left(\frac{1054}{9275} + \theta\left(-\frac{4682}{27825} + \theta\frac{379}{5565}\right)\right)$$

$$b_4(\theta) = -\frac{5}{2}\theta^2\left(\frac{27}{40} + \theta\left(-\frac{9}{5} + \theta\frac{83}{96}\right)\right)$$

$$b_5(\theta) = \frac{18225}{848}\theta^2\left(-\frac{3}{250} + \theta\left(\frac{22}{375} - \theta\frac{37}{600}\right)\right)$$

$$b_6(\theta) = -\frac{22}{7}\theta^2\left(-\frac{3}{10} + \theta\left(\frac{29}{30} - \theta\frac{17}{24}\right)\right).$$

These are fourth order polynomials in $\theta$.

**Implicit Runge–Kutta Methods**

For implicit Runge–Kutta methods based on collocation polynomials these polynomials can serve as continuous representation of the solution. Unfortunately for many collocation methods the order of the continuous representation is $q < p - 1$, so that the requirement at the beginning of this section is not met.

**Theorem 4.5.2** *The continuous representation based on a collocation polynomial $u$ of an $s$-stage implicit Runge–Kutta method is of order $s$, i.e.*

$$\|X_n(\theta) - \hat{x}(t_n + \theta h)\| \leq Ch^{s+1}$$

*with $X_n(\theta) = u_n(t_n + \theta h)$, $\hat{x}$ being the local exact solution and $u_n$ being the collocation polynomial of step $n$ [HNW87].*

By this theorem the continuous representation of the fifth order Radau IIa method has order three.

# 5 Implicit Ordinary Differential Equations

## 5.1 Implicit ODEs

### 5.1.1 Types of Implicit ODEs

In the last chapter we discussed the numerical treatment of explicit ordinary differential equations. Here, we will consider the more general case, *implicit ordinary differential equations*.

The equations of motion of unconstrained multibody systems are normally given as implicit ODE

$$E(x)\dot{x} = f(t, x) \tag{5.1.1}$$

with $E$ being a square, regular matrix of the form

$$E(x) = \begin{pmatrix} I & 0 \\ 0 & M(x) \end{pmatrix}$$

with the regular mass matrix $M(x)$.

Due to the regularity of $E$, Eq. (5.1.1) can be transformed into explicit form, though computationally this might not always be advantageous.

When dealing with constrained multibody systems, the situation is different, because $E$ is singular. Here we have

$$E(x) = \begin{pmatrix} I & 0 & 0 \\ 0 & M(p) & 0 \\ 0 & 0 & 0 \end{pmatrix} \quad \text{and } x = (p^T, v^T, \lambda^T)^T$$

(see Eq. (1.3.11)). Eq. (5.1.1) turns out to be a *singularly implicit ODE*, which cannot be transformed into an explicit form. A special case occurs, when the singular part of $E$ is given by some zero rows and the non zero rows form a nonsingular submatrix of $E$ as in the case of multibody systems.

In that particular case the system consists of differential equations and some non-linear equations, often called in this context (somehow sloppy) algebraic equations. Consequently, this type of singularly implicit ODEs is also called a system of *differential algebraic equations* (DAEs).

By setting $f^T = (f_D^T, f_C^T)$ and $x^T = (y^T, \lambda^T)$ we can write such a system as

$$E_D(y)\dot{y} \;=\; f_D(t, y, \lambda) \qquad\qquad (5.1.2a)$$
$$0 \;=\; f_C(t, y, \lambda) \qquad\qquad (5.1.2b)$$

with a square, regular matrix $E_D(y)$. If this matrix is the identity, the system is called a *semi-explicit DAE*, as one set of variables is given in terms of an explicit ODE, the other set is defined implicitly.

Sometimes it might be appropriate to write DAEs in a more compact form

$$F(t, x, \dot{x}) = 0$$

with the partial derivative $\frac{\partial F}{\partial \dot{x}}$ being singular. Although in mechanical systems like in most other applications the highest derivative occurs only linearly, we sometimes use this notation for simplicity.

After presenting some properties of DAEs we will devote this chapter to numerical integration methods for this class of problems.

Some facts and concepts on DAEs like consistency of initial conditions, index of the system were already scoped in Ch. 2. They will be reintroduced in this section from a different aspect.

## 5.1.2   On the Existence and Uniqueness of Solutions of Multi-body System Equations

In this section we give necessary and sufficient conditions for the existence and uniqueness of solutions of the initial value problem

$$\dot{p} \;=\; v \qquad\qquad (5.1.3a)$$
$$M(p)\dot{v} \;=\; f_a(p, v) - G(p)^T \lambda \qquad\qquad (5.1.3b)$$
$$0 \;=\; g(p) \qquad\qquad (5.1.3c)$$

with the initial conditions $p(t_0) = p_0$, $v(t_0) = v_0$, $\lambda(t_0) = \lambda_0$.

**Definition 5.1.1 (Solution, consistent initial values)**
*A function $x : [t_0, t_e] \to \mathbb{R}^n$, $x(t) = \left(p(t)^T, v(t)^T, \lambda(t)^T\right)^T$ fulfilling the smoothness requirement: $p \in C^2[t_0, t_e]$, $v \in C^1[t_0, t_e]$, $\lambda \in C^0[t_0, t_e]$ is called a solution of Eq. (5.1.3) if it satisfies Eq. (5.1.3). The value $x(t_0)$ is then called* consistent initial value *of (5.1.3).*

Note, that this definition requires that the initial values $x(t_0) = x_0$ fulfill the constraint (5.1.3c). Some authors extend the definition of a solution and include functions with an initial impulse due to inconsistent initial values.

Differentiation of the algebraic equations yields

$$0 \quad = \quad G(p)\dot{p} = G(p)v \qquad (5.1.4a)$$

$$0 \quad = \quad G(p)\dot{v} + \left(\frac{d}{dt}G(p)\right)v$$

$$=: \quad G(p)\dot{v} + \zeta(p,v). \qquad (5.1.4b)$$

A solution in the sense of Def. 5.1.1 only exists if the initial values satisfy these two equations.

From (5.1.3b) and (5.1.4b) $\dot{v}$ and $\lambda$ can be determined as the solution of

$$\begin{pmatrix} M(p) & G(p)^T \\ G(p) & 0 \end{pmatrix} \begin{pmatrix} \dot{v} \\ \lambda \end{pmatrix} = \begin{pmatrix} f_a(p,v) \\ -\zeta(p,v) \end{pmatrix}.$$

We get

$$\dot{v} \quad = \quad M(p)^{-1}(f_a(p,v) - G(p)^T\lambda) \qquad (5.1.5a)$$

$$\lambda \quad = \quad \left(G(p)M(p)^{-1}G(p)^T\right)^{-1}\left(G(p)M(p)^{-1}f_a(p,v) + \zeta(p,v)\right). \qquad (5.1.5b)$$

Note that $\left(G(p)M(p)^{-1}G(p)^T\right)$ is regular as we assumed $G$ having full rank.
By inserting $\lambda$ into the differential equation for $v$ an explicit ODE for $p$ and $v$ is obtained

$$\dot{p} \quad = \quad v \qquad (5.1.6a)$$

$$\dot{v} \quad = \quad M(p)^{-1}\left(P(p)f_a(p,v) + Q(p)\zeta(p,v)\right) \qquad (5.1.6b)$$

with

$$P(p) = I - G(p)^T \left(G(p)M(p)^{-1}G(p)^T\right)^{-1} G(p)M(p)^{-1}$$

and

$$Q(p) = G(p)^T \left(G(p)M(p)^{-1}G(p)^T\right)^{-1}.$$

The solvability of Eq. (5.1.5) is guaranteed by standard theorems if the functions are sufficiently smooth.

Consistency of the initial values for problem (5.1.3) requires that $x_0 = (p_0^T, v_0^T, \lambda_0^T)^T$ obeys equations (5.1.3b), (5.1.4a), and (5.1.5b). Consequently, unlike the explicit ODE case, initial conditions cannot be chosen arbitrarily. They have to meet the constraint equations and their time derivatives. This is one major difference between DAEs and explicit ODEs. It can be seen as one source of difficulties in numerically solving DAEs.

Let us consider two classical examples, [BCP89]:

**Example 5.1.2**

$$\dot{x}_2 = x_1$$
$$0 = x_1 - f_2$$

with $f_2$ being a sufficiently smooth function. The solution of this problem is given by

$$x_1(t) = f_2(t) \quad and \quad x_2(t) = x_2(t_0) + \int_{t_0}^{t} f_2(s)ds. \qquad (5.1.7)$$

If $x_1(t_0) \neq f_2(t_0)$, there exists no solution in the sense of Definition 5.1.1.

This example shows what one might have expected. The freedom in selecting initial conditions is limited by the constraint. Clearly, the initial conditions must meet the constraints.

The second example shows that there are types of equations where the initial values must be restricted even more:

**Example 5.1.3**

$$\dot{x}_2 = x_1 - f_1$$
$$0 = x_2 - f_2$$

with $f_1$ and $f_2$ being sufficiently smooth functions. The solution of this problem is given by

$$x_2(t) = f_2(t) \quad and \quad x_1(t) = \dot{f}_2(t) + f_1(t). \qquad (5.1.8)$$

In that case not only the constraint must be met by the initial condition but also its derivative reduces the freedom in the choice of initial conditions.

For mechanical systems (5.1.3) the consistency requirements allow to choose $2(n_p - n_\lambda)$ initial values freely while the remaining are determined by (5.1.3c), (5.1.4a), and (5.1.5b).

The question arises, which of the values can be freely chosen. This question is coupled to initially selecting appropriate state variables, see Sec. 2.1 and Sec. 5.3.2. Let $R : {I\!\!R}^{2n_p} \to {I\!\!R}^{2(n_p - n_\lambda)}$ describe the $2(n_p - n_\lambda)$ freely chosen initial values for the position and velocity variables in the form

$$R(p_0, v_0) = 0. \qquad (5.1.9)$$

Then, solvability of (5.1.3c), (5.1.4a), (5.1.9) with respect to the remaining variables requires that

$$\begin{pmatrix} \frac{\partial R}{\partial p}(p, v) & \frac{\partial R}{\partial v}(p, v) \\ G(p) & 0 \\ v^T \frac{\partial G}{\partial p}(p) & G(p) \end{pmatrix}$$

is nonsingular in $(p_0, v_0)$.

**Example 5.1.4** *For the constrained truck we can chose* $(R(p,v))_i = p_i - p_{i,0}$, $i = 1, 2, 4, 5, 6, 8, 9$ *and* $(R(p,v))_{7+i} = v_i - v_{i,0}$, $i = 1, 2, 4, 5, 6, 8, 9$. *This corresponds to the matrix V calculated in Ex. 2.2.1.*
*One may also use* $(R(p,v))_i = p_i - p_{i,0}$, $i = 1, 2, 3, 4, 5, 6, 9$ *and* $(R(p,v))_{7+i} = v_i - v_{i,0}$, $i = 1, 2, 3, 4, 5, 6, 9$.

### 5.1.3 Sensitivity under Perturbations

DAEs can be classified by their index. We defined in Sec. 2.6.3 the index of a linear DAE by the index of nilpotency of the corresponding matrix pencil. For the general, nonlinear case there are many different index definitions, which might lead to different classifications in special cases. Here we will introduce the so-called *perturbation index*, which allows to categorize the equations by means of their sensitivity under perturbations, [HLR89].

**Definition 5.1.5 (Perturbation index)** *The DAE*

$$F(t, x, \dot{x}) = 0$$

*has index $k$ along a solution $x$ in $[t_0, t_e]$, if $k$ is the smallest number such that for all functions $\hat{x}$ with*

$$F(t, \hat{x}, \dot{\hat{x}}) = \delta(t)$$

*the following inequality holds*

$$\|\hat{x}(t) - x(t)\| \quad \leq \quad C\Big(\|\hat{x}(t_0) - x(t_0)\| + \max_{t_0 \leq \hat{t} \leq t} \|\int_{t_0}^{\hat{t}} \delta(\tau)d\tau\| + $$
$$+ \max_{t_0 \leq \hat{t} \leq t} \|\delta(\hat{t})\| + \cdots + \max_{t_0 \leq \hat{t} \leq t} \|\delta^{(k-1)}(\hat{t})\|\Big)$$

*as long as the expression on the right hand side is sufficiently small. $C$ is a constant which is independent of the perturbation $\delta$ and depends only on $F$ and the length of the interval. [HLR89]*

Def. 5.1.5 also includes the case $k = 0$, if the integral of $\delta$ is denoted by $\delta^{(-1)}$. This case characterizes the behavior under small perturbations of explicit ODEs:

**Lemma 5.1.6** *Let $x$ be the solution of the explicit initial value problem*

$$\dot{x} = f(t, x), \quad x(t_0) = x_0,$$

*and let $\hat{x}$ be the solution of the perturbed problem*

$$\dot{\hat{x}} = f(t, \hat{x}) + \delta(t), \quad \hat{x}(t_0) = \hat{x}_0.$$

*Assume that $f$ has a Lipschitz constant $L$. Then, the difference between the two solutions can be bounded by*

$$\|\hat{x}(t) - x(t)\| \leq e^{L(t-t_0)} \left( \|\hat{x}(t_0) - x(t_0)\| + \max_{t_0 \leq \tilde{t} \leq t} \| \int_{t_0}^{\tilde{t}} \delta(\tau) d\tau \| \right).$$

*Proof:* see e.g. [WA86].□

The dependency on higher derivatives of the perturbations may cause problems when attempting to solve higher index problems. Roundoff-errors, discretization errors and truncation errors of iterative processes can be viewed as perturbations of the original problem.

Next, we show that multibody system equations have the perturbation index 3. Let $p, v, \lambda$ denote the solutions of the linear unperturbed system (5.1.3) and $\hat{p}, \hat{v}, \hat{\lambda}$ the solution of the corresponding perturbed system

$$\dot{\hat{p}} = \hat{v} + \delta_1(t) \tag{5.1.10a}$$
$$M(\hat{p})\dot{\hat{v}} = f_a(\hat{p}, \hat{v}) - G(\hat{p})^T \hat{\lambda} + \delta_2(t) \tag{5.1.10b}$$
$$0 = g(\hat{p}) + \delta_3(t) \tag{5.1.10c}$$

In analogy to (5.1.5b) we obtain by differentiating (5.1.10c) twice and inserting the result into (5.1.10b)

$$\hat{\lambda} = \left( G(\hat{p})M(\hat{p})^{-1}G(\hat{p})^T \right)^{-1} \left( G(\hat{p})M(\hat{p})^{-1}(f_a(\hat{p}, \hat{v}) + \delta_2(t)) + \zeta(\hat{p}, \hat{v}) + \right.$$
$$\left. + G(\hat{p})\dot{\delta}_1(t) + \Gamma(\hat{p}, \dot{\hat{p}})\delta_1(t) + \ddot{\delta}_3(t) \right) \tag{5.1.11}$$

with $\Gamma(p, \dot{p}) = \frac{d}{dt}G(p)$.

Inserting $\hat{\lambda}$ into the differential equation for $\hat{v}$ finally leads to the following system of explicit ODEs

$$\dot{\hat{p}} = \hat{v} + \delta_1(t) \tag{5.1.12a}$$
$$\dot{\hat{v}} = M(\hat{p})^{-1}(P(\hat{p})(f_a(\hat{p}, \hat{v}) + \delta_2(t)) +$$
$$+ Q(\hat{p})(\zeta(\hat{p}, \hat{v}) + G(\hat{p})\dot{\delta}_1(t) + \Gamma(\hat{p}, \dot{\hat{p}})\delta_1(t) + \ddot{\delta}_3(t))). \tag{5.1.12b}$$

This is just Eq.(5.1.6) perturbed by

$$\delta = \left( \begin{array}{c} \delta_1 \\ M^{-1}\left( P\delta_2 + Q(G\dot{\delta}_1 + \Gamma\delta_1 + \ddot{\delta}_3) \right) \end{array} \right).$$

Assuming for simplicity that the perturbed and unperturbed problem are initialized by the same initial conditions, we obtain from Lemma 5.1.6

$$\left\| \left( \begin{array}{c} \hat{p}(t) \\ \hat{v}(t) \end{array} \right) - \left( \begin{array}{c} p(t) \\ v(t) \end{array} \right) \right\| \leq$$

$$C \max_{t_0 \leq \tau \leq t} \int_{t_0}^{\tau} \left\| \left( \begin{array}{c} \delta_1 \\ M^{-1}\left( P\delta_2 + Q(G\dot{\delta}_1 + \Gamma\delta_1 + \ddot{\delta}_3) \right) \end{array} \right) \right\| ds.$$

Thus mechanical systems have index 3.

A sharper bound is given in [Arn95] where it has been shown, that $\ddot{\delta}_3$ affects only the perturbations in $\lambda$ but not those in $p$ and $v$. That motivates to distinguish also the *index of variables*: $\lambda$ has perturbation index-3, while $p$ and $v$ have perturbation index-2.

There is another way to formally determine the index of a DAE, sometimes called the *differentiation index*, as the index is determined by successively differentiating the constraints:

Let us consider Eqs. (5.1.3a), (5.1.3b) and (5.1.5b). This set of equations was obtained by twice differentiating the position constraint with respect to time and some algebraic manipulations of the equations. A further time differentiation of the algebraic equation leads to an ODE in all variables $p, v$ and $\lambda$:

$$
\begin{aligned}
\dot{p} &= v \\
M(p)\dot{v} &= f_{\rm a}(p,v) - G(p)^T \lambda \\
\dot{\lambda} &= \frac{d}{dt}\left( \left( G(p)M(p)^{-1}G(p)^T \right)^{-1} \left( G(p)M(p)^{-1}f_{\rm a}(p,v) + \zeta(p,v) \right) \right).
\end{aligned}
$$

The *differentiation index* is defined as the number of differentiations required to obtain an ODE in all unknowns. Thus, the equations of motion with the constraint given on position level are an index-3 system conformal to their perturbation index. Analogously, the velocity constrained equations are an index-2 problem and the acceleration constrained equations an index-1 system.

**Remark 5.1.7** *For general systems of the form (5.1.2) with $E_D(y)$ being regular it can be easily checked that (5.1.2) is an*

- *index-1 system, if*

$$
\frac{\partial f_C}{\partial \lambda}(t, y, \lambda) \quad \text{is regular} \tag{5.1.13}
$$

  *in a neighborhood of the solution and*

- *index-2 system, if*

$$
\frac{\partial f_C}{\partial y}(t, y, \lambda) E_D(y)^{-1} \frac{\partial f_D}{\partial \lambda}(t, y, \lambda) \quad \text{is regular and} \quad \frac{\partial f_C}{\partial \lambda} = 0 \tag{5.1.14}
$$

  *in a neighborhood of the solution.*

*In multibody dynamics there are often higher index systems with some constraints being index-1. In order to precise this statement for index-2 systems we partition the constraints and the constraint forces into two corresponding groups $f_{C_1}, \lambda_1$ and $f_{C_2}, \lambda_2$. Then index-2 systems with index-1 equations are such that the DAE (5.1.2) satisfies for $f_{C_1}, \lambda_1$ the index-1 condition (5.1.13) and for $f_{C_2}, \lambda_2$ the index-2 condition (5.1.14). Examples for this type of systems can be found in contact mechanics, see Sec. 5.5.2.*

We consider in this book exclusively DAEs of the form (5.1.2) with a regular matrix $E_D(y)$. For these systems the differentiation index and the perturbation index are the same, [HLR89].

**Remark 5.1.8** *If $G$ is singular we have redundant constraints. $\lambda$ cannot be determined uniquely. In this case the index is not defined but formally set to infinity. In mechanics such systems are called statically undetermined. It can be shown that also in that case $G^T\lambda$ is well-defined, so that the solution components $p$ and $v$ are uniquely given by the equations of motion even in this redundant case.*

### 5.1.4 ODEs with Invariants

The three formulations of the equations of motion are equivalent in the sense that when initialized with the same data, the index-3 formulation, the index-2 formulation and the index-1 formulation have the same analytic solution. For example, the solution $p(t), v(t), \lambda(t)$ of the index-1 problem (5.1.4b)

$$\dot{p} = v \tag{5.1.15a}$$
$$M(p)\dot{v} = f_a(p, v) - G(p)^T\lambda \tag{5.1.15b}$$
$$0 = G(p)\dot{v} + \zeta(p, v) \tag{5.1.15c}$$

also satisfies the position and velocity constraint if the initial values $p(t_0) = p_0$, $v(t_0) = v_0$, $\lambda(t_0) = \lambda_0$ satisfy these constraints, i.e.

$$g(p(t)) = g(p_0) \text{ and } G(p(t))v(t) = G(p_0)v_0.$$

Due to the relation (5.1.5) and the regularity of $M(p)$ this system is equivalent to an explicit ODE of the form

$$\dot{x} = f(x), \quad x(0) = x_0$$

where

$$\begin{pmatrix} g(p) - g(p_0) \\ G(p)v(p) - G(p_0)v_0 \end{pmatrix} = 0 \tag{5.1.16}$$

is an a priori known property of the solution.
This motivates the following definition.

**Definition 5.1.9** *Consider the overdetermined system*

$$\dot{x} = f(x), \quad x(0) = x_0 \tag{5.1.17a}$$
$$0 = \varphi(x) - \varphi(x_0). \tag{5.1.17b}$$

*(5.1.17b) is called an* integral invariant *of (5.1.17a) if it holds for all initial values $x_0$.*

Necessary and sufficient for $\varphi$ to be an integral invariant is $\varphi'(x)f(x) = 0$.
Note that invariants are not constraining the solution, they are additional information on properties of the solution, which are automatically fulfilled. We therefore call the system (5.1.17) an *overdetermined system of ODEs*.
Similarly, (5.1.15a), (5.1.15b), (5.1.15c) form together with (5.1.3c), (5.1.4a) an *overdetermined system of DAEs*. It is often advantageous to consider overdetermined systems as the additional information can be used to stabilize a numerical method, see Sec. 5.3.

**Example 5.1.10** *As pointed out in the introduction (see Sec. 1.3.11) a three parametric description of rotations may lead to singularities in the equations of motion. Therefore, a description in terms of four parameters $q_1, q_2, q_3, q_4$, so-called quaternions or Euler parameters, is used. Quaternions have the property*

$$q_1^2 + q_2^2 + q_3^2 + q_4^2 = 1. \tag{5.1.18}$$

*The following differential equations describe an idealized gyroscope in terms of quaternions*

$$
\begin{aligned}
\dot{q}_1 &= (q_4\omega_1 - q_3\omega_2 + q_2\omega_3)/2 & \text{(5.1.19a)} \\
\dot{q}_2 &= (q_3\omega_1 + q_4\omega_2 - q_1\omega_3)/2 & \text{(5.1.19b)} \\
\dot{q}_3 &= (-q_2\omega_1 + q_1\omega_2 + q_4\omega_3)/2 & \text{(5.1.19c)} \\
\dot{q}_4 &= -(q_1\omega_1 + q_2\omega_2 + q_3\omega_3)/2 & \text{(5.1.19d)} \\
\dot{\omega}_1 &= (b - c)/a\ \omega_2\omega_3 & \text{(5.1.19e)} \\
\dot{\omega}_2 &= (c - a)/b\ \omega_1\omega_3 & \text{(5.1.19f)} \\
\dot{\omega}_3 &= (a - b)/c\ \omega_1\omega_2 & \text{(5.1.19g)}
\end{aligned}
$$

*with the moments of inertia $a = b = 3$ and $c = 1$. These equations together with (5.1.18) form an overdetermined system of ODEs with (5.1.18) being an integral invariant.*

**State Space Form for ODEs with Invariants**
Integral invariants can be used to reduce the number of coordinates. The system in reduced coordinates is called the *state space form* of the ODE with invariants.
We will give here a formal definition of a state space form and relate it to the state space form of constrained mechanical systems, which we already studied in the linear case, see Sec. 2.1.
Let $\varphi : \mathbb{R}^{n_x} \to \mathbb{R}^{n_\varphi}$ and assume that its derivative $H = \frac{d\varphi}{dx}$ has full rank.
We fix a point $x_P \in \mathbb{R}^{n_x}$ with $\varphi(x_P) = 0$ and choose an $n_x \times (n_x - n_\varphi)$ matrix $\tilde{V}(x_P)$ such that

$$
\begin{pmatrix}
\tilde{V}(x_P)^T \\
H(x_P)
\end{pmatrix}
$$

is regular. Then, by applying the Inverse Function Theorem 3.2.1, the existence of a neighborhood $U_1(x_P) \subset \mathbb{R}^{n_x}$ can be shown, such that

$$
\begin{aligned}
y &= \tilde{V}(x_P)^T(x - x_P) \\
0 &= \varphi(x)
\end{aligned}
$$

has a unique solution $x \in U_1(x_P)$. Furthermore, there exists a neighborhood $U_2(0) \subset \mathbb{R}^{n_y}$ with $n_y := n_x - n_\varphi$ and an inverse mapping $\Phi_{x_P} : U_2 \to U_1,\, y \mapsto x$ with $\varphi(\Phi_{x_P}(y)) = 0$.

Differentiating the first equation leads to $\dot{y} = \tilde{V}^T \dot{x} = \tilde{V}^T f$.

**Definition 5.1.11 (State Space Form)**
*The differential equation*

$$\dot{y} = \tilde{V}(x_P)^T f(\Phi_{x_P}(y)) \tag{5.1.20}$$

*is called the* local state space form *of (5.1.17) in $U_1$ with respect to a given parameterization matrix $\tilde{V}(x_P)^T$. $x_P$ is called the* origin of parameterization.*[EFLR90]*

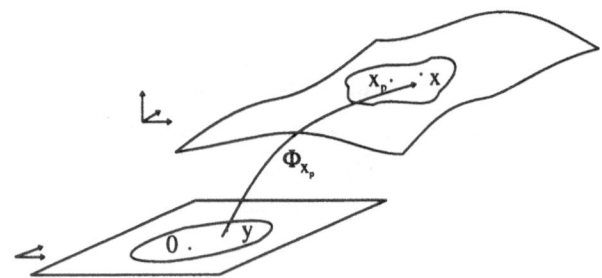

Figure 5.1: Parameterization

State space forms are not unique. They depend locally on the particular choice of $\tilde{V}$. Nevertheless we will speak about *the* state space form, assuming that a particular choice of $\tilde{V}$ was made.

If $\tilde{V}(x_P)$ is chosen such that its columns span the null space of $H(x_P)$, then we refer to the parameterization as a *tangent space parameterization*.

**Example 5.1.12**
*For linear systems of the form*

$$
\begin{aligned}
\dot{x} &= Ax \\
0 &= Hx - z(t)
\end{aligned}
$$

*a state space form can be derived by selecting $\tilde{V}$ such that $H\tilde{V} = 0$ and $\tilde{V}^T\tilde{V} = I_{n_y}$. Setting $\Phi_0(y) := \tilde{V}y + H^+z$ leads to*

$$\dot{y} = \tilde{V}^T A \tilde{V} y + \tilde{V}^T A H^+ z(t). \tag{5.1.21}$$

**Example 5.1.13** *We consider the mechanical system in its index-1 formulation (5.1.15) with its invariants (5.1.16).*
*For H we get*

$$H = \varphi'(x) = \begin{pmatrix} G(p) & 0 & 0 \\ \Gamma(p,v) & G(p) & 0 \end{pmatrix},$$

*with $\Gamma(p,v) = \frac{\partial}{\partial p}(G(p)v)$ and $x^T = (p^T, v^T, \lambda^T)$. Consequently a possible choice for $\tilde{V}(x)$ is*

$$\tilde{V}(x) = \begin{pmatrix} V(p)M(p) & 0 & 0 \\ -G^\dagger(p)\Gamma(p,v)V(p)M(p) & V(p) & 0 \\ 0 & 0 & I \end{pmatrix}$$

*with $V(p)$ such that $G(p)V(p) = 0$ and $G^\dagger = M^{-1}G^T(GM^{-1}G^T)^{-1}$, see Eq. (2.1.7). We will use this parameterization later when considering the so-called implicit state space method.*

## 5.2   Linear Multistep Methods for DAEs

Historically, multistep methods were the first methods applied to DAEs. They were originally based on the observation that BDF methods can be applied to DAEs in essentially the same way as to stiff explicit ODEs, [Gear71a, Gear71b]. The first applications were electrical networks, which are mostly index-1 DAEs. Later, BDF methods were also applied to mechanical systems, [Orl73]. There, numerical problems occurred due to the fact, that mechanical systems have a higher index than electrical networks. This led to the investigation of index reduction techniques and stabilizations, regularization methods and the transformation to explicit ODEs by coordinate partitioning and reduction to state space form. We will describe some of these approaches in the sequel.
First, we demonstrate numerical problems of the various formulations when using Euler's method and the trapezoidal rule.

### 5.2.1   Euler's Method and Adams-Moulton for DAEs

When applying the implicit Euler method to the DAE (5.1.2) the derivatives $\dot{y}(t_n)$ are replaced by a backward difference $(y_n - y_{n-1})/h$. This results in a nonlinear system of equations which has to be solved for obtaining $y_n$ and $\lambda_n$:

$$E_D(y_n)(y_n - y_{n-1}) = h f_D(t_n, y_n, \lambda_n) \qquad (5.2.1a)$$
$$0 = f_C(y_n, \lambda_n). \qquad (5.2.1b)$$

We saw that the equations of motion can be formulated as an index-3 system by using constraints on position level. By using constraints on velocity or $\lambda$ level the equations are obtained in index-2 or index-1 form, respectively. The scheme (5.2.1) is formally applicable to each of these formulations though there are significant differences in performance and accuracy.    First, we will demonstrate this by integrating the mathematical pendulum formulated in Cartesian coordinates as DAE.

The pendulum equations read

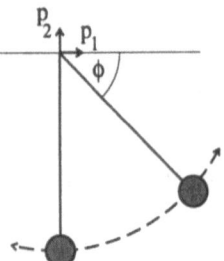

$$\dot{p}_1 = v_1 \qquad\qquad (5.2.2a)$$
$$\dot{p}_2 = v_2 \qquad\qquad (5.2.2b)$$
$$\dot{v}_1 = -\lambda p_1 \qquad\qquad (5.2.2c)$$
$$\dot{v}_2 = -\lambda p_2 - g_{gr} \qquad\qquad (5.2.2d)$$
$$0 = p_1^2 + p_2^2 - 1 \qquad\qquad (5.2.2e)$$
$$0 = p_1 v_1 + p_2 v_2 \qquad\qquad (5.2.2f)$$
$$0 = v_1^2 + v_2^2 - \lambda(p_1^2 + p_2^2) - p_2 g_{gr}. \qquad (5.2.2g)$$

Herein we assume the mass of the pendulum and its length being one. Furthermore, if we set the gravitational constant $g_{gr} = 13.7503671\,\mathrm{m/s^2}$, the period of the pendulum is $2\,\mathrm{s}$, when started with the initial values $p_1(0) = 1, p_2(0) = v_1(0) = v_2(0) = 0$. The function $f_D$ is defined by Eqs. (5.2.2a) - (5.2.2d). We take as constraint $f_C$ either the index-3 formulation (5.2.2e), or the index-2 formulation (5.2.2f) or the index-1 formulation (5.2.2g).

With the implicit Euler method the results given in Tab. 5.1 were obtained.

This table also includes results for the problem formulated in its state space form (ssf)

$$\dot{\phi}_1 = \phi_2$$
$$\dot{\phi}_2 = -g_{gr} \cos \phi_1.$$

Comparing after 2 periods, at $t = 4.0\,\mathrm{s}$, the absolute error in position $e_p$, in velocity $e_v$ and in the Lagrange multipliers $e_\lambda$, we see that for all step sizes the index-2 formulation and the explicit ODE formulation (ssf) give the best results while the index-3 and index-1 approach may even fail. It can also be seen from this table that the effort for solving the nonlinear system varies enormously with the index of the problem. In this experiment the nonlinear system was solved by Newton's method, which was iterated until the norm of the increment became less than $10^{-8}$. If this margin was not reached within 10 iterates the Jacobian was updated. The process was regarded as failed, if even with an updated Jacobian this bound was not met. We see, that the index-3 formulation requires the largest amount of re-evaluations of the Jacobian (NJC) and fails for small step sizes. It is a typical property of

|        | NFE   | NJC  | NIT  | $e_p$        | $e_v$        | $e_\lambda$  |
|--------|-------|------|------|--------------|--------------|--------------|
| $h = 10^{-2}$ | | | | | | |
| Ind. 1 | fail  | fail | fail | fail         | fail         | fail         |
| Ind. 2 | 4009  | 80   | 7.   | $8.10^{-1}$  | $3.10^{+0}$  | $3.10^{+1}$  |
| Ind. 3 | 4175  | 95   | 7.   | $8.10^{-1}$  | $2.10^{+0}$  | $2.10^{+1}$  |
| ssf    | 1946  | 1    | 4.   | $4.10^{-1}$  | $2.10^{+0}$  | $1.10^{+1}$  |
| $h = 10^{-3}$ | | | | | | |
| Ind. 1 | 25990 | 1    | 5.   | $5.10^{-2}$  | $7.10^{-1}$  | $8.10^{-1}$  |
| Ind. 2 | 32499 | 77   | 7.   | $7.10^{-2}$  | $7.10^{-1}$  | $1.10^{+0}$  |
| Ind. 3 | 34856 | 133  | 7.   | $1.10^{-1}$  | $6.10^{-1}$  | $2.10^{+0}$  |
| ssf    | 15346 | 1    | 3.   | $3.10^{-2}$  | $2.10^{-1}$  | $5.10^{-1}$  |
| $h = 10^{-4}$ | | | | | | |
| Ind. 1 | 155365 | 1   | 3.   | $5.10^{-3}$  | $7.10^{-2}$  | $7.10^{-3}$  |
| Ind. 2 | 305978 | 61  | 7.   | $6.10^{-3}$  | $7.10^{-2}$  | $8.10^{-2}$  |
| Ind. 3 | fail   | fail | fail | fail         | fail         | fail         |
| ssf    | 120001 | 1   | 2.   | $4.10^{-3}$  | $2.10^{-2}$  | $5.10^{-2}$  |

Table 5.1: Comparison of performance results for the implicit Euler scheme applied to different formulations of the pendulum problem. NFE: number of function evaluations, NJC: number of Jacobian evaluations, NIT: average number of Newton iterations, $e_p, e_v$ and $e_\lambda$: absolute errors in $p_1(4), v_1(4), \lambda(4)$, resp.

higher index systems, that the Newton iteration fails to converge due to problems in solving the linear systems accurately.

**The Drift-off Effect**
In the index-2 and index-1 formulations the position constraint is no longer incorporated. This constraint is violated when using one of these formulations. The result is given in Fig. 5.2, where the development of the residual of the position constraint is visualized for the pendulum example.

It can be seen that the residuals on position level are growing with increasing time leading to physically meaningless solutions because the invariant is not fulfilled. This instability is called *drift-off* effect. For linear differential equations with constant coefficients this effect can be explained by Theorem 2.7.3, which states that index reduction introduces multiple zero eigenvalues into the system. Multiple zero eigenvalues cause a polynomial instability, i.e. errors due to perturbations of the initial conditions grow like a polynomial. Errors due to discretization and due truncation of the corrector iteration can be viewed as perturbations of the system. This is reflected by the fact, that the choice of smaller step sizes diminishes also the violation of the constraint.

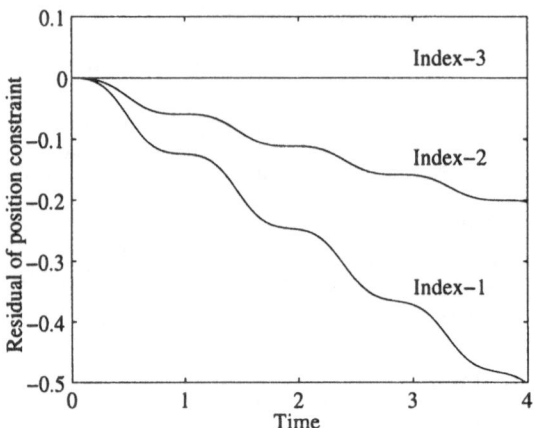

Figure 5.2: Residual of the position constraint, when integrating the pendulum with the implicit Euler method and $h = 0.005$.

## Adams Methods and DAEs

Also methods using past values of the right hand side like Adams methods can at least formally be applied to DAEs. We will start here with the trapezoidal rule and will treat a general multistep formulation in the next section.

We will rewrite the DAE (5.1.2) in *semi-explicit form* be introducing an additional variable $w$

$$\dot{y} = w$$
$$0 = E_D(y)w - f_D(t, y, \lambda)$$
$$0 = f_C(y, \lambda)$$

Discretizing this equation with the Adams–Moulton two step method (see Ex. 4.1.2) then results in

$$y_n = y_{n-1} + \frac{h}{12}(5w_n + 8w_{n-1} - w_{n-2}) \qquad (5.2.4a)$$
$$0 = E_D(y_n)w_n - f_D(t_n, y_n, \lambda_n) \qquad (5.2.4b)$$
$$0 = f_C(y_n, \lambda_n). \qquad (5.2.4c)$$

Applying this method to the pendulum problem we observe an unstable behavior for the index-2 and index-3 formulation, see Fig. 5.3.

By decreasing the step size this instability is not removed. The two step Adams–Moulton method is like all higher order methods in this class not zero-stable when applied to index-2 or index-3 systems. We will give in Sec. 5.2.3 criteria for zero stability of multistep methods applied to index-2 systems.

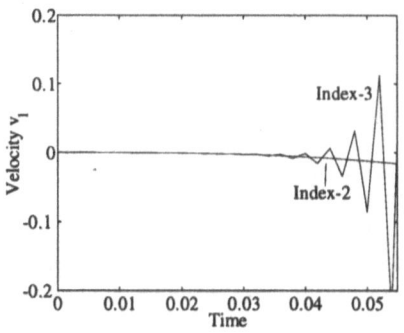 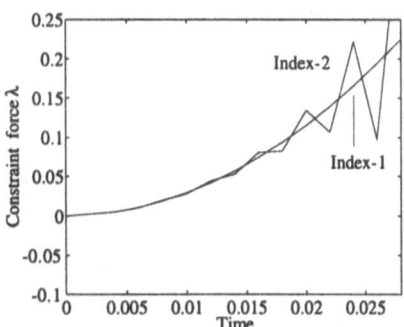

Figure 5.3: The solution of the pendulum problem generated with the two step Adams–Moulton method and $h = 0.01$

### Start Phase in Index-3 Problems

Discretized index-3 problems have an initial phase (boundary layer), where the method's order differs from subsequent steps. The same holds for the situation right after step size changes. In order to demonstrate this effect we consider the semi-explicit index-3 system,

$$
\begin{aligned}
\dot{y}_1 &= y_2 \\
\dot{y}_2 &= \lambda \\
0 &= y_1 - g(t)
\end{aligned}
$$

[BCP89, Füh88]. Consistent initial values are $y_1(0) = g(0)$, $y_2(0) = \dot{g}(0)$ and $\lambda(0) = \ddot{g}(0)$.

Discretizing this system with the implicit Euler method yields

$$
\begin{aligned}
y_{1,1} &= g(h) \\
y_{2,1} &= \frac{g(h) - g(0)}{h} \\
\lambda_1 &= \frac{g(h) - g(0) - h\dot{g}(0)}{h^2},
\end{aligned}
$$

$$
\begin{aligned}
y_{1,2} &= g(2h) \\
y_{2,2} &= \frac{g(2h) - g(h)}{h} \\
\lambda_2 &= \frac{g(2h) - 2g(h) + g(0)}{h^2}
\end{aligned}
$$

and for a general step, $n > 2$, we get

$$
\begin{aligned}
y_{1,n} &= g(nh) \\
y_{2,n} &= \frac{g(nh) - g((n-1)h)}{h} \\
\lambda_2 &= \frac{g(nh) - 2g((n-1)h) + g((n-2)h)}{h^2}.
\end{aligned}
$$

We note that in this example of a system with no degrees of freedom, the influence of the initial conditions vanishes completely after two steps. Furthermore, we observe in the first step

$$
\lambda(h) - \lambda_1 = \frac{g(h)}{2} + \mathcal{O}(h) = \mathcal{O}(1)
$$

while in a general step the global error is

$$
\lambda(nh) - \lambda_n = \ddot{g}(nh)h + \mathcal{O}(h^2),
$$

which is exactly what we expect from the situation for explicit ODEs. If, instead, we would have started the process with initial conditions

$$
\begin{pmatrix}
g(0) \\
\frac{g(0) - g(-h)}{h} \\
\frac{g(0) - 2g(-h) + g(-2h)}{h^2}
\end{pmatrix}
\tag{5.2.5}
$$

we would have observed the "right" order of convergence right from the beginning. These initial conditions are in that sense *consistent with the discretization*.
A step size change can be viewed as a change of the discretization and we can expect the same phenomenon again. Indeed, if the step size in step $n$ is changed from $h$ to $h/r$ we get

$$
\begin{aligned}
y_{1_{n+1}} &= g(t_{n+1}) \\
y_{2_{n+1}} &= \frac{g(t_{n+1}) - g(t_n)}{h_{n+1}} = \frac{r(g(t_{n+1}) - g(t_n))}{h_n} \\
\lambda_{n+1} &= \frac{1}{h_{n+1}} \left( \frac{g(t_{n+1}) - g(t_n)}{h_{n+1}} - \frac{g(t_n) - g(t_{n-1})}{h_n} \right) \\
&= r \frac{r\,(g(t_{n+1}) - g(t_n)) - (g(t_n) - g(t_{n-1}))}{h_n^2} \\
&= \mathcal{O}(1) \text{ if } r \neq 1.
\end{aligned}
$$

Especially, if the step size is decreased very much, the error in $\lambda$ can become large.

Again, after two more steps with step size $h_{n+1}$ this effect vanishes. Due to these effects, error estimation and step size control become complicated for index-3 problems and presently no implementation is known that handles these problems. In [Are93] a projection step is suggested to improve the results in these phases after step size changes.

This example shows that the influence of inconsistencies in the input data vanishes completely after a fixed number of steps. We will see in Sec. 5.2.3 that this "forgiving" property is shared by all BDF methods. It can be viewed as a reaction of a discrete *dead-beat controller* on disturbances, [ÅW90, AFS97].

The observations of this section are summarized in Table 5.2. We see that each formulation has its own flaws and we will later in this chapter see how a combination of all formulations leading to an overdetermined DAEs (ODAE) will overcome these problems.

| Formulation | Implicit Euler Method | Adams–M. 2 step |
|---|---|---|
| Index-3 (Position constraint) | problems when solving the linear system in corrector iteration step size changes | unstable in $v$ and $\lambda$ |
| Index-2 (Velocity constraint) | (slow) drift-off | unstable in $\lambda$ |
| Index-1 ($\lambda$- constraint) | drift-off | drift-off |
| State space form (ssf) | - | - |

Table 5.2: Problems of Euler's method and the trapezoidal rule for equations of constrained motion

## 5.2.2  Multistep Discretization Schemes for DAEs

We consider system (5.1.2) and an equidistant partition of the time axis with the step size $h := t_n - t_{n-1}$. Corresponding to Sec. 4 we denote the numerical approximation to the solution at time $t_n$ by $x_n$. We use the notation $\dot{x}_n$ for the numerical approximation to $\dot{x}(t_n)$. When we want to express its dependency on $x_n$ we write $\delta_h(x_n)$ instead.

After normalizing $\beta_k = 1$ we rewrite the implicit multistep discretization scheme (4.1.12) as

$$\delta_h(x_n) := \dot{x}_n := \frac{\alpha_k}{h} x_n + \frac{1}{h} \sum_{i=1}^{k} \alpha_{k-i} x_{n-i} - \sum_{i=1}^{k} \beta_{k-i} \dot{x}_{n-i}. \tag{5.2.6}$$

Especially in case of BDF this reads

$$\delta_h(x_n) = \frac{\alpha_k}{h} x_n + \frac{1}{h} \sum_{i=1}^{k} \alpha_{k-i} x_{n-i}. \tag{5.2.7}$$

This leads to the discretized form of (5.1.2)

$$E_D(y_n)\delta_h(y_n) \;=\; f_D(t_n, y_n, \lambda_n) \tag{5.2.8a}$$
$$0 \;=\; f_C(t_n, y_n, \lambda_n). \tag{5.2.8b}$$

This is a set of nonlinear equations for the unknowns $y_n$ and $\lambda_n$, which can be solved by *Newton's method.*
Viewing DAEs as the limiting case of *singularly perturbed ODEs* of the form

$$E_D(y)\dot{y} \;=\; f_D(t, y, \lambda) \tag{5.2.9a}$$
$$\varepsilon \dot{\lambda} \;=\; f_C(t, y, \lambda) \tag{5.2.9b}$$

leads to an alternative approach. After discretizing this problem in the usual way, the parameter $\varepsilon$ is set to zero, which leads to the method

$$E_D(y_n)\delta_h(y_n) \;=\; f_D(t_n, y_n, \lambda_n) \tag{5.2.10a}$$

$$0 \;=\; \sum_{i=0}^{k} \beta_{k-i} f_C(t_{n-i}, y_{n-i}, \lambda_{n-i}). \tag{5.2.10b}$$

We see that this so-called *direct approach* is only in the BDF-case identical to method (5.2.8). In the general case the numerical values do not satisfy the constraints and it can be shown, that the Adams-Moulton methods fail in this implementation due to stability problems, see [HW96, GM86]. (For an extensive discussion of the direct approach and some generalizations, see [Are93].) So, it might not always be a good idea viewing DAEs as a limiting case of singular perturbed or stiff ODEs.

## 5.2.3   Convergence of Multistep Methods

In this section we will cite three results concerning convergence of multistep methods applied to index-1, index-2 and index-3 problems of the form (5.1.2). To illustrate the assumptions in these theorems we will investigate a multistep discretization of the linearized, unconstrained truck. There, we will see how discretization can be discussed in terms of mapping eigenvalues from the left half plane into (or not into) the unit circle. This will extend the discussion of stability for explicit ODEs by looking at extreme cases: eigenvalues at infinity or multiple eigenvalues in zero. We will see, the image of these eigenvalues under discretization cannot be influenced by changing the step size and we have to introduce the notion of *stability at* $\infty$.

**Index-1 Systems**

**Theorem 5.2.1** *Let (5.1.2) have index 1 and let the starting values $x_i = (y_i^T, \lambda_i^T)^T$ satisfy*

$$x_i - x(t_i) = \mathcal{O}(h^p), \quad i = 0, \dots, k-1.$$

*If the multistep discretization defined by (4.1.12) is consistent of order p and zero stable, then the method (5.2.6) converges with order p, i.e. the global error is*

$$y_n - y(t_n) = \mathcal{O}(h^p), \quad \lambda_n - \lambda(t_n) = \mathcal{O}(h^p)$$

*for $t_n = nh \le t_e$. [HW96]*

Consequently, problems which are index-1 "per se" pose not more problems than explicit ODEs. Unfortunately mechanical problems can only be transformed into an index-1 formulation by differentiation, which results in instabilities in the problem, the drift-off effect. Thus, it is the stability of the *problem* and not of the *method* which causes problems.

**Index-2 Systems**

The index-2 case is the situation where the equations of motion are set up together with constraints on velocity level. We will see that the negative observation concerning the two step Adams-Moulton method holds in general for all higher order Adams-Moulton methods. The central convergence theorem requires $\infty$-stability of the method.

**Definition 5.2.2** *A method is called* stable at infinity, *if $\infty$ is in its stability region, i.e. if the roots of the $\sigma$ polynomial lie in the unit circle and those on its boundary are simple. If there are no roots on the boundary of the unit circle the method is called strictly stable at infinity.*

In light of the linear test equation this means that such a method is stable for *all* step sizes in the ultimately stiff case, where the stiffness parameter tends to infinity. Clearly, BDF methods are strictly stable at infinity as the generating polynomial $\sigma(\xi) = \xi^k$ has all its roots in the origin.

In contrast, Adams-Moulton methods other than the implicit Euler are not strictly stable at infinity. The two step Adams-Moulton method is not stable at infinity as its $\sigma$ polynomial has a root at -1.76.

**Theorem 5.2.3** *Assume that (5.1.2) has index-2 and let k starting values $x_i = (y_i^T, \lambda_i^T)^T$ satisfy*

$$y_i - y(t_i) = \mathcal{O}(h^{p+1}), \quad \lambda_i - \lambda(t_i) = \mathcal{O}(h^p), \qquad i = 0, \dots, k-1.$$

*If the multistep discretization defined by (4.1.12) is consistent of order $p$, zero stable and strictly stable at infinity, then the method (5.2.6) converges with order $p$, i.e. the global error is*

$$y_n - y(t_n) = \mathcal{O}(h^p), \quad \lambda_n - \lambda(t_n) = \mathcal{O}(h^p)$$

*for $t_n = nh \leq t_e$.[HW96, Are93]*

Note that higher accuracy has to be required for the initial values. This theorem is formulated for the idealized case, where it is assumed that the nonlinear corrector system is solved exactly.

### Index-3 Systems

For the index three case the boundary layer effect (see Sec. 5.2.4) has to be taken into account. We cite the following convergence results for BDF methods and systems, where $f_C$ is independent of $\lambda$:

**Theorem 5.2.4** *Assume that (5.1.2) has index-3. Let this equation be discretized by a $p$-th order BDF method using $k$ starting values $x_i = (y_i^T, \lambda_i^T)^T$ satisfying*

$$y_i - y(t_i) = \mathcal{O}(h^{p+1}), \quad f_C(t_i, y_i) = \mathcal{O}(h^{p+2}) \quad i = 0, \ldots, k-1.$$

*Then there exists a numerical solution which converges after $k+1$ steps with order $p$, i.e. the global error is*

$$y_n - y(t_n) = \mathcal{O}(h^p), \quad \lambda_n - \lambda(t_n) = \mathcal{O}(h^p)$$

*for $t_n = nh \leq t_e$.[BCP89]*

### Stability Problems*

We defined consistency exclusively as a property of a discretization scheme which is independent of a given differential equations. So it is stability which causes convergence problems for higher index DAEs and which led to ask for stability at infinity.

This can be demonstrated by considering different discretizations for linearized mechanical systems. The presentation here follows the lines of [AFS97] where also proofs of the following theorems can be found.

We consider systems of the generic form

$$\begin{aligned} \dot{x} &= Ax - G_2^T \lambda & \text{(5.2.11a)} \\ 0 &= G_1 x, & \text{(5.2.11b)} \end{aligned}$$

with

$$x = \begin{pmatrix} p \\ v \end{pmatrix} \text{ and } A = \begin{pmatrix} O & I \\ -M^{-1}K & -M^{-1}D \end{pmatrix}$$

(see (2.1.1)). By setting $G_2 = G_1 = (0 \quad G)$ this form describes the index-2 case (velocity constraint), while $G_2 = (0 \quad G)$, $G_1 = (G \quad 0)$ describes the index-3 case (position constraint).

After discretizing this system and writing it in explicit one-step form, it takes the form

$$\begin{pmatrix} X \\ \Lambda \end{pmatrix}_n = \Phi(h) \begin{pmatrix} X \\ \Lambda \end{pmatrix}_{n-1}. \tag{5.2.12}$$

Stability of the discretization can be described in terms of the eigenvalues of the discrete transfer matrix $\Phi(h)$ and like in the discrete case, the eigenvalues should lie inside the unit circle, and those of unit modulus must have a full set of linearly independent eigenvalues.

The system can be first transformed into its state space form (2.1.12). The state space form describes the system's dynamic behavior in minimal coordinates. Discretizing this system by the same method leads to a discrete transfer matrix $S(h)$ of the state space form.

In Sec. 2.6 we discussed generalized eigenvalues of linear constant coefficient DAEs and found three groups of eigenvalues

- the eigenvalues of the state space form, related to the dynamics of the system (dynamic eigenvalues),

- multiple eigenvalues at infinity, related to the constraints and their derivatives (structural eigenvalues due to constraints),

- defective zero eigenvalues introduced by index reduction. These may be related to solution invariants, see Sec. 5.1.4 (structural eigenvalues due to invariants).

It is interesting to see what happens with these eigenvalues when considering the discretized system. We will see that the image of the structural eigenvalues is $h$-independent, while the image of the dynamic eigenvalues depends on $h$.

Given a sequence of square matrices $\{A_j\}_{j=0}^{k-1}$ we introduce the notation for companion matrices

$$\operatorname*{comp}_{j=0:k-1} (A_j) := \begin{pmatrix} -A_{k-1} & \cdots & & -A_0 \\ I & & & \\ & \ddots & & \\ & & I & 0 \end{pmatrix}.$$

A straightforward calculation of the discrete-time transfer matrix $\Phi(h)$ yields the structure

$$\Phi(h) = \left( \begin{array}{cc} \mathrm{comp}_{j=0:k-1}(PA_k^{-1}A_j) & 0 \\ \mathrm{comp}_{j=0:k-1}(QA_k^{-1}A_j) & \mathrm{comp}_{j=0:k-1}(\beta_j I) \end{array} \right) =: \left( \begin{array}{cc} \mathcal{A} & 0 \\ \mathcal{D} & \mathcal{B} \end{array} \right),$$

$$(5.2.13)$$

where the matrices $A_j$ are defined by

$$A_j = \alpha_j I - h\beta_j A, \qquad j = 0 \ldots k. \qquad (5.2.14)$$

The step size $h$ is assumed to be sufficiently small so that $A_k$ is invertible. We have also used the projector

$$P = I - A_k^{-1}G_2^T(G_1 A_k^{-1} G_2^T)^{-1} G_1$$

and the $n_\lambda \times n_x$ matrix

$$Q = (G_1 A_k^{-1} G_2^T)^{-1} G_1$$

(see also Sec. 2.6.3). Note that $QP = 0$.

The submatrix $\mathcal{B} := \mathrm{comp}_{j=0:k-1}(\beta_j I)$ is called the $\beta$-*block* of $\Phi(h)$. It is the companion matrix of the $\sigma$ polynomial of the discretization method and the eigenvalues of this matrix are a subset of $h$-independent eigenvalues of $\Phi(h)$.

Thus, a necessary condition for the stability of the discretization is that these eigenvalues are inside the unit circle, which is just the assumption we made in Theorem 5.2.3.

For BDF methods we get

$$\mathcal{B}_{\mathrm{BDF}} = \left( \begin{array}{cccc} 0 & 0 & \cdots & 0 \\ I & & & \\ & \ddots & & \\ & & I & 0 \end{array} \right).$$

This matrix has the property $\mathcal{B}^k = 0$ and all its eigenvalues are in the origin. A perturbation in $\lambda$ vanishes after $k$ steps.

On the other hand, for Adams-Moulton methods with order $p > 2$ this matrix has eigenvalues larger than one in modulus and the discretization becomes unstable. The discretization scheme is no longer stable at infinity and Theorem 5.2.3 can no longer be applied.

Recently, a special family of multistep methods has been developed which shifts the eigenvalues of the $\beta$-block $\mathcal{B}$ into the unit circle. Therefore, these methods are called $\beta$-*blocked multistep methods*, [AFS96].

Also the submatrix $\mathcal{A}$ gives an interesting insight in the nature of the discretized linear system. We cite the following theorem from [AFS97].

**Theorem 5.2.5** *The $kn_x \times kn_x$ submatrix $A$ is similar to a matrix of the structure*

$$\begin{pmatrix} A_1 & 0 \\ * & S \end{pmatrix} \tag{5.2.15}$$

*where $S$ is the transfer matrix of the discrete state space form.*
*For index 3 systems the matrix $A_1$ has $k \cdot n_\lambda$ structural eigenvalues (independent of $h$, $\sigma$ and $\rho$) equal to zero, and $k \cdot n_\lambda$ eigenvalues equal to the $k$ roots of $\sigma$, each repeated $n_\lambda$ times (the second $\beta$-block).*

By this theorem we see that the discretized state space form and the discretized linear DAE share the same (dynamic) eigenvalues.

**Example 5.2.6** *We consider the equations of motion of the linearized constrained constrained truck in its index-2 formulation. Discretizing these equations by the two step Adams-Moulton method (see Sec. 4.1.2) leads to the eigenvalues of the discrete transfer matrix which are given in Table 5.3. One clearly identifies structural and dynamic eigenvalues and the source of instability of AM3 when applied to an index-2 problem, see also Tab. 2.2.*

The analysis so far was related to linear systems with constant system matrices. It suggests that the dynamics of discretized systems is entirely determined by the discretized state space form (see middle part of Tab. 5.3). Having the linear test equation in mind, one might think of stiff DAEs as those systems having a stiff state space form.
Unfortunately, the situation is much complicater than that. We consider the following example from [HIM97].

**Example 5.2.7** *Consider the index-2 system*

$$\begin{aligned} \dot{x}_1 &= ax_1 + x_2 + x_3 \\ \dot{x}_2 &= (b^2 t^2 - bt + b)\, x_1 - ax_2 + bt\, x_3 \\ 0 &= (1 - bt)\, x_1 + x_2. \end{aligned} \tag{5.2.16}$$

*A state space form of this system is just the linear test equation*

$$\dot{x}_1 = ax_1,$$

*while the other components are determined by the algebraic relations*

$$x_2 = (bt - 1)x_1 \ and \ x_3 = -x_2.$$

*Consequently, the stability of the problem is determined by the size of $a$. We discretize this system with the five step BDF method*

$$\frac{137}{60} x_{n+1} - 5x_n + 5x_{n-1} - \frac{10}{3} x_{n-2} + \frac{5}{4} x_{n-3} - \frac{1}{5} x_{n-4} = hf(t_{n+1}, x_{n+1})$$

*and set a := −10 so that the discretized state space form is stable for all step sizes. In Fig. 5.4 the norm of the numerical solution is shown for different values of b and a step size h = 0.1. One clearly sees the unstable behavior of the numerical solution for b ≥ 0. For b = 5 there is a stability bound at h ≈ 0.07, while due to the A(α) stability of the method there is no stability bound for b = 0. For b = 0*

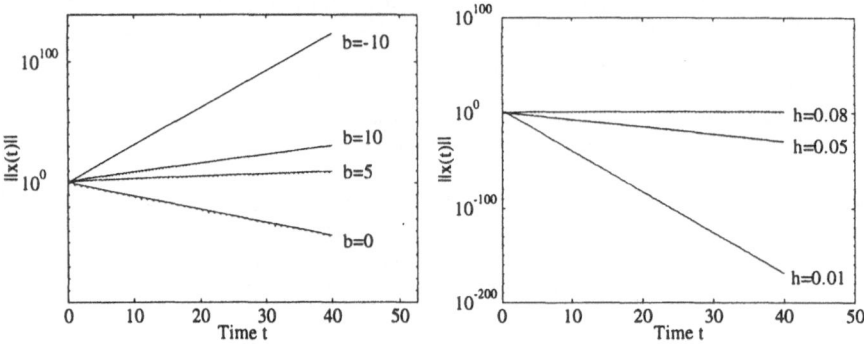

Figure 5.4: $\|x(t)\|$ for various values $b$ and $h = 0.1$ (left) and $\|x(t)\|$ for various step sizes and $b = 5$ (right)

*the system gets constant coefficients and the stability of the numerical method is just the one of the linearized state space form. Thus, there are examples were the numerical treatment of the state space form can be more efficient than the direct treatment of the DAE. This field is a topic of current research and we refer to [HIM97, WWS94, AP93] for details.*

### 5.2.4 Solving Corrector Equations in Discretized DAEs

#### 5.2.4.1 Exploiting the Multibody System Structure

In Sec. 4.1.6 it has been shown how the second order nature of the multibody equations can be exploited to reduce the linear algebra effort [Eich91]. These techniques will be extended here to the index-1 DAE case.

For the index-1 formulation of the multibody equations we have to solve the non-linear system

$$p_n = \xi_p + \gamma v_n \tag{5.2.17a}$$

$$v_n = \xi_v + \gamma w_n \tag{5.2.17b}$$

$$M(p_n)w_n = f_a(p_n, v_n) - G(p_n)^T \lambda_n \tag{5.2.17c}$$

$$0 = G(p_n)w_n + \zeta(p_n, v_n) \tag{5.2.17d}$$

with $\gamma = h_n \beta_k^c$ and $\xi$ being the "old" data as in Sec. 4.1.6. For ease of notation we omit in the sequel the index $n$. Applying Newton's method and inserting the expressions for $p, v$ leads to (cf. System (4.1.32))

$$
\begin{pmatrix} M + \gamma^2 \frac{\partial}{\partial p} (Mw - f_a + G^T \lambda) - \gamma \frac{\partial}{\partial v} f_a & G^T \\ G + \gamma^2 \frac{\partial}{\partial p} (Gw + \zeta) + \gamma \frac{\partial}{\partial v} \zeta & 0 \end{pmatrix}^{(i)} \begin{pmatrix} \Delta w \\ \Delta \lambda \end{pmatrix}^{(i)} = - \begin{pmatrix} r_1 \\ r_2 \end{pmatrix}^{(i)}
$$

and $w^{(i+1)} = w^{(i)} + \Delta w^{(i)}$, $\lambda^{(i+1)} = \lambda^{(i)} + \Delta \lambda^{(i)}$, $v^{(i+1)} = \gamma w^{(i+1)} + \xi_v$, and $p^{(i+1)} = \gamma v^{(i+1)} + \xi_p$. Herein we set

$$
\begin{pmatrix} r_1(p(w), v(w), w, \lambda) \\ r_2(p(w), v(w), w, \lambda) \end{pmatrix} := \begin{pmatrix} M(p(w))w - f_a(p(w), v(w)) + G(p(w))^T \lambda \\ G(p(w))w + \zeta(p(w), v(w)) \end{pmatrix}
$$

This system has only $n_p + n_\lambda$ equations and variables, whereas (5.2.17) has the dimension $3n_p + n_\lambda$. Thus the effort for solving the linear equations is reduced significantly.

Taking this structure into consideration, the iteration matrix can be computed efficiently using finite differences:

$$
\left( M + \gamma^2 \frac{\partial}{\partial p} (Mw - f_a + G^T \lambda) - \gamma \frac{\partial}{\partial v} f_a \right)_{j,i}
$$
$$
= \frac{r_{1,j}(p + \gamma^2 \eta e_i, v + \gamma \eta e_i, w + \eta e_i, \lambda) - r_{1,j}(p, v, w, \lambda)}{\eta} + \mathcal{O}(\eta)
$$

and

$$
\left( G + \gamma^2 \frac{\partial}{\partial p} (Gw + \zeta) + \gamma \frac{\partial}{\partial v} \zeta \right)_{j,i}
$$
$$
= \frac{r_{2,j}(p + \gamma^2 \eta e_i, v + \gamma \eta e_i, w + \eta e_i, \lambda) - r_{2,j}(p, v, w, \lambda)}{\eta} + \mathcal{O}(\eta)
$$

Note that the predictor values for $p$ and $v$ must be computed using (5.2.17a) and (5.2.17b).

If the partial derivatives in these expressions are small with respect to $M$ and $G$, respectively, the above system can be simplified to

$$
\begin{pmatrix} M & G^T \\ G & 0 \end{pmatrix}^{(i)} \begin{pmatrix} \Delta w \\ \Delta \lambda \end{pmatrix}^{(i)} = - \begin{pmatrix} r_1 \\ r_2 \end{pmatrix}^{(i)}.
$$

The leading matrix in this equation has the same structure as matrices occuring in constrained least squares problems and solution techniques discussed in Sec. 2.3 can be applied. This leads to an additional gain by a factor two in the number of operations for decomposing the Jacobian.

## 5.3  Stabilization and Projection Methods

The index reduced system and the original system have the same solutions if the initial values are consistent. In the case of semi-explicit DAEs this means that the algebraic equations and their derivatives are fulfilled. Then the solution lies in all manifolds defined by the algebraic equations and the overdetermined system

$$
\begin{array}{rcll}
\dot{p} &=& v & (5.3.1a) \\
M(p)\dot{v} &=& f_{\mathrm{a}}(p,v) - G^T(p)\lambda & (5.3.1b) \\
0 &=& g(p) & (5.3.1c) \\
0 &=& G(p)v & (5.3.1d) \\
0 &=& G(p)M(p)^{-1}(f_{\mathrm{a}}(p,v) - G(p)^T\lambda) - \zeta(p,v) & (5.3.1e)
\end{array}
$$

has a solution.

The situation is different for the discretized system. Due to discretization and roundoff errors the discretized version of (5.3.1), the overdetermined system

$$
\begin{array}{rcll}
\delta_h(p_n) &=& v_n & (5.3.2a) \\
M(p_n)\delta_h(v_n) &=& f_{\mathrm{a}}(p_n,v_n) - G(p_n)^T\lambda_n & (5.3.2b) \\
0 &=& g(p_n) & (5.3.2c) \\
0 &=& G(p_n)v_n & (5.3.2d) \\
0 &=& G(p_n)M(p_n)^{-1}(f_{\mathrm{a}}(p_n,v_n) - G(p_n)^T\lambda_n) - \zeta_n, & (5.3.2e)
\end{array}
$$

is in general not solvable in the classical sense. $\delta_h(x_n)$ denotes the multistep approximation to $\dot{x}(t_n)$, see (5.2.6).

We present two classes of methods to overcome this difficulty:

- the *coordinate projection* and

- the *implicit state space form*.

The coordinate projection method is motivated by avoiding the drift-off effect: Eqs. (5.3.1a), (5.3.1b), (5.3.1e) are integrated one step and the solution is projected onto the manifold given by (5.3.1d), (5.3.1c).

The implicit state space form is motivated by a transformation to state space form, discretization of the state space form equations and back transformation. Integrating constrained mechanical systems by first transforming them to state space form is used frequently, [HY90, Yen93]. This leads to approaches known as *coordinate splitting methods*. The method we discuss here does not carry out the transformation to state space form explicitly. It embeds the choice of state variables in the iterative solution of the nonlinear system (5.3.2), while the integration method always treats the full set of coordinates.

## 5.3.1    Coordinate Projection Method

Coordinate projection is applied after an integration step is completed. Considering a multibody system in its index-1 form (5.1.3a),(5.1.3b),(5.1.4b), a typical integration step results in values $\tilde{p}_n$, $\tilde{v}_n$, and $\tilde{\lambda}_n$ being the solution of the nonlinear system

$$\delta_h(\tilde{p}_n) = \tilde{v}_n \tag{5.3.3a}$$
$$M(\tilde{p}_n)\delta_h(\tilde{v}_n) = f_a(\tilde{p}_n,\tilde{v}_n) - G(\tilde{p}_n)^T\lambda_n \tag{5.3.3b}$$
$$0 = G(\tilde{p}_n)\delta_h(\tilde{v}_n) + \zeta(\tilde{p}_n,\tilde{v}_n). \tag{5.3.3c}$$

These values are then projected orthogonally back onto the manifolds given by the constraints on position and velocity level, i.e. the projected values are defined as the solution of the *nonlinear constrained least squares problem*

$$\|p_n - \tilde{p}_n\|_2 = \min_{p_n} \tag{5.3.4a}$$
$$g(p_n) = 0 \tag{5.3.4b}$$
$$\tag{5.3.4c}$$
$$\|v_n - \tilde{v}_n\|_2 = \min_{v_n} \tag{5.3.4d}$$
$$G(p_n)v_n = 0 \tag{5.3.4e}$$

The projected values, which now fulfill also position and velocity constraints, are then used to advance the solution.

This projection method can be applied to any discretization method. It has first been proposed for differential equations with invariants in [Sha86], where also convergence for one step methods has been proven. Convergence for multistep methods was shown in [Eich93].

**Computation of the Projection Step**
The numerical solution of the projection step can be computed iteratively by a Gauß–Newton method, see also Ch. 7.2.2. There, a nonlinear constrained least squares problem is solved iteratively. The nonlinear functions are linearized about the current iterate and the problem is reduced to a sequence of linear constrained least squares problems.

In a typical step the current iterate $p_n^{(i)}$ is updated by

$$p_n^{(i+1)} = p_n^{(i)} + \Delta p_n^{(i)}$$

where the increment $\Delta p_n^{(i)}$ is the solution of the problem linearized about $p_n^{(i)}$

$$\|p_n^{(i)} - \tilde{p}_n + \Delta p_n^{(i)}\|_2 = \min_{\Delta p_n^{(i)}} \tag{5.3.5a}$$
$$g(p_n^{(i)}) + G(p_n^{(i)})\Delta p_n^{(i)} = 0. \tag{5.3.5b}$$

Similarly, $v_n^{(i)}$ is updated by

$$v_n^{(i+1)} = v_n^{(i)} + \Delta v_n^{(i)}$$

with

$$\|v_n^{(i)} - \tilde{v}_n + \Delta v_n^{(i)}\|_2 \quad = \quad \min_{\Delta v_n^{(i)}} \tag{5.3.6a}$$

$$G(p_n)(v^{(i)}) + G(p_n)\Delta v_n^{(i)} \quad = \quad 0 \tag{5.3.6b}$$

and $p_n$ being the solution of the first iteration process (5.3.5).
To initialize the iteration $p_n^{(0)} := \tilde{p}_n$ and $v_n^{(0)} := \tilde{v}_n$ can be used.
The solution of the linear constrained least squares problems (5.3.5) and (5.3.6) are according to (2.3.4a)

$$\Delta p_n^{(i)} \quad = \quad -G^+(p_n^{(i)})g(p_n^{(i)}) + (I - G^+(p_n^{(i)})G(p_n^{(i)}))(\tilde{p}_n - p_n^{(i)})$$
$$\Delta v^{(i)} \quad = \quad -G^+(p_n)G(p_n)v_n^{(i)} + (I - G^+(p_n)G(p_n))(\tilde{v}_n - v_n^{(i)}).$$

$G^+$ denotes the pseudo-inverse defined in (2.3.11).
We refer to Sec. 2.3.3 for a description of various methods to efficiently compute $\Delta p_n^{(i)}$ and $\Delta v_n^{(i)}$.

## 5.3.2   Implicit State Space Form*

We  will first motivate this approach by considering the linearized form of the multibody equations (5.3.1)

$$\dot{p} \quad = \quad v \tag{5.3.7a}$$
$$M\dot{v} \quad = \quad -Kp - Dv - G^T\lambda \tag{5.3.7b}$$
$$Gp \quad = \quad z \tag{5.3.7c}$$
$$G\dot{p} \quad = \quad \dot{z} \tag{5.3.7d}$$
$$G\ddot{p} \quad = \quad \ddot{z}, \tag{5.3.7e}$$

with an excitation term $z(t)$, cf. (1.5.1).
We will define a solution of the corresponding discretized linear system, cf. (5.3.2),

$$\delta_h(p_n) \quad = \quad v_n \tag{5.3.8a}$$
$$M\delta_h(v_n) \quad = \quad -Kp_n - Dv_n - G^T\lambda_n \tag{5.3.8b}$$
$$0 \quad = \quad Gp_n - z(t_n) \tag{5.3.8c}$$
$$0 \quad = \quad Gv_n - \dot{z}(t_n) \tag{5.3.8d}$$
$$0 \quad = \quad GM^{-1}(-Kp_n - Dv_n - G^T\lambda_n) - \ddot{z}(t_n) \tag{5.3.8e}$$

by the following steps:

1. Transformation to the state space form of (5.3.7), which in that case is globally defined.

2. Discretization of the state space form by an implicit BDF method. The resulting system is – in contrast to the discretized form of (5.3.7) – a square linear system with a well defined solution.

3. Back transformation to the full set of variables $p_n, v_n, \lambda_n$. These values will be taken to define a generalized solution of (5.3.8).

As shown in Sec. 2.1 a state space form of (5.3.7) reads

$$V^T M V \ddot{y} = -V^T K V y - V^T D V \dot{y} - V^T K G^\dagger z - V^T D G^\dagger \dot{z},$$

cf. (2.1.11), with the matrix $V$ spanning the nullspace of $G$.
Transforming it to a first order differential equation by setting $w := \dot{y}$ and discretizing yields

$$
\begin{aligned}
\delta_h(y_n) &= w_n \\
V^T M V \delta_h(w_n) &= -V^T K V y_n - V^T D V w_n \\
&\quad -V^T K G^\dagger z(t_n) - V^T D G^\dagger \dot{z}(t_n).
\end{aligned}
$$

Multiplying the first equation by $V^T M V$, adding on the left hand side the vanishing term $V^T M G^\dagger \delta_h(z(t_n))$, on the right hand side $V^T M G^\dagger \dot{z}(t_n)$, and using Eqs. (2.1.3), (2.1.4), (2.1.8), (2.1.9) yields

$$V^T M \delta_h(p_n) = V^T M v_n. \tag{5.3.9}$$

The second equation can be transformed back by formally adding the vanishing terms $V^T M G^\dagger \delta_h(\dot{z}(t_n))$ on the left hand side and $-V^T G^T \lambda$ on the right hand side. Using the derivatives of Eqs. (2.1.3), (2.1.4) and (2.1.9) finally results in

$$V^T M \delta_h(v_n) = -V^T K p_n - V^T D v_n - V^T G^T \lambda. \tag{5.3.10}$$

By introducing the residuals

$$
\begin{aligned}
r_1 &:= \delta_h(p_n) - v_n & \text{(5.3.11a)} \\
r_2 &:= M \delta_h(v_n) + K p_n + D v_n + G^T \lambda_n & \text{(5.3.11b)}
\end{aligned}
$$

we can summarize these results as

$$
\begin{aligned}
V^T M r_1(p_n, v_n) &= 0 & \text{(5.3.12a)} \\
V^T r_2(p_n, v_n, \lambda_n) &= 0. & \text{(5.3.12b)}
\end{aligned}
$$

This can be viewed as a characterization of the solution of the discretized state space form in terms of the original variables.
We take this characterization to define a generalized solution of the discretized overdetermined system:

**Definition 5.3.1** *The solution* $p_n, v_n, \lambda_n$ *of the system*

$$V^T M r_1(p_n, v_n) = 0 \qquad (5.3.13\text{a})$$
$$V^T r_2(p_n, v_n) = 0 \qquad (5.3.13\text{b})$$
$$r_p(p_n) = 0 \qquad (5.3.13\text{c})$$
$$r_v(v_n) = 0 \qquad (5.3.13\text{d})$$
$$r_\lambda(p_n, v_n, \lambda_n) = 0 \qquad (5.3.13\text{e})$$

*with*

$$
\begin{aligned}
r_p &:= G p_n - z(t_n) \\
r_v &:= G v_n - \dot{z}(t_n) \\
r_\lambda &:= G M^{-1} \left( -K p_n - D v_n - G^T \lambda_n - \ddot{z}(t_n) \right).
\end{aligned}
$$

*is called the* implicit state space solution *of the overdetermined problem (5.3.2).*

We derive now an expression for the implicit state space solution of a linear multibody system. To simplify notation, we will restrict ourselves to BDF methods. The results can easily be extended to the general case.

The linear system (5.3.13) consisting of $2n_p + n_\lambda$ equations in the $2n_p + n_\lambda$ unknowns $p_n, v_n, \lambda_n$ can be rewritten as

$$
\begin{pmatrix}
\frac{\alpha_k}{h} V^T M & -V^T M & 0 \\
V^T K & V^T (\frac{\alpha_k}{h} M + D) & 0 \\
-G M^{-1} K & -G M^{-1} D & -G M^{-1} G^T \\
G & 0 & 0 \\
0 & G & 0
\end{pmatrix}
\begin{pmatrix}
p_n \\
v_n \\
\lambda_n
\end{pmatrix}
=
\begin{pmatrix}
-\frac{1}{h} V^T M \xi_p \\
-\frac{1}{h} V^T M \xi_v \\
\ddot{z}(t_n) \\
z(t_n) \\
\dot{z}(t_n)
\end{pmatrix}
\tag{5.3.14}
$$

with $\xi_p$ and $\xi_v$ being "old" data, see Sec. 4.1.6.
Defining

$$
A_1 :=
\begin{pmatrix}
\frac{\alpha_k}{h} M & -M & 0 \\
K & \frac{\alpha_k}{h} M + D & G^T \\
-G M^{-1} K & -G M^{-1} D & -G M^{-1} G^T
\end{pmatrix}
, \qquad
a_1 :=
\begin{pmatrix}
M \xi_p \\
M \xi_v \\
\ddot{z}(t_n)
\end{pmatrix}
\tag{5.3.15}
$$

and

$$
A_2 :=
\begin{pmatrix}
G & 0 & 0 \\
0 & G & 0
\end{pmatrix}
\quad \text{and} \quad
a_2 :=
\begin{pmatrix}
z(t_n) \\
\dot{z}(t_n)
\end{pmatrix}
,
\tag{5.3.16}
$$

Eq. (5.3.14) can be rewritten as

$$
\begin{pmatrix}
\tilde{V}^T A_1 \\
A_2
\end{pmatrix}
x_n =
\begin{pmatrix}
\tilde{V}^T a_1 \\
a_2
\end{pmatrix}
$$

with

$$\tilde{V}^T := \begin{pmatrix} V^T & 0 & 0 \\ 0 & V^T & 0 \\ 0 & 0 & I \end{pmatrix}, \quad x_n = \begin{pmatrix} p_n \\ v_n \\ \lambda_n \end{pmatrix}.$$

The solution of this system can be expressed as

$$x_n := A_2^+ a_2 + \tilde{V}(\tilde{V}^T A_1 \tilde{V})^{-1}\tilde{V}^T (a_1 - A_1 A_2^+ a_2) \qquad (5.3.17)$$

whenever the matrix $\tilde{V}^T A_1 \tilde{V}$ is invertible. The matrix

$$\begin{pmatrix} A_1 \\ A_2 \end{pmatrix}^{ssf+} := \begin{pmatrix} \tilde{V}(\tilde{V}^T A_1 \tilde{V})^{-1}\tilde{V}^T & (I - \tilde{V}(\tilde{V}^T A_1 \tilde{V})^{-1}\tilde{V}^T A_1)A_2^+ \end{pmatrix} \quad (5.3.18)$$

is called *ssf inverse* of $\begin{pmatrix} A_1 \\ A_2 \end{pmatrix}$. It defines a solution of the overdetermined system

$$\begin{pmatrix} A_1 \\ A_2 \end{pmatrix} x = \begin{pmatrix} a_1 \\ a_2 \end{pmatrix}$$

in a generalized sense based on the characterization of the solution of the discretized state space form.
The ssf inverse is independent of the particular basis $V$ of the null space of $G$.
It can be shown that it is an $(1, 2, 4)$ generalized inverse in the sense of Def. 2.3.1.
Alternatively, one may consider a *range-space representation* of the ssf-generalized solution

$$x_n = A_1^{-1} \left(I - A_2^T(A_2 A_1^{-1} A_2^T)^{-1}A_2 A_1^{-1}\right) a_1 + A_1^{-1} A_2^T \left(A_2 A_1^{-1} A_2^T\right)^{-1} a_2. \quad (5.3.19)$$

obtained from writing the system in its augmented form

$$\begin{aligned} A_1 x_n &= a_1 + A_2^T \mu \\ A_2 x_n &= a_2, \end{aligned}$$

see Sec. 2.3.
In that representation the ssf inverse takes the form

$$\begin{pmatrix} A_1 \\ A_2 \end{pmatrix}^{ssf+} := A_1^{-1} \left( I - A_1^{-1} A_2^T(A_2 A_1^{-1} A_2^T)^{-1}A_2 A_1^{-1} \quad A_2^T(A_2 A_1^{-1} A_2^T)^{-1} \right).$$

$$(5.3.20)$$

The definition of the ssf-solution can be applied to the corrector step (5.3.2) in non-linear problems. In every iteration we have to solve the linearized overdetermined system which reads after scaling the kinematic equations by $M$

$$
\begin{pmatrix}
\frac{\alpha_k}{h}M & -M & 0 \\
K & \frac{\alpha_k}{h}M+D & G^T \\
\Gamma_1 & \Gamma_2 & GM^{-1}G^T \\
G & 0 & 0 \\
\Gamma & G & 0
\end{pmatrix}^{(i)}
\begin{pmatrix}
\Delta p_n \\
\Delta v_n \\
\Delta \lambda_n
\end{pmatrix}^{(i)}
= -
\begin{pmatrix}
Mr_1(p_n,v_n) \\
r_2(p_n,v_n,\lambda_n) \\
r_p(p_n) \\
r_v(p_n,v_n) \\
r_\lambda(p_n,v_n,\lambda_n)
\end{pmatrix}^{(i)}
$$

with

$$
\begin{aligned}
r_1(p,v) &:= \delta_h(p)-v \\
r_2(p,v,\lambda) &:= M(p)\delta_h(v)-f_a(p,v)+G(p)^T\lambda \\
r_p(p) = &:= g(p) \\
r_v(p,v) &:= G(p)v \\
r_\lambda(p,v,\lambda) &:= \zeta(p,v)+G(p)M(p)^{-1}\left(f_a(p,v)-G(p)^T\lambda\right),
\end{aligned}
$$

and

$$
\begin{aligned}
K &:= \tfrac{\partial r_2}{\partial p}(p,v,\lambda) & \tfrac{\alpha_k}{h}M+D &= \tfrac{\partial r_2}{\partial v}(p,v,\lambda) \\
D &:= -\tfrac{\partial f_a}{\partial v}(p,v) & \Gamma &:= \tfrac{\partial r_v}{\partial p}(p,v)=\zeta(p,v) \\
\Gamma_1 &= \tfrac{\partial r_\lambda}{\partial p}(p,v,\lambda) & \Gamma_2 &:= \tfrac{\partial r_\lambda}{\partial v}(p,v,\lambda).
\end{aligned}
$$

(We omitted the obvious arguments of the matrices for notational simplicity).
We split into sub-matrices

$$
A_1 := 
\begin{pmatrix}
\frac{\alpha_k}{h}M & -M & 0 \\
K & \frac{\alpha_k}{h}M+D & G^T \\
\Gamma_1 & \Gamma_2 & GM^{-1}G^T
\end{pmatrix}, \quad
A_2 :=
\begin{pmatrix}
G & 0 & 0 \\
\Gamma & G & 0
\end{pmatrix}
\tag{5.3.21}
$$

so that the linear system within the corrector iteration becomes

$$
\begin{pmatrix} A_1 \\ A_2 \end{pmatrix}
\begin{pmatrix} p_n \\ v_n \\ \lambda_n \end{pmatrix}
=
\begin{pmatrix} a_1 \\ a_2 \end{pmatrix}
$$

with $a_1$ composed of $Mr_1,r_2,r_\lambda$ and $a_2$ of $r_p$ and $r_v$.
For this system the ssf-generalized solution can be computed via (5.3.17) or (5.3.19),
where in the first case $\tilde{V}$ with $A_2\tilde{V}=0$ may be taken as

$$
\tilde{V} =
\begin{pmatrix}
V & 0 & 0 \\
-G^\dagger\Gamma V & V & 0 \\
0 & 0 & I
\end{pmatrix}
$$

and an arbitrary matrix $V$ with columns spanning the null space of $G$.

We summarize this approach by writing the corrector equation formally as

$$r(x) = 0 \quad \text{with} \quad r : I\!\!R^{2n_p+n_\lambda} \longrightarrow I\!\!R^{2n_p+3n_\lambda}.$$

The implicit state space approach is based on the iteration

$$x^{(i+1)} := x^{(i)} - \left(r'(x^{(i)})\right)^{ssf+} r(x^{(i)}).$$

As this iteration uses an $(1, 2, 4)$ generalized inverse we can apply Th. 3.4.1 which states the conditions for $x^{(i)}$ converging to $x^*$ with

$$\left(r'(x^*)\right)^{ssf+} r(x^*) = 0. \tag{5.3.22}$$

An immediate consequence is that by (5.3.17) the conditions given in Def. 5.3.1 hold in $x^*$. Thus, this iteration gives an implicit state space solution also in the nonlinear case.

We note that every integration method which is convergent for index-1 DAEs can be applied when treating overdetermined DAEs together with the implicit state space method, in particular also Adams–Moulton methods.

**Example 5.3.2** *We consider Example 5.2.7 again, but rewrite it now in overdetermined form*

$$
\begin{align}
\dot{x}_1 &= ax_1 + x_2 + x_3 & \text{(5.3.23a)} \\
\dot{x}_2 &= (b^2 t^2 - bt + b)\, x_1 - ax_2 + bt\, x_3 & \text{(5.3.23b)} \\
0 &= x_2 + x_3 & \text{(5.3.23c)} \\
0 &= (1 - bt)\, x_1 + x_2, & \text{(5.3.23d)}
\end{align}
$$

*where equation (5.3.23c) was obtained by taking the time derivative of the constraint (5.3.23d) and by inserting (5.3.23a),(5.3.23b). Discretizing this system as above with the five step BDF and solving the corrector equations using the ssf inverse results in a scheme which has for all values of b the same stability. As it can be expected from the construction of the approach, the stability of the numerical method is just the stability of the discretized state space form, see Fig. 5.5.*

## 5.3.3  Projection Methods for ODEs with Invariants

In this chapter we apply the two projection methods to ordinary differential equations with invariants, see Sec. 5.1.4.

First we show that *linear* invariants are preserved by linear multistep methods, so that in that case no projections need to be applied. There will be no drift off from the manifold given by the invariant.

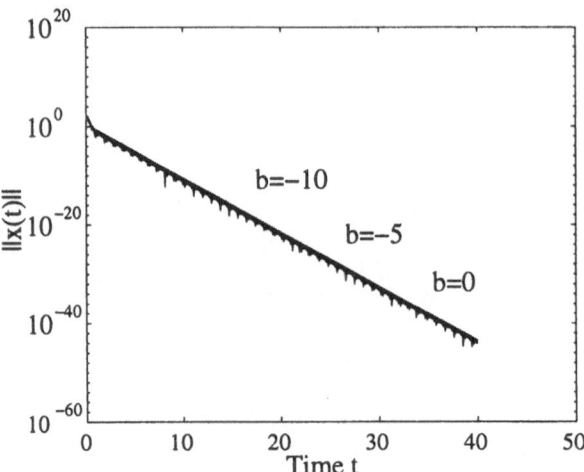

Figure 5.5: $\|x(t)\|$ for various values $b$ when solved with the implicit state space method and $h = 0.1$.

**Theorem 5.3.3** *Assume that the differential equation $\dot{x} = f(t,x)$ with $x(0) := x_0$ has a linear integral invariant $Hx - b = 0$. Given $k$ starting values $x_i$, $i = 0, \dots , k-1$ with $Hx_i - b = 0$, the solution $x_n$ of the problem discretized by a consistent linear $k$ step method satisfies this invariant too, i.e. $Hx_n - b = 0$.*

The proof is straightforward and uses the property $Hf = 0$, Def. 5.1.9, [Gear86]. This result no longer holds for linear invariants of the type

$$Hx(t) + z(t) = 0$$

with $z \neq$ const. and general nonlinear invariants. Therefore we study projections onto the manifold defined by the invariant.

We pointed out earlier that for equations of motion of constrained mechanical systems written in index-1 form the position and velocity constraints form integral invariants, see (5.1.16). Thus the coordinate projection and the implicit state space method introduced in the previous section can be viewed as numerical methods for ensuring that the numerical solution satisfies these invariants.

We will now consider these methods in a more general setting. As before we use $\delta_h(x_n)$ to denote the numerical approximation of $\dot{x}(t_n)$ by a discretization method depending on the step size $h$.

**Coordinate Projection for ODEs with Invariants**
This method is the most frequently used approach. As above, first a numerical

solution is computed which normally does not lie in the manifold $\mathcal{M} = \{x | \varphi(x) = \varphi(x_0)\}$ so that in a second step it is corrected by projection:

1. The numerical solution $\tilde{x}_n$ of the ODE is computed by a discretization method:

$$\delta_h(\tilde{x}_n) = f(\tilde{x}_n)$$

2. The solution $\tilde{x}_n$ is then corrected by

$$\|x_n - \tilde{x}_n\| \;=\; \min_{x_n} \tag{5.3.24a}$$

$$\varphi(x_n) \;=\; \varphi(x_0). \tag{5.3.24b}$$

These values $x_n$ and $f(x_n)$ are then used to advance the solution.

Due to the nonlinearity of the invariant the projection has to be performed by iteratively solving linearly constrained least squares problems, cf. Sec. 5.3.1. It has been shown in [Sha86, Eich93] that this projection does not disturb the convergence properties of the discretization scheme. Additionally, the global error increment is smaller, because only errors in the manifold are present, whereas the error components orthogonal to the manifold vanish due to the projection. For implementational details of this method see [Sha86, Eich91].

**Example 5.3.4** *We solve the equations of motion of the gyroscope example, Ex. 5.1.10, by using the Runge-Kutta method ODE45 in MATLAB with the initial values*

$$y(0)^T = (\; \sqrt{(1 - \cos(0.1))/2} \;\; 0 \;\; 0 \;\; \sqrt{(1 + \cos(0.1))/2} \;\; 0 \;\; 1 \;\; 20 \;).$$

*One clearly sees in Fig. 5.7 the growth of the residuals $q_1^2 + q_2^2 + q_3^2 + q_4^2 - 1$ when the method without the projection step is used. On the other hand projecting onto the manifold removes this instability. In Fig. 5.6 the influence of the projection on the solution is shown.*

**Derivative Projection for ODEs with Invariants**
For explaining this approach we first note that the expanded system

$$\dot{x} \;=\; f(x) - H(x)^T \mu \tag{5.3.25a}$$

$$0 \;=\; \varphi(x) - \varphi(x_0) \tag{5.3.25b}$$

with $H := \varphi'$ is equivalent to (5.1.17) in the sense that it has the same solution $x$ iff (5.3.25b) is an integral invariant. In that case $\mu \equiv 0$. Due to roundoff and discretization errors Eqs. (5.1.17) and (5.3.25) are no longer equivalent under

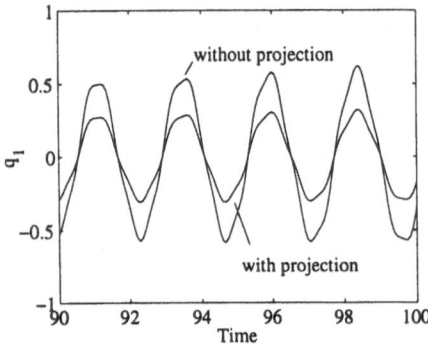

Figure 5.6: Solution of the gyroscope

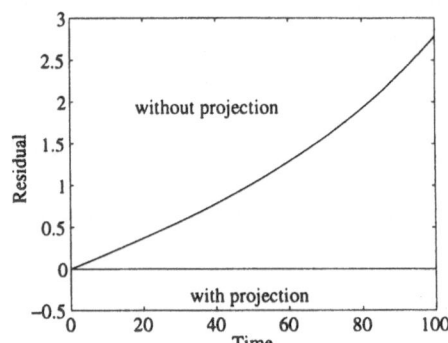

Figure 5.7: Residual of the gyroscope with and without projection

discretization.

Discretizing (5.3.25) leads to

$$\delta_h(x_n) = f(x_n) - H(x_n)^T \mu_n \qquad (5.3.26a)$$
$$0 = \varphi(x_n) - \varphi(x_0). \qquad (5.3.26b)$$

Herein, $\mu_n$ plays the role of a control variable for keeping the numerical solution $x_n$ on the manifold determined by the invariant. This variable is no longer zero when solving the discretized equation. The drift-off effect is removed and (5.3.25) can be viewed as a way to stabilize the original problem. This stabilization has been suggested by Gear, [Gear86]. Its application to index reduced DAEs was discussed in [GGL85] and became well known under the name *Gear-Gupta-Leimkuhler approach*.

A shortcoming of this stabilized formulation is that it consists of an index-2 formulation which can only be solved by numerical methods which are convergent for this class of DAEs. In particular, higher order Adams-Moulton methods are no longer applicable, see Th. 5.2.3. They introduce an instability in $\mu$.

To avoid this disadvantage we relate this formulation to the implicit state space method. First we note, that the discretized differential equation in (5.1.17) is solved up to a non vanishing residual

$$r_\mu := H(x)^T \mu.$$

Let $V(x_n)$ be an $n_x \times n_\mu$-matrix which spans the nullspace of $H(x_n)$. Then $V^T r_\mu = 0$ holds. This corresponds to the condition characterizing the ssf solution of overdetermined DAEs, see Def. 5.3.1.

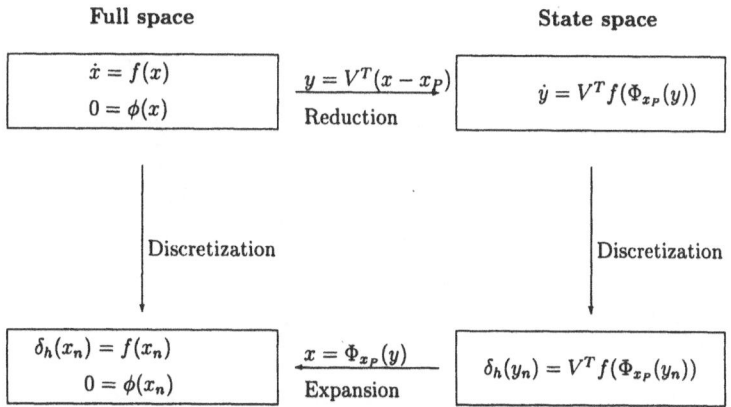

Figure 5.8: Commutativity relationship for the implicit state space method

Instead of solving (5.3.25) one might consider the discrete overdetermined system

$$\delta_h(x_n) = f(x_n)$$
$$0 = \varphi(x_n) - \varphi(x_0)$$

where the nonlinear equation is solved by an iterative process which uses the ssf inverse. The iteration matrix reads

$$A := \begin{pmatrix} A_1 \\ A_2 \end{pmatrix} = \begin{pmatrix} \frac{\alpha_k}{h} I - f'(x_n) \\ \varphi'(x_n) \end{pmatrix},$$

and $A^{ssf+}$ is given by (5.3.20).

By similar considerations as in the DAE case above this approach can be related to the state space form (5.1.20) of the ODE with invariant. Choosing the ssf inverse for solving the overdetermined systems corresponds to the requirement that the diagram in Fig. 5.8 commutes.

### Relationship Between the Two Projection Methods*
Both projection methods differ by the quantities which are projected. The coordinate projection method projects the solution of the discretized equation orthogonally onto the manifold $\mathcal{M} := \{x | \varphi(x) - \varphi(x_0) = 0\}$ while the derivative projection determines $x_n$ in such a way that the residual $\delta_h(x_n) - f_n$ is orthogonal to the tangent on $\mathcal{M}$ in $x_0$.

Both approaches can be modified by introducing appropriate weights in the inner product which defines orthogonality. Other projections have been suggested in the literature, e.g. [Bar89]. For an overview we refer to [EFLR90].

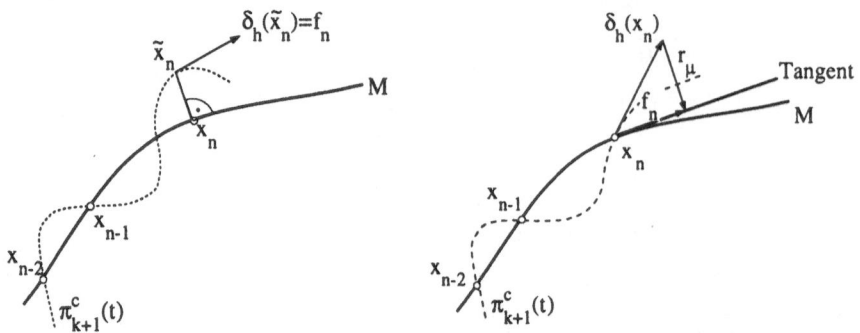

Figure 5.9: Projection methods: coordinate projection (left) and derivative projection (right)

## 5.4   One Step Methods

In this section we will present one-step methods for DAEs. Though there are successful implementations of extrapolation methods for differential algebraic equations [DHZ87] and mainly for constrained mechanical systems [Lub91] we restrict our discussion of one-step methods to Runge–Kutta methods.

### 5.4.1   Runge-Kutta Methods

We consider the equations of motion (5.1.2) in their semi-explicit form

$$\dot{y} \;=\; E_D(y)^{-1} f_D(t, y, \lambda) \tag{5.4.1a}$$
$$0 \;=\; f_C(t, y, \lambda). \tag{5.4.1b}$$

Note, that the inversion of the regular matrix $E_D(y)$ can be performed in $\mathcal{O}(n_y)$ operations by applying the techniques described in Sec. 1.4.1. In order to simplify notation we will assume in the sequel $E_D(y) = I$. Furthermore we assume that (5.4.1) has an index of at most 2, i.e. we assume that either condition (5.1.13) or (5.1.14) hold.

First, we consider implicit methods and restrict ourselves to those related to collocation polynomials. For this end we have to extend Definition 4.3.1 to semi-explicit DAEs.

**Definition 5.4.1** *The polynomials $u$ and $v$ of degree $s$ defined by the conditions*

$$
\begin{aligned}
u(t_n) &= y_n & v(t_n) &= \lambda_n \\
\dot{u}(t_n + c_i h) &= f_D(t_n + c_i h, u(t_n + c_i h), v(t_n + c_i h)), \quad i = 1, \dots, s \\
0 &= f_C(t_n + c_i h, u(t_n + c_i h), v(t_n + c_i h))
\end{aligned}
$$

are called a pair of collocation polynomials *of the semi-explicit DAE (5.4.1). The distinct values* $c_i \in [0,1], i = 1, \ldots, s$ *are called* collocation points.

A *collocation method* then uses this pair of polynomials to determine the numerical solution of the problem at $t = t_{n+1}$, i.e.

$$y_{n+1} := u(t_n + h) \text{ and } \lambda_{n+1} := v(t_n + h).$$

Note, that in general $f_C(t_{n+1}, y_{n+1}, \lambda_{n+1}) \neq 0$ unless $c_s = 1$.
Like in the explicit ODE case (see Sec. 4.3.1) the choice of collocation points defines an $s$-stage Runge–Kutta method. We describe $\dot{u}$ and $\dot{v}$ in terms of Lagrange polynomials and obtain by integrating

$$u(t_n + c_i h) \quad = \quad y_n + h \sum_{j=1}^{s} a_{ij} f_D(t_n + c_j h, u(t_n + c_j h), v(t_n + c_j h)) \quad (5.4.2)$$

$$v(t_n + c_i h) \quad = \quad \lambda_n + h \sum_{j=1}^{s} a_{ij} l_j \qquad\qquad\qquad\qquad (5.4.3)$$

with the $l_j$ determined by requiring consistency at the interpolation points,

$$f_C(t_n + c_i h, u(t_n + c_i h), v(t_n + c_i h)) = 0.$$

The coefficients $a_{ij}$ are defined by (4.3.2).
By introducing *stage values* $Y_i := u(t_n + c_i h)$ and $\Lambda_i = v(t_n + c_i h)$ we can write the collocation method as an implicit Runge–Kutta method,

$$Y_i \quad = \quad y_n + h \sum_{j=1}^{s} a_{ij} f_D(t_n + c_j h, Y_j, \Lambda_j) \qquad\qquad (5.4.4a)$$

$$\Lambda_i \quad = \quad \lambda_n + h \sum_{j=1}^{s} a_{ij} l_j, \qquad\qquad i = 1, \ldots, s \qquad (5.4.4b)$$

$$0 \quad = \quad f_C(t_n + c_i h, Y_i, \Lambda_i) \qquad\qquad\qquad (5.4.4c)$$

$$y_{n+1} \quad = \quad y_n + h \sum_{i=1}^{s} b_i f_D(t_n + c_i h, Y_i, \Lambda_i) \qquad\qquad (5.4.4d)$$

$$\lambda_{n+1} \quad = \quad \lambda_n + h \sum_{i=1}^{s} b_i l_i \qquad\qquad\qquad (5.4.4e)$$

and $b_i$ defined by (4.3.4).
For the treatment of mechanical systems Radau IIA methods and Gauß methods proved to be advantageous. Gauß methods are of special interest due to their symmetry especially when treating boundary value problems, cf. Sec. 7. As already pointed out when discussing explicit ODEs, Radau IIA methods are favorable in

multibody dynamics due their stability properties when treating highly oscillatory problems.

In general only the stage values lie on the manifold $\{(y, \lambda) | f_C(t_n + c_i h, y, \lambda) = 0\}$. For methods with $b_i = a_{si}$ for $i = 1, \ldots, s$ also the solution values are on the manifold. Methods with this property are called *stiffly accurate* Runge–Kutta methods. Note, collocation methods with the interval end point being a collocation point, $c_s = 1$, are stiffly accurate by construction.

On the other hand for not stiffly accurate methods, like Gauß methods, one can modify the method by projecting the numerical solution onto

$$\mathcal{M}_{n+1} := \{(y, \lambda) | f_C(t_{n+1}, y, \lambda) = 0\}$$

after every integration step. This leads to so-called projected Runge–Kutta methods introduced in [AP91]. The projection is identical to the coordinate projection described in Sec. 5.3.3.

### Stability of Implicit Runge–Kutta Methods for DAEs

We saw in Sec. 4.2.3 that every one-step method is stable for $h$ small enough. For this we considered the method's stability function $R(z) := (1 + zb^T(I_s - zA)^{-1}\mathbf{1})$, for which $R(0) = 1$ holds. Zero stability is connected to the discrete dynamics of the process. In DAEs there are in the continuous process dynamic components (state variables) and non dynamic components, which are defined by algebraic relationships. We saw already when considering multistep methods that even for zero-stable methods these algebraic components might cause stability problems. To see this in the Runge–Kutta case, we first look at the simplest index-1 problem $\lambda = 0$ and "discretize" it by the implicit Runge–Kutta method (5.4.4). This gives

$$\Lambda_i = \lambda_n + h \sum_{j=1}^{s} a_{ij} l_j, \qquad\qquad i = 1, \ldots, s$$

$$0 = \Lambda_i$$

$$\lambda_{n+1} = \lambda_n + h \sum_{i=1}^{s} b_i l_i.$$

This results in the recursion

$$\lambda_{n+1} = (1 - b^T A^{-1} \mathbf{1}) \lambda_n = R(\infty) \lambda_n.$$

Starting with inconsistent initial values, i.e. $\lambda_0 \neq 0$, leads to an increasing error and an unstable growth of $\lambda_i$ if $R(\infty) > 1$. The error is damped when $R(\infty) < 1$. For $R(\infty) = 1$ errors in the individual steps and the initial error are just summed up. Thus, we have at least to require *stability at infinity* for a method being stable in the DAE case.

As for stiffly accurate methods, like Radau IIA methods,

$$a_{si} = b_i \Rightarrow A^T e_s = b \Rightarrow b^T A^{-1} = 1$$

these methods are stable at infinity with optimal damping, i.e. $R(\infty) = 0$. This property together with A-stability is called *L-stability*.

We have indeed the same requirements as in the multistep case where stability at infinity plays a central role (see Def. 5.2.2 and Th. 5.2.3). There, the method with optimal damping at infinity is the BDF.

**Example 5.4.2** *We discretize the linearized constrained truck with a 3-stage Gauß method and a 3-stage Radau IIA method. The eigenvalues of the discrete-time transfer matrix $\Phi$ with*

$$\begin{pmatrix} y_{n+1} \\ \lambda_{n+1} \end{pmatrix} = \Phi(h) \begin{pmatrix} y_n \\ \lambda_n \end{pmatrix}$$

*reflect the structure of the problem. There are 14 eigenvalues depending on the step size h corresponding to the 14 degrees of freedom and eigenvalues independent of h corresponding to 6 algebraic variables. The latter correspond to the four infinity eigenvalues and the two zero generalized eigenvalues of the corresponding index-2 continuous system (see Tab. 2.2 in Sec. 2.2). We see in Table 5.5 that these structural eigenvalues are mapped in the discrete case to eigenvalues of the size $R(\infty)$ and $R(0)$ respectively. Note, that the discrete system is stable if all eigenvalues are inside the unit circle and those on its boundary are simple. Table 5.5 shows the eigenvalues for $h = 0.1$. Stability is not affected by changing h as both methods are A-stable.*

**Accuracy of Collocation Methods for DAEs**

The order of implicit Runge-Kutta methods is different for $y$ and $\lambda$ components in index-2 problems. For the two most important collocation methods we give order results in Tab. 5.4. Applying projection to the Gauß method in order to force that the numerical result is on the constraint manifold increases the order, so that the accuracy of the $y$ variable corresponds to the order of the method in the explicit ODE case.

## 5.4.2   Half-Explicit Methods

All methods for DAEs we discussed so far were implicit. A nonlinear constraint $f_C$ defines the algebraic variables in an implicit way. Therefore, iteration methods, mainly of Newton type are inevitable when treating DAEs. On the other hand, the differential part of the problem may be discretized by an explicit scheme. Such a method is called a *half-explicit method*.

| No. | eigenvalue | $h$ dep. | comment |
|:---:|:---:|:---:|:---:|
| Eigenvalues related to the constraints | | | |
| 1 - 2 | 0.0 | - | velocity |
| 3 - 4 | 0.0 | - | constraint |
| 5 - 6 | 0.1165 | - | roots of |
| 7 - 8 | $-1.7165$ | - | $\sigma$ polynomial |
| Eigenvalues related to the "dynamics" | | | |
| 9 - 10 | $-0.0026 \pm 0.0051i$ | $+$ | |
| 11 - 12 | $-0.0003 \pm 0.0066i$ | $+$ | |
| 13 | $-0.0112$ | $+$ | |
| 14 - 15 | $-0.0074 \pm 0.0102i$ | $+$ | eigenvalues of |
| 16 - 17 | $-0.0063 \pm 0.0129i$ | $+$ | the discrete |
| 18 - 19 | $-0.0053 \pm 0.0480i$ | $+$ | state space form |
| 20 - 21 | $0.0117 \pm 0.0497i$ | $+$ | |
| 22 | $-0.1065$ | $+$ | |
| 23 | 0.4641 | $+$ | |
| 24 - 25 | $0.7209 \pm 0.3756i$ | $+$ | |
| 26 | 0.8825 | $+$ | |
| 27 - 28 | $0.9066 \pm 0.0999i$ | $+$ | eigenvalues of |
| 29 - 30 | $0.9097 \pm 0.1293i$ | $+$ | the discrete |
| 31 - 32 | $0.7660 \pm 0.5244i$ | $+$ | state space form |
| 33 - 34 | $0.9664 \pm 0.0575i$ | $+$ | |
| 35 - 36 | $0.9904 \pm 0.0775i$ | $+$ | |
| Eigenvalues related to the invariants | | | |
| 37 - 38 | 0.0 | - | roots of |
| 39 - 40 | 1.0000 | - | $\rho$ polynomial |

Table 5.3: Eigenvalues of the discrete linear truck example in its index 2 formulation for the two step Adams–Moulton method with step size $h = 0.005$.

| | Index-1 DAE | | Index-2 DAE | | $R(\infty)$ |
|:---:|:---:|:---:|:---:|:---:|:---:|
| | $y$ | $\lambda$ | $y$ | $\lambda$ | |
| Radau IIA | $2s - 1$ | $2s - 1$ | $2s - 1$ | $s$ | 0 |
| Gauß $s$ odd | $2s$ | $s + 1$ | $s + 1$ | $s - 1$ | -1 |
| $s$ even | $2s$ | $s$ | $s$ | $s - 2$ | 1 |

Table 5.4: Order results for one step $s$-stage collocation methods applied to DAEs

| No. | Gauß-m. eigenvalue | Radau IIA-m. eigenvalue | $h$ dep. | comment |
|---|---|---|---|---|
| Eigenvalues related to the constraints | | | | |
| 1 | $-1.0$ | $0.0$ | - | velocity |
| 2 | $-1.0$ | $0.0$ | - | constraint |
| 3 | $-1.0$ | $0.0$ | - | acc. constraint |
| 4 | $-1.0$ | $0.0$ | - | ("hidden" constraint) |
| Eigenvalues related to the "dynamics" | | | | |
| 5 | $-0.0346$ | $-0.0345$ | + | |
| 6/7 | $0.0445 \pm 0.1648\,i$ | $0.0691 \pm 0.1792\,i$ | + | |
| 8 | $0.2065$ | $0.2070$ | + | |
| 9/10 | $0.1805 \pm 0.3556\,i$ | $0.1802 \pm 0.3560\,i$ | + | discrete |
| 11/12 | $0.0609 \pm 0.3982\,i$ | $0.0701 \pm 0.3944\,i$ | + | dynamics |
| 13/14 | $0.3919 \pm 0.4613\,i$ | $0.0607 \pm 0.3993\,i$ | + | |
| 15/16 | $0.6279 \pm 0.5000\,i$ | $0.6279 \pm 0.4999\,i$ | + | |
| 17/18 | $0.5443 \pm 0.7182\,i$ | $0.5443 \pm 0.7181\,i$ | + | |
| Eigenvalues related to the invariants | | | | |
| 19 | $1.0$ | $1.0$ | - | position constraint |
| 10 | $1.0$ | $1.0$ | - | (invariant) |

Table 5.5: Eigenvalues of the discrete linear truck example in its index-2 formulation for the three stage Gauß and Radau IIa methods, $h = 0.1$.

We introduce half-explicit methods by considering the explicit Euler method for index-1 and index-2 problems.

For an index-1 problem the *half-explicit Euler method* reads

$$y_{n+1} = y_n + h f_D(t_n, y_n, \lambda_n) \qquad (5.4.5a)$$
$$0 = f_C(t_n, y_n, \lambda_n) \qquad (5.4.5b)$$

while for an index-2 problem the half-explicit Euler methods takes the form

$$y_{n+1} = y_n + h f_D(t_n, y_n, \lambda_n) \qquad (5.4.6a)$$
$$0 = f_C(t_{n+1}, y_{n+1}). \qquad (5.4.6b)$$

The procedure of solving these equations is divided into two parts. First we insert the first equation into the second one to get

$$f_C(t_{n+1}, y_n + h f_D(t_n, y_n, \lambda_n)) = 0.$$

Due to the index-2 condition this equation is solvable for $\lambda_n$. Once $\lambda_n$ is computed, $y_{n+1}$ can be obtained by simply evaluating the first equation. Note, that here the functions $f_D$ and $f_C$ have to be evaluated at different points, which causes practical problems in most of the current multibody simulation codes.

Multistep generalizations of this approach have first been introduced and discussed by Arévalo [Are93], extrapolation methods based on this idea were introduced by Lubich [Lub91] and resulted in the code MEXX [LENP95]. Ostermann investigated half-explicit extrapolation methods for index-3 problems [Ost90].

We will restrict us here to the presentation of half-explicit one-step methods. They have been first introduced in [HLR89].

**Half-Explicit Runge–Kutta Methods for Index-1 Problems**
The half-explicit Euler method can be extended in various ways to get higher order methods. In the index-1 case such an extension is

$$Y_1 = y_n \tag{5.4.7a}$$

$$Y_i = y_n + h \sum_{j=1}^{i-1} a_{ij} f_D(t_n + c_j h, Y_j, \Lambda_j), \quad i = 2, \ldots, s \tag{5.4.7b}$$

$$0 = f_C(t_n + c_i h, Y_i, \Lambda_i), \quad i = 1, \ldots, s \tag{5.4.7c}$$

$$\tag{5.4.7d}$$

$$y_{n+1} = y_n + h \sum_{i=1}^{s} b_i f_D(t_n + c_i h, Y_i, \Lambda_i) \tag{5.4.7e}$$

$$0 = f_C(t_{n+1}, y_{n+1}, \Lambda_s) \tag{5.4.7f}$$

$$\lambda_{n+1} = \Lambda_s \ . \tag{5.4.7g}$$

All explicit Runge–Kutta methods which are convergent with order $q$ for explicit ODEs can be applied in this way to index-1 ODEs too. The order of the error in $y(t)$ and $\lambda(t)$ is $q$.

**Half-Explicit Runge–Kutta Methods for Index-2 Problems**
For the index-2 a half-explicit Runge–Kutta method can be defined by

$$Y_1 = y_n \tag{5.4.8a}$$

$$Y_i = y_n + h \sum_{j=1}^{i-1} a_{ij} f_D(t_n + c_j h, Y_j, \Lambda_j) \tag{5.4.8b}$$

$$0 = f_C(t_n + c_i h, Y_i), \qquad i = 1, \dots, s \tag{5.4.8c}$$

$$y_{n+1} = y_n + h \sum_{i=1}^{s} b_i f_D(t_n + c_i h, Y_i, \Lambda_i) \tag{5.4.8d}$$

$$0 = f_C(t_{n+1}, y_{n+1}) \tag{5.4.8e}$$

$$\lambda_{n+1} = \Lambda_s \tag{5.4.8f}$$

with $c_s = 1$, cf. [HLR89].

Unfortunately, most explicit Runge–Kutta formulas have very poor convergence properties when used for half-explicit methods in the above way. In general they do not exceed order two. Exceptions are those sets of coefficients developed in [BH93]. We cite the coefficients of their third order method

$$\frac{c \;|\; A}{\;\;|\; b^T} \;=\; \begin{array}{c|ccc} 0 & & & \\ 1/3 & 1/3 & & \\ 1 & -1 & 2 & \\ \hline & 0 & 3/4 & 1/4 \end{array} \;.$$

Methods of order 4 and 5 (HEM4 and HEM5) as well as numerical results can be found in [Bra92].

It can be shown, that the convergence properties of half-explicit Runge–Kutta methods can be improved significantly by introducing an additional stage to approximate $\lambda_{n+1}$ and by refraining from requiring $f_C(t_n + c_2 h, Y_2) = 0$. This leads to the so-called *type B class of half-explicit methods* introduced in [Arn98, Mur97,

AM97]. We write them in the following form:

$$Y_1 = y_n \tag{5.4.9a}$$

$$Y_2 = y_n + h a_{21} f_D(t_n + c_1 h, Y_1, \Lambda_1) \tag{5.4.9b}$$
$$\Lambda_1 = \lambda_n \tag{5.4.9c}$$

$$Y_i = y_n + h \sum_{j=1}^{i-1} a_{ij} f_D(t_n + c_j h, Y_j, \Lambda_j) \tag{5.4.9d}$$

$$0 = f_C(t_n + c_i h, Y_i), \qquad\qquad i = 3, \ldots, s+2 \tag{5.4.9e}$$

$$y_{n+1} = y_n + h \sum_{i=1}^{s} b_i f_D(t_n + c_i h, Y_i, \Lambda_i) \tag{5.4.9f}$$

$$\lambda_{n+1} = \Lambda_{s+1}. \tag{5.4.9g}$$

with $a_{s+1,i} = b_i$ and $c_{s+1} = c_s = 1$. This and the fact $f_D(t_{n+1}, Y_1, \Lambda_1) = f_D(t_n + c_s h, Y_s, \Lambda_s)$ allows to implement this method with $s$ evaluations of $f_D$ and $f_C$ per step.

The Butcher-tableau takes the following form

| | | | | | | |
|---|---|---|---|---|---|---|
| $c_1$ | $0$ | | | | | |
| $c_2$ | $a_{21}$ | $0$ | | | | |
| $c_3$ | $a_{31}$ | $a_{32}$ | | | | |
| $\vdots$ | $\vdots$ | $\vdots$ | $\ddots$ | $\ddots$ | | |
| $1$ | $a_{s1}$ | $a_{s2}$ | $\cdots$ | $a_{s,s-1}$ | $0$ | |
| $1$ | $b_1$ | $b_2$ | $\cdots$ | $b_{s-1}$ | $b_s$ | $0$ |
| $c_{s+2}$ | $a_{s+2,1}$ | $a_{s+2,2}$ | $\cdots$ | $a_{s+2,s-1}$ | $a_{s+2,s}$ | $a_{s+2,s+1}$ |
| | $b_1$ | $b_2$ | $\cdots$ | $b_{s-1}$ | $b_s$ | $0$ |

where the $c_i, a_{ij}$, and $b_i$ for $i = 1, \ldots, s$ are the coefficients of an explicit method for explicit ODEs. The additional coefficients $c_{s+2}, a_{s+2,i}, i = 1, \ldots, s+1$ have to be chosen so to achieve highest possible order.

**Example 5.4.3** *The* 5th *order Dormand and Prince method can be extended to an half-explicit method. For this method* $s = 6$ *and the coefficients* $a_{ij}, b_i,$ $i, j = 1, \ldots 6$ *are given in (4.2.11). The additional coefficients can be selected in various ways. Arnold ([Arn98]) sets them so, that the residual of the hidden constraint ($\lambda$-constraint) is damped in an optimal way. This leads to the coefficients of the code*

*HEDOPRI5 which are*

$$a_{81} = -\frac{18611506045861}{19738176307200} \quad a_{82} = \frac{59332529}{893904224850} \quad a_{83} = -\frac{2509441598627}{893904224850}$$

$$a_{84} = \frac{2763523204159}{3289696051200} \quad a_{85} = -\frac{41262869588913}{116235927142400} \quad a_{86} = \frac{46310205821}{287848404480}$$

$$a_{87} = -\frac{3280}{75413}$$

*The code HEDOPRI5 has been implemented among other special codes for multibody systems in the collection MBSPACK [Sim95].*

## 5.5   Contact Problems as DAEs

In this section we discuss the modeling of contact problems between two rigid bodies. In contrast to others joints, contact joints allow that the contact point or "point of attachment" moves along the surface of the body.

We give a DAE formulation of this problem, discuss the nature of the equations and point out open problems.

### 5.5.1   DAE formulation for Single Point Contact of Planar Bodies

We consider two planar bodies with arbitrary shapes, Body 0 and Body 1. For simplicity we assume that the body fixed reference system of Body 0 is the reference system in which all motions are expressed. With respect to that coordinate system the reference point of Body 1 is given by the two coordinates $x_1$ and $y_1$ (cf. Fig. 5.10).

Assume that the surfaces of the bodies are described by the functions $r_0, r_1$, respectively: a surface point of Body 0 is given by its two coordinates $r_0(\eta_0), \eta_0$ and one of Body 1 by $r_1(\eta_1), \eta_1$ .

The distance vector $d$ between two arbitrary points of the surfaces is given as

$$d = \begin{pmatrix} x_1 \\ y_1 \end{pmatrix} + \begin{pmatrix} r_1(\eta_1) \\ \eta_1 \end{pmatrix} - \begin{pmatrix} r_0(\eta_0) \\ \eta_0 \end{pmatrix}.$$

If $d$ satisfies

$$d^T \tau_i = 0, \quad i = 0, 1$$

with the normalized tangential vectors $\tau_i = \frac{1}{\sqrt{1+r_i'^2}} \begin{pmatrix} r_i'(\eta_i) \\ 1 \end{pmatrix}$ then $d$ is the vector of shortest distance between the two bodies. Thus

$$r_1'(\eta_1) = r_0'(\eta_0). \tag{5.5.1}$$

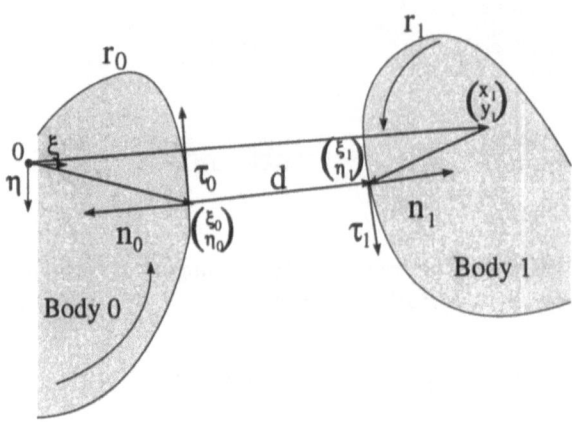

Figure 5.10: Planar Contact Problem

In a contact point we obtain from $d = 0$ by eliminating $\eta_0 = \eta_1 + y_1$ the *contact condition*

$$g_{con}(x_1, y_1, \eta_1) = x_1 + r_1(\eta_1) - r_0(y_1 + \eta_1) = 0,$$

and from (5.5.1)

$$g_{npe}(x_1, y_1, \eta_1) = r_1'(\eta_1) - r_0'(y_1 + \eta_1) = 0$$

the *non-penetrating condition.*

The equations contain two types of coordinates

- coordinates $p^T = (x_1^T, y_1^T)$ describing the motion of the system. For these coordinates we have equations of motion in form of ODEs.

- Auxiliary coordinates $p_r = \eta_1$ used for the description of the contact point.

Summarizing the equations of motion one obtains

$$\dot{p} = v \tag{5.5.2a}$$
$$M(p)\dot{v} = f(p, p_r, v) - G^T(p, p_r)\lambda \tag{5.5.2b}$$
$$0 = g_{con}(p, p_r) \tag{5.5.2c}$$
$$0 = g_{npe}(p, p_r). \tag{5.5.2d}$$

Classically, the non-penetrating equations are solved for $p_r = p_r(p)$ and this result is inserted into the contact condition [DJ82]. Even for tree structured systems, the state space form cannot be obtained explicitly, because the algebraic condition $\tilde{g}_{con}(p) = g_{con}(p, p_r(p)) = 0$ is a nonlinear equation which in general cannot be solved explicitly for a set of independent coordinates. Therefore in [DJ82] the solution is obtained iteratively and the result is stored in tables. During the integration table-look-ups are used to compute the state space form. This leads to discontinuities because the intermediate values have to be interpolated by e.g. piecewise linear interpolation. These discontinuities are artificial because they are not present in the original system.

In contrast, a formulation of the equations of motion in DAE form avoids these difficulties.

The non-penetrating condition can be solved for $p_r = \eta_1$ as function of $y_1$ if $r_1''(\eta_1) \neq r_0''(y_1 + \eta_1)$ holds, i.e. if (5.5.2d) is an index-1 equation. This will be assumed in the sequel.

The contact condition (5.5.2c) is then a constraint on the position coordinates and has therefore the same structure as the constraint equations discussed before. Thus, (5.5.2) is an index-3 DAE with an additional index-1 constraint. This is the typical form of the equations of motion also in case of contact of three dimensional bodies.

The constraint matrix is given by

$$G = \frac{dg_{con}}{dp} = \frac{\partial g_{con}}{\partial p} + \frac{\partial g_{con}}{\partial p_r} \frac{\partial p_r}{\partial p}.$$

$\frac{\partial p_r}{\partial p}$ can be obtained by differentiating the non-penetrating condition:

$$\frac{\partial p_r}{\partial p} = -\left(\frac{\partial g_{npe}}{\partial p_r}\right)^{-1} \frac{\partial g_{npe}}{\partial p}.$$

One can show that

$$\frac{\partial g_{con}}{\partial p_r} = r_1'(\eta_1) - r_0'(y_1 + \eta_1) = 0$$

holds. Hence,

$$G = \frac{\partial g_{con}}{\partial p}.$$

**Example 5.5.1 (Wheelset running on a track)**
*Figure 5.11 shows a wheelset. The multibody model is given in Figure 5.12. The frame moves with a constant speed $v$. We only consider relative motion with respect to the frame. This is described by the coordinates $p = (x, y, z, \theta, \phi)^T$, where $x$ is the lateral, $y$ the vertical and $z$ the longitudinal displacement. $\theta$ and $\phi$ are the yaw and roll angle. The deviation from the nominal angular velocity is described by $\beta$.*

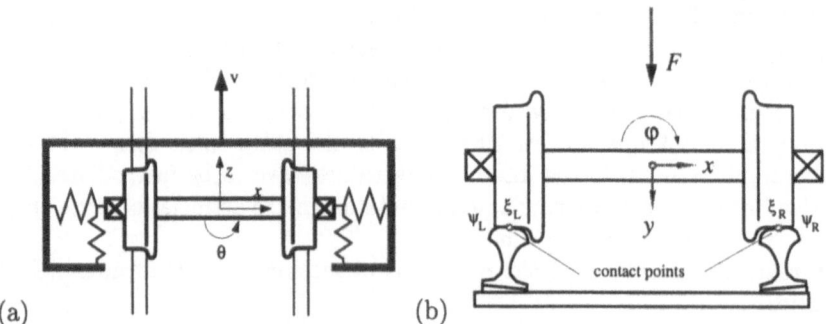

Figure 5.11: The wheelset and the track. (a) View from above, (b) lateral cross section.

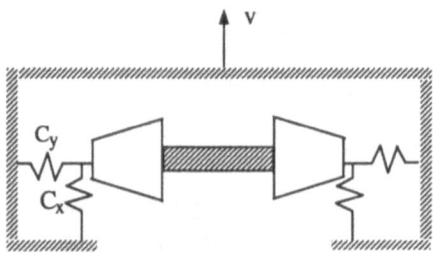

Figure 5.12: Multibody model

*Let $P_W$ and $P_R$ be the profile functions of the wheel and the rail depending on the contact variables. Then, the contact and non-penetrating conditions between wheel and rail have the form*

$$P_R(\hat{\xi}) - y - \xi \sin \phi + P_W(\xi) \cos \phi \cos \psi = 0.$$

$$P_R'(\hat{\xi})\,(P_W'(\xi) \sin \phi + \cos \phi \cos \psi) - \cos\phi \sin \phi \cos \psi + \sin \theta \sin \psi \;=\; 0$$
$$P_W'(\xi) \sin \theta \cos \theta - \sin \theta \sin \phi \cos \psi - \cos\theta \sin \psi \;=\; 0$$

*with*

$$\hat{\xi} = x + \xi \cos \theta \cos \phi + R(\xi)\,(\cos \theta \sin \phi \cos \psi - \sin \theta \sin \psi)$$

*and $\xi$, $\psi$ being auxiliary variables describing the contact points of the left and right wheel respectively: $\psi_{L/R}$ being the left (L) and right (R) shift angle and $\xi_{L/R}$ the left (L) and right (R) contact point coordinate left/right with respect to the wheelset fixed reference frame, see Fig. 5.13.*

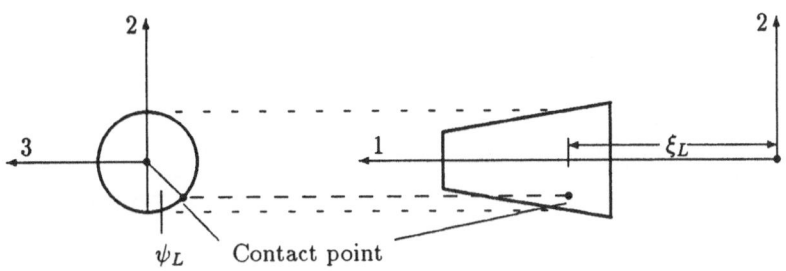

Longitudinal cross section          Lateral cross section

Figure 5.13: Shift angle and coordinate of contact point on the left side

*Thus, there are two contact equations and four non-penetrating equations. The four auxiliary variables are $p_r^T := (\psi_L, \xi_L, \psi_R, \xi_R)$. As equations of motion we then obtain*

$$\dot{p} = v$$

$$M(p) \begin{pmatrix} \dot{v} \\ \dot{\beta} \end{pmatrix} = \begin{pmatrix} f(p, v, \beta, p_r, t, N(p, p_r, \lambda)) - G^T(p, p_r)\lambda \\ h(p, v, \beta, p_r, t, N(p, p_r, \lambda)) \end{pmatrix}$$

$$0 = g_{con}(p, p_r)$$

$$0 = g_{npe}(p, p_r),$$

*The forces $f$ and $h$ include*

- *gravity and centrifugal forces,*

- *creep forces in the contact point,*

- *forces of the car body and*

- *constraint forces.*

*The creepage forces depend on the normal forces $N$ and thus on $\lambda$. Wheel and track are modeled as rigid bodies, but for the computation of the creepages local elasticity and the existence of a contact ellipse is assumed.*

*The following figures show simulation results obtained using data given in [SFR91][1].*

---

[1] See also the testset for initial value problem solvers, where subroutines describing the wheelset problem are given: "http://www.cwi.nl/cwi/projects/IVPtestset.shtml".

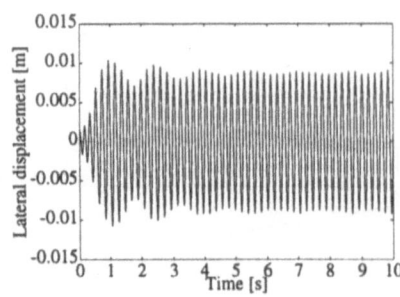

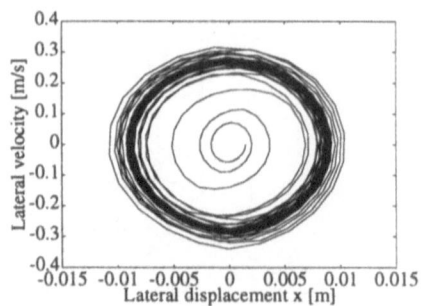

Figure 5.14: Lateral displacement                Figure 5.15: Phase diagram

## 5.5.2  Problems

We now point out some problems, especially when contact point jumps can occur. We demonstrate this type of problems by first considering a circle with radius $d$ moving along a sinusoidal curve, see Fig. 5.16:

$$r_0(\eta_0) = \sin \omega \eta_0$$
$$r_1(\eta_1) = -\sqrt{d^2 - \eta_1^2}.$$

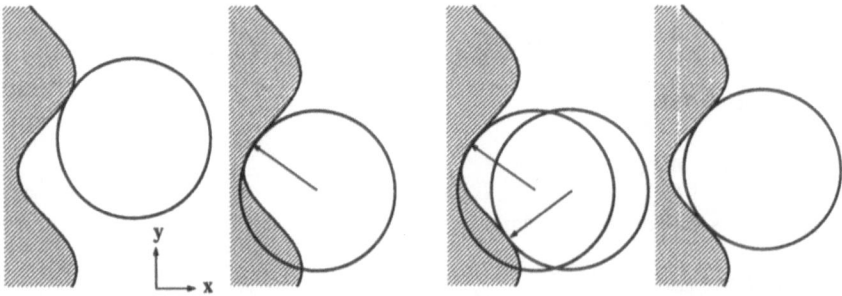

Figure 5.16:          Figure 5.17:          Figure 5.18:          Figure 5.19:

If $d > \omega^2$, there exist points where the curvature of both curves coincide, see Fig. 5.17. In these points does $g_{npe}$ not satisfy the conditions of the Implicit Function Theorem and Eq. (5.5.2d) cannot be solved for $p_r = \eta_1$. If the circle would move beyond such a point both bodies would locally penetrate. There exists no physical solution beyond this point. The solution can be continued if one admits *contact point jumps*.

Plotting $\eta_1$ as a function of $y_1$ one
obtains a curve looking like hysteresis. At the end points of the branches
the Jacobian $\frac{\partial}{\partial p_r} g_{\mathrm{npe}}$ becomes singular. Between these points there is an
interval where two solutions exist (see
Fig. 5.18). Taking a closer look at these
solutions one sees that only one solution
is physically possible, the one presented
in Fig. 5.19.

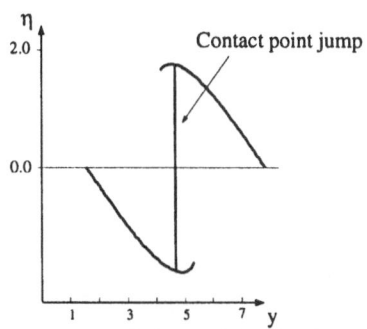

The exact localization of this jump requires the investigation of the problem's global
geometry. When numerically solving a DAE with contact conditions only local
properties of the functions $g_{\mathrm{npe}}$ and $g_{\mathrm{con}}$ come into play and the numerical solution
approaches the singularity until the method fails to converge. By monitoring the
condition number of the iteration matrix during numerical integration with an
implicit method one gets indicators that the numerical solution is in such a critical
region. We will demonstrate this by considering a more realistic example.

**Example 5.5.2 (Simulation for German railroad profiles)**
*The German railroad company (Deutsche Bahn AG) uses wheel and rail models
for simulation which are defined by piecewise smooth functions consisting of polynomials and circles. They are given in Figs. 5.20 and 5.21. When using these*

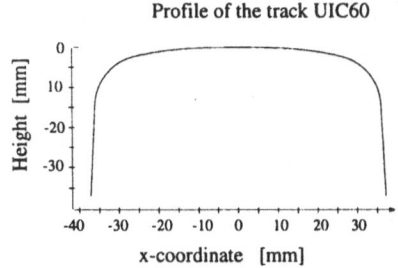

Figure 5.20: Profile of the track            Figure 5.21: Profile of the wheel

*profiles in Ex. 5.5.1 one encounters frequent contact point jumps. Monitoring the
condition number of the iteration matrix during a numerical simulation typically
gives results like those depicted in Fig. 5.22. The condition of the iteration matrix
increases when approaching $t = 1.29523$s and can be considered being numerically
singular at that point. When plotting the contact variable $\xi_L$ as a function of the
lateral displacement $x$ one obtains Fig. 5.23. Again, there are regions where the
location of the contact point mathematically is not uniquely defined.
The lateral displacement at the singularity is $x(\hat{t}) \approx 0.491$mm, which corresponds
to the "nose" in the left upper corner of Fig. 5.23.*

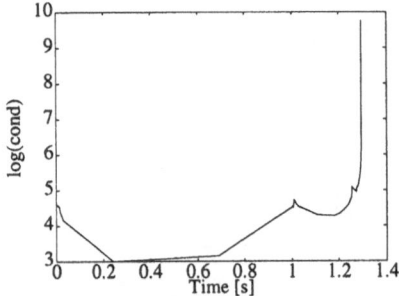

Figure 5.22: Condition of the itera-    Figure 5.23: Contact variable
tion matrix

*The contact point jump which corresponds to this critical region cannot be deter-
mined from this picture and as pointed out before not from standard DAE solvers.*

This kind of contact problems may be handled, e.g. by some of the following
techniques:

- If the integrator encounters an ill-conditioned iteration matrix a critical region with a potential contact point jump is assumed. The minima of the distance function between the two profiles are computed together with the corresponding distance vectors. Then a backward integration is performed with
  monitoring these distance vectors until one of those becomes zero. There, a
  contact point jump has to be performed.

- A contact point candidate is introduced from the beginning. This candidate
  is defined as the point where the distance to the curve which potentially can
  be penetrated is minimal.

- One uses a sort of preprocessing of the contact planes and establishes where
  contact point jumps may occur.

The first variant has the disadvantage that the physically correct contact point
jump may not be found. The second variant has the disadvantage that such a
point may not exist and if it exists its computation can be a problem. The third
variant can be very costly because the whole surface has to be analyzed, though
the solution is only a curve on it.

Points beyond which the solution cannot be continued are also present when using
the classical approach by table-look-ups.

For a more detailed analysis of contact point jumps and their numerical treatment
in the context of wheel/rail dynamics we refer to [AN96].

# 6 Integration of ODEs with Discontinuities

The equations of motion of multibody systems often contain *discontinuities* in the right hand side, in the solution or in its higher derivatives. In the sequel we summarize discontinuities in the function itself as well as in its higher derivatives under the expression "discontinuity".

In multibody systems typical elements introducing discontinuities are:

- *Friction:* Dry friction leads to jump discontinuities in the right hand side of the differential equations.

- *Approximation of characteristic lines by piecewise smooth functions:* Forces are often described in terms of data obtained from measurements and given in tables. Approximating or interpolating these data introduces discontinuities into the problem depending on the particular interpolation method.

- *Impact phenomena:* Impacts lead to jump discontinuities in the velocities.

- *Computer-controlled systems:* The right hand side of the differential equation depends on the state at explicitly given points.

- *Hysteresis:* The right hand side depends on the history of motion.

- *Structure varying systems:* The number of the degrees of freedom changes.

- *Time dependent input functions:* Excitations to the system are often modeled using time dependent input functions, which may have discontinuities at certain previously known time points (*time events*).

The aim of this section is to give some insight into the treatment of discontinuous systems: Which problems may occur when integrating discontinuous equations? How must a model be formulated for an integration method with a switching algorithm? How do switching algorithms work? How can integration be accelerated?

We saw in Ch. 4 that numerical integration methods require continuity also of higher derivatives of the solution depending on the order of the method and the error control mechanism. Ignoring this fact causes wrong error estimates and a failure of step size and error control. Thus, standard numerical integration software shows unreliable and inefficient behavior when applied to non-smooth systems. We will start this chapter by demonstrating these effects. Then, we will introduce the concept of differential equations with switching conditions.

Sometimes discontinuities occur at previously known time points and the integration can be stopped and restarted at these time events after a re-initialization. In other cases, the occurrence of a discontinuity is state dependent and one attempts to describe it implicitly in terms of roots of so-called *switching functions*. Integration methods must be extended to localize these roots in order to permit a re-initialization also at these points. For the numerical solution of discontinuous systems we present a *switching algorithm*, the basic principles of which are given in Sec. 6.3.

This method is of a particular importance in the context of optimization methods, where the computation of sensitivity matrices with prescribed accuracy is required, see Sec. 7.

We will discuss differential equations with switching conditions and then present techniques for localizing roots of these functions.

Systems with *dry friction (Coulomb-friction)*, or more general systems with a jump discontinuity in the right hand side may have no classical solution in certain intervals. The numerical solution tends to oscillate with high frequency around a so-called switching manifold and is very inefficient to compute. We will introduce Filipov's concept of a generalized solution [Fil64], give its physical interpretation and show how numerical methods can take advantage from this approach.

We will conclude this chapter by classifying discontinuities and by discussing switching techniques for these classes individually. This might ease the use of these methods, encourage a careful discontinuity handling and reduce the effort.

## 6.1   Difficulties with Discontinuities

In practice, discontinuous systems are often solved without localizing the points of discontinuity explicitly. This may lead to an inefficient behavior or even to a failure of the integration caused by an order reduction of the method and wrong error estimates.

Standard integration methods with automatic step size and order selection estimate the global error increment, see Sec. 4.1.4. By comparing this with a given error tolerance a decision is made to accept or reject a step and which step size and order to choose for the next step. If a discontinuity is present in an integration interval $(t_n, t_n + h]$ the error estimate becomes large, causing a rejection of the step and a drastic reduction of the step size. After such a reduction the discontinuity is often located outside of the new integration interval and this time the step is accepted. Then, some steps are taken with increasing step sizes until a new attempt is made to pass the discontinuity. This can occur repeatedly and thus requires a high computational effort until eventually the discontinuity is passed.

This effect is demonstrated by means of the truck driving over a curb:

**Example 6.1.1 (Truck driving over a curb)** *We assume that the truck drives with $v = 15$ m/s after 2 m over a curb with a height of 0.05 m. This leads to discontinuities at $t_1 := \frac{2m}{v} \approx 0.1333$ m and at $t_2 := t_1 + (a_{23} - a_{12})/v \approx 0.4333$ s due to the time delay between the wheels.*
*Integrating with MATLAB's integrator ODE45 leads to the critical steps given in Tab. 6.1. ERR/TOL is the estimated local error increment divided by the scaled required tolerance and thus should be less than one. Fig. 6.1 shows some simulation results.*

| $t$ | rejected step | | | accepted step | | |
|---|---|---|---|---|---|---|
| | $h$ | $t+h$ | ERR/TOL | $h$ | $t+h$ | ERR/TOL |
| 0.0948 | 0.0424 | 0.1372 | 56.43 | 0.0151 | 0.1099 | $3.7 \cdot 10^{-4}$ |
| 0.1099 | 0.0589 | 0.1689 | 262.55 | 0.0155 | 0.1254 | $1.4 \cdot 10^{-3}$ |
| 0.1254 | 0.0461 | 0.1715 | 2286.7 | 0.0079 | 0.1332 | $5.1 \cdot 10^{-5}$ |
| 0.1332 | 0.0454 | 0.1786 | 2143.9 | | | |
| 0.1332 | 0.0078 | 0.1411 | 1.4053 | | | |
| 0.1332 | 0.0059 | 0.1391 | 1.3472 | | | |
| 0.1332 | 0.0044 | 0.1376 | 1.2917 | | | |
| 0.1332 | 0.0034 | 0.1366 | 1.1100 | 0.0026 | 0.1359 | $9.3 \cdot 10^{-1}$ |
| | | | $\vdots$ | | | |
| 0.4200 | 0.0341 | 0.4541 | 73.86 | 0.0115 | 0.4315 | $5.2 \cdot 10^{-4}$ |
| 0.4315 | 0.0419 | 0.4734 | 507.923 | | | |
| 0.4315 | 0.0096 | 0.4412 | 7.3766 | 0.0052 | 0.4367 | $1.0 \cdot 10^{-1}$ |

Table 6.1: Critical steps integrating the truck driving over a curb

In addition to the high effort necessary to obtain a solution one cannot trust in the results: The estimation of the error as well as the method itself is based on the assumption that the solution and its derivatives are continuous up to a certain order. If this assumption is not fulfilled the error estimation is not valid.
This is even more serious for multistep methods as in $k$-step methods $k$ subsequent steps may be dubious.
If the integration routine is used in the context of optimization procedures an efficient and reliable computation of sensitivity matrices is necessary (see Ch. 7). This can only be done by localizing switching points.
The difficulties arising in the integration of ODEs with discontinuities are often circumvented by using integration methods with fixed step size or low order. The first approach has the drawback that the error is not controlled. Both approaches have the drawback that they may be very inefficient.
Another way out is to smooth the discontinuity. Smoothing approaches lead to artificially stiff systems (see Ex. 6.4.2) and to large, often discontinuous higher

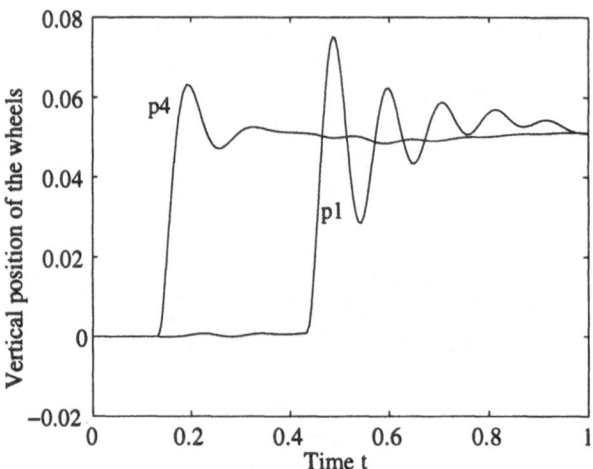

Figure 6.1: Truck driving over a curb

derivatives.

To summarize, all approaches which do not localize the points of discontinuity remain unsatisfactory and must use several heuristics. Thus we present another approach which is based on localizing the discontinuities as roots of switching functions.

## 6.2   Differential Equations with Switching Conditions

We first discuss the case of an explicit ODE, $\dot{x} = f(t, x)$, $x(t_0) = x_0$, where $f$ is piecewise continuous. The extension to the case of differential-algebraic equations is given in Sec. 6.3.8.

The conditions under which a discontinuity occurs are in general known and can be formulated as roots of a vector-valued, time and state dependent *switching function*

$$q(t, x(t)) = (q_1(t, x(t)), \ldots, q_{n_q}(t, x(t)))^T.$$

The differential equation can then be written as

$$\dot{x} = f(t, x, \operatorname{sign} q),$$

where the expression "$\operatorname{sign} q$" has to be understood componentwise. $f$ depends on a combination of the signs of the components of $q$.

In Example 6.1.1 the switching functions are simple time functions:

$$q(t) = \begin{pmatrix} q_1(t) \\ q_2(t) \end{pmatrix} = \begin{pmatrix} t - t_1 \\ t - t_2 \end{pmatrix}.$$

**Example 6.2.1** *The problem*

$$\dot{x} = |x|$$

*can be formulated using switching functions as*

$$\dot{x} = \begin{cases} x & \text{for } \operatorname{sign} q = 1 \\ -x & \text{for } \operatorname{sign} q = -1 \end{cases} \quad \text{with } q(t, x) = x.$$

For the numerical treatment of ODEs with discontinuities it is necessary to rewrite the differential equation as

$$\dot{x} = f(t, x, s) \text{ with } s = \operatorname{sign} q.$$

We make the following assumption.

**Assumption 6.2.2** *For fixed $s$, $f$ is assumed to be smooth:*

$$f(., ., s) \in C^l(\mathbb{R} \times \mathbb{R}^n, \mathbb{R}^n)$$

*with $l$ being sufficiently large.*

Discontinuities can therefore occur only if a component of $q$ changes its sign. The solution itself may have jump discontinuities, e.g. modeling the impact between two bodies may lead to jumps in the velocity variables. The time $\hat{t}$ of the occurrence of such a jump discontinuity is assumed to be implicitly defined as root of a switching function, too:

$$q_i(\hat{t}, x^-(\hat{t})) = 0.$$

The right limit $x^+(\hat{t})$ is a function of the left limit $x^-(\hat{t})$:

$$x^+(\hat{t}) = x^-(\hat{t}) + z(\hat{t}, x^-(\hat{t})). \qquad (6.2.1)$$

$x^+(\hat{t})$ is the initial value for the new integration interval.
Then, the solution is composed of the two branches[1]

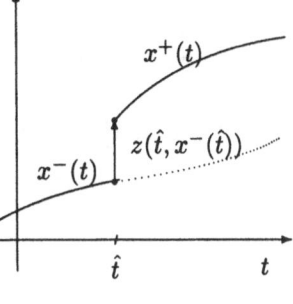

$$x(t) = \begin{cases} x^-(t) & t \le \hat{t} \\ x^+(t) & t > \hat{t}. \end{cases}$$

---

[1] Note that the figure implies that the solution for the old right hand side can be continued, at least for a small interval. This assumption is necessary for switching algorithms.

## 6.3   Switching Algorithm

We are now ready to discuss the treatment of differential equations with switching conditions. The main task is the localization of the switching points as roots of the switching functions. Note, the complexity of this task depends on the form of the switching function. Switching functions may in general have the same complexity as the right hand side function. For example, an impact requires the computation of relative distances for both functions.

In general, switching algorithms work in the following way:

1. After every integration step check if a discontinuity has occurred by testing the sign of the switching function.

2. If no sign changed, continue the integration otherwise localize the discontinuity as a root of the switching function.

3. Change to the new right hand side, modify variables according to Eq. (6.2.1).

4. Adapt the discretization, normally by restarting the integration.

These steps are described in detail in the following sections.

### 6.3.1   Test for Occurrence of a Discontinuity

After every integration step the sign of the switching functions is checked. If a sign change occurs, the localization of the switching point is started.

Note, this foregoing is based on the assumption that a solution of the differential equation with the "old" right hand side exists until $t = t_{n+1}$. Especially the "old" right hand side function is not allowed to have a singularity at the switching point.

**Remark 6.3.1** *For parallel processing it may be advantageous to decouple sign check and discretization: The integration can be performed and in parallel with a small time lag the switching functions can be evaluated to check their sign. If a sign change occurs, the integration process has to be stopped and has to be reset to the point of discontinuity. This is especially interesting for problems with few discontinuities and many or computationally complex switching functions.*

### 6.3.2   Localization of Switching Points

In order to localize a discontinuity, a root $t^* \in [t_n, t_{n+1}]$ of the switching function $q$ must be found:

$$q(t^*, x(t^*)) = 0.$$

Since the exact solution $x(t)$ is not known, the root of the switching function is computed for the numerical solution. The localization of this switching point involves a one-dimensional root finding. This is in general a nonlinear problem which must be solved iteratively. For this, values of the switching functions and of the states $x$ are not only required at discretization points, but also in between. They can be approximated using a *continuous representation* of the solution, as shown below.

The determination of the switching point can be done, by using Newton's method, the secant rule or by *inverse interpolation*. We prefer derivative-free methods, because the numerical computation of the time derivative of $q$ involves an extra evaluation of the right hand side function due to

$$
\begin{aligned}
\frac{d}{dt}q(t, x(t)) \;&=\; q_t + q_x \dot{x} = q_t + q_x f \\
&=\; \frac{q(t + \varepsilon, x(t) + \varepsilon f(t, x(t))) - q(t, x(t))}{\varepsilon} + \mathcal{O}(\varepsilon). \quad (6.3.1)
\end{aligned}
$$

This evaluation may require an unnecessary additional computational effort. Another reason to use derivative-free methods is that in the DAE case derivatives for the algebraic variables are not available and may even not exist, see Sec. 5.1.2.

Thus, we focus here only on the method of inverse interpolation. The secant rule is a special case of this approach.

Later we combine this determination of the iterates with *safeguard techniques*.

### 6.3.2.1  Inverse Interpolation

The idea of inverse interpolation is quite simple: instead of looking for a root of $Q(t) = q(t, x(t))$ we evaluate the inverse function $t = T(Q)$ at $Q = 0$.

Since $T(Q)$ is unknown, we construct a polynomial $T_k$ of order $k$ interpolating $(Q(\tau_i), \tau_i)$, $i = 0, \ldots, k$, i.e. $T_k(Q(\tau_i)) = \tau_i$, $i = 0, \ldots, k$. Then, $\tau_{k+1} := T_k(0)$ is a good approximation to the switching point.

Starting with $\tau_0 := t_n, \tau_1 := t_{n+1}$, we can compute $T_k(0)$ by using the Neville-Aitken scheme [SB93]:

$$
\begin{aligned}
T_{i0} \;&:=\; \tau_i, \quad i = 0, \ldots, k \\
T_{ij} \;&:=\; T_{i,j-1} + \frac{T_{i,j-1} - T_{i-1,j-1}}{\dfrac{q_{i-j}}{q_i} - 1}, \quad 1 \le j \le i \le k
\end{aligned}
$$

with $q_l = Q(\tau_l) = q(\tau_l, x(\tau_l))$. The desired value $\tau_{k+1}$ is then obtained from $T_k(0) = T_{kk}$. For $k = 1$ we obtain the secant rule.

### 6.3.2.2  Safeguard Methods for Root-Finding

In order to avoid that after the restart of the integration a localized switching point results in a sign change again, care must be taken to ensure that the restart is performed at a point $\tilde{t}$ with $\tilde{t} > t^*$. This is the reason to determine not only a sequence of approximations $\tau_j$ to $t^*$ but a sequence of intervals $I_j$ containing $t^*$:

$$I_j = [a_j, b_j] \subset I_{j-1}, \quad t^* \in I_j.$$

$t^* \in I_j$ can be assured by $Q(a_j)Q(b_j) \leq 0$. The restart is then executed at $\tilde{t} = b_j$ if $b_j - a_j$ is sufficiently small. Due to $q(t_n, x(t_n)) \cdot q(t_{n+1}, x(t_{n+1})) \leq 0$ we can take $I_0 = [t_n, t_{n+1}]$ as initial value for the interval, where $t_n$ and $t_{n+1}$ are the discretization points of the interval enclosing a sign change.

The sequence of intervals is constructed such that the length of the intervals becomes arbitrarily small, $|I_j| \to 0$.

The idea of safeguard methods consists of combining a method which is convergent for an arbitrary initial interval $I_0$ with a rapidly convergent method. Such a combination may consist of bisection and inverse interpolation. Fig. 6.2 shows the principles of the algorithm; each step is discussed in the sequel.

First, an iterate $\tau_{j+1}$ is determined by a rapidly convergent method and then checked using the following criteria, [GMW81] (the numbering refers to Fig. 6.2):

1. We require the new iterate $\tau_{j+1}$ to be inside the interval $[a_j, b_j]$ obtained so far. Otherwise $\tau_{j+1}$ is determined by bisection: $\tau_{j+1} = \frac{a_j + b_j}{2}$.

2. $\tau_{j+1}$ should be closer to the "better" one of the points $a_j, b_j$. That is the one with the smaller absolute value of $Q$. $\tau_{j+1}$ is determined by bisection if the distance to the better point is larger than half of the interval length, because the bisection step is expected to result in a larger reduction of the interval length.

3. Care must be taken to ensure that two iterates have a minimal distance: it may happen that $\tau_{j+1}$ cannot be distinguished numerically from the best point so far. Roundoff errors may then disturb the interpolation process and lead to useless values for the next iterate. This is the reason for requiring a minimal distance $\delta$.

   Furthermore, this strategy avoids that one boundary point of the interval converges to the root and the other remains constant. Thus, $|I_j| \to 0$ is enforced.

4. $a_{j+1}, b_{j+1}$ are chosen such that

   $$\tau_{j+1} \in \{a_{j+1}, b_{j+1}\}, \quad a_{j+1}, b_{j+1} \in \{a_j, b_j, \tau_{j+1}\}, \quad a_{j+1} < b_{j+1}$$

   and that the enclosing property remains true: $Q(a_{j+1}) \cdot Q(b_{j+1}) \leq 0$ ("regula falsi"). Then $j$ is increased by 1.

The interval length converges globally to zero and the convergence is at least linear with a rate less than $\frac{1}{2}$, i.e.

$$\frac{|I_j|}{|I_{j+1}|} \leq \frac{1}{2} \quad .$$

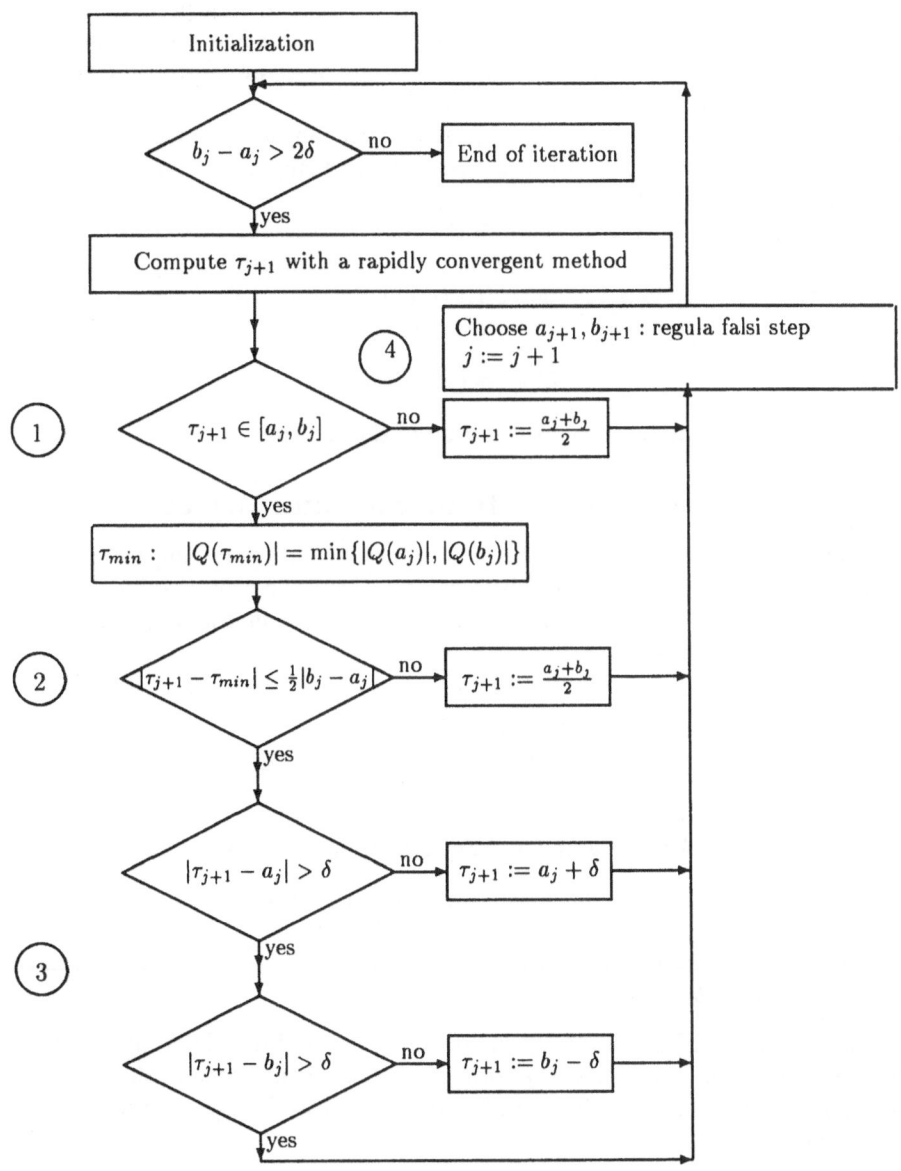

Figure 6.2: Safeguard-method

### 6.3.2.3   Evaluation of Switching Functions

In order to evaluate the switching function in the root-finding process, values of
the numerical solution are not only required at discretization points but also in
between them. These values can be computed using an error controlled continuous
representation of the solution, see Sec. 4.5.

For BDF or Adams methods such a representation can be obtained directly from
the construction idea of the discretization method using the polynomial already
computed for the discretization.

In [SG84] it is shown for Adams methods that this representation is error controlled,
for BDF methods the corresponding result can be found in [Eich91].

The values of the switching function $q(t, x(t))$ at the exact solution are substituted
by its values at the continuous representation

$$Q(t) := q(t, P(t)),$$

where $P(t)$ is the continuous representation of the solution.

## 6.3.3   Changing the Right Hand Side and Restarting

As result of the root finding algorithm an interval $[a, b]$ containing the switching
point is obtained. Now, it must be checked if there exists a root, maybe of an other
switching function in $[t_n, a]$. If this is the case, the root finding algorithm has to
be restarted in this interval.

Thus we propose the following foregoing:

1. Estimate the location of all roots using the secant rule. Let the index of the
   switching function with the first estimated root be $i_0$.

2. Localize the root of $q_{i_0}$ as discussed above. As result of the root finding
   algorithm an interval $[a, b]$ is obtained containing the switching point.

3. Test if there exists a root located left of $a$, i.e. in $[t_n, a]$. If this is the case
   one has to restart the root finding algorithm in this interval. Otherwise the
   signs of all components of $s$ corresponding to all switching functions having
   a root in $[t_n, b]$ are changed.

The determination of the new right hand side to be used may be a difficult task if
the switching of one function causes other variables or functions in the right hand
side to have jump discontinuities. In Sec 6.5.2 this case is discussed for structure
varying systems.

## 6.3.4   Adaptation of the Discretization

One-step methods can be restarted at point $b$. In this case, changing to a lower
order method is not necessary as no past values are used. This is different for

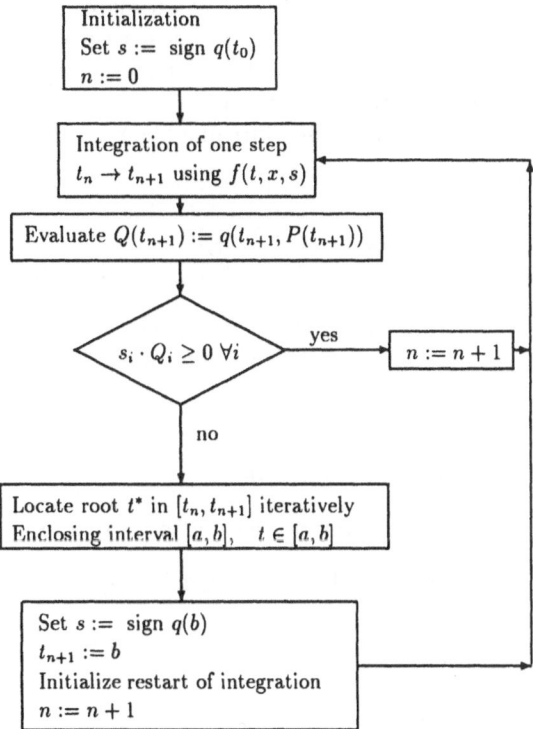

Figure 6.3: Switching algorithm

multistep methods: If the discontinuity is of order $m$ in $x$, i.e. the $m$-th derivative of $x$ has a jump

$$
\begin{aligned}
x^{(m)}(t+) &= x^{(m)}(t-) + z(t-, x(t-)) \\
x^{(i)}(t+) &= x^{(i)}(t-), \quad i = 0, \ldots, m-1,
\end{aligned}
$$

the order of the discretization method has to be reduced in order to meet the smoothness requirements of the method, the error estimation and step size control. For the $k$-step BDF method this is: $y \in C^{k+1}$, such that the order has to be reduced to $m-2$.

This involves the knowledge of the order of the discontinuity. Numerical approaches to determine this order work only reliable for orders 0 and 1 [GO84]. We therefore suggest to restart the integration method for safety reasons. When restarting a multistep method, much time is spent for regaining the appropriate integration step size and order. The best one can do is to use the information available from the last time interval before the discontinuity: in the case of the BDF method for example one can use the backward differences to estimate the appropriate step size for the new reduced order $k$ (which is $k = 1$ in most cases).

To accelerate the restart for multistep methods Runge–Kutta methods can be used to generate the necessary starting values, [SB95].

Now we have discussed all steps of the switching algorithm. The overall algorithm is summarized in Fig. 6.3.

### 6.3.5   Setting up a Switching Logic –
### A Model Problem: The Woodpecker Toy

We are going to demonstrate the above switching algorithm using the woodpecker toy. This example has a sophisticated switching logic. The modeling originates from [Pfei84].

The woodpecker toy consists of

- a rod where the woodpecker glides down

- a sleeve which glides down the rod with some play and

- a woodpecker connected to the sleeve by a spring.

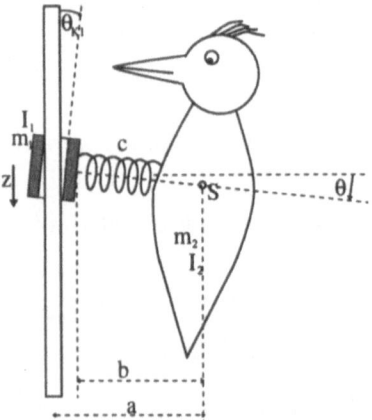

Figure 6.4: The woodpecker toy

The woodpecker has two degrees of freedom: one rotational with respect to the rod (angle $\theta$) and one translational down the rod (height $z$ of point $S$).

The most essential part is the sleeve which hinders the woodpecker from moving down for angles $|\theta| > |\theta_{K1}|$. Thus, in this situation there is no translational degree of freedom.

The motion can be subdivided into five phases:

1. $\theta > \theta_{K1}$:
   The woodpecker swings to the right until it reaches the maximal amplitude and then back to the left. It does not move down.

2. $-\theta_{K1} < \theta < \theta_{K1}$:
   Phase 1 ends for $\theta = \theta_{K1}$ and the woodpecker moves down due to gravitation. The rotation is anti-clockwise ($\dot{\theta} < 0$).

3. $-\theta_{K2} < \theta < -\theta_{K1}$:
   For $\theta = -\theta_{K1}$ an impact due to the stop of the sleeve occurs, afterwards the translational degree of freedom is again jammed.

4. $-\theta_{K2} < \theta < -\theta_{K1}$:
   For $\theta = -\theta_{K2}$ the woodpecker touches the rod with its bill and the direction of rotation changes. The translational degree of freedom remains blocked.

5. $-\theta_{K1} < \theta < \theta_{K1}$:
   For $\theta = -\theta_{K1}$ the self blocking phase ends and the woodpecker moves again down the rod, the rotation is in clockwise sense until the sleeve stops for $\theta = \theta_{K1}$ and phase 1 starts again.

The resulting switching logic is shown in Fig. 6.5.

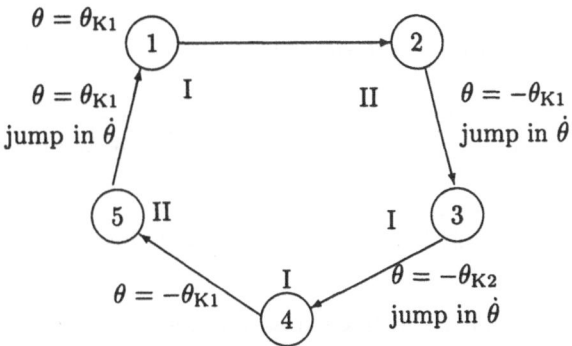

Figure 6.5: Switching diagram for the woodpecker toy

The sleeve is modeled as a massless joint [Pfei84]. Only small oscillations are considered. Then the following system of differential equations is obtained:

- **Eq. I:** ( $\theta$ is the only degree of freedom, $z = $ const)

$$(I_2 + m_2 b^2)\,\ddot{\theta} = -c\theta + m_2 b g_{gr}$$

- **Eq. II:** (two degrees of freedom)

$$\left(I_2 + m_2 b^2 \left(1 - \frac{m_2}{m_1 + m_2}\right)\right)\ddot{\theta} = -c\,\theta$$

$$(m_1 + m_2)\ddot{z} + m_2 b\,\ddot{\theta} = (m_1 + m_2)g_{gr}.$$

All impacts but the impact of the bill are non-elastic and lead to jumps in the angular velocities $\dot{\theta}$:

- impact of the sleeve at the top:

$$\dot{\theta}^+ = (1 - d_3)\left(\dot{\theta}^- + \frac{m_2 b}{I_2 + m_2 b^2}\dot{z}^-\right)$$

- impact of the sleeve at the bottom:

$$\dot{\theta}^+ = (1 - d_1)\left(\dot{\theta}^- + \frac{m_2 b}{I_2 + m_2 b^2}\dot{z}^-\right)$$

- impact of the bill:

$$\dot{\theta}^+ = -\dot{\theta}^-.$$

$\dot{\theta}^-, \dot{z}^-$ are the velocities before the impact and $\dot{\theta}^+, \dot{z}^+$ the velocities after the impact.

The constants are taken from [Pfei84]: $b = 0.015$ m, $a = 0.025$ m, $m_1 = 0.003$ kg, $m_2 = 0.0045$ kg, $I_2 = 7 \cdot 10^{-7}$ kgm$^2$, $c = 0.0056$ Nm, $d_3 = 0.04766$, $d_1 = 0.18335$, $g_{gr} = 9.81$ m/sec$^2$, $\theta_{K1} = 10°$, $\theta_{K2} = 12°$.

The simulated solution is given in Figs. 6.6 – 6.8.

This example could only be solved using switching functions due to the occurrence of

- a change of type of the differential equation (one and two degrees of freedom) and

- the jumps in $\dot{\theta}$.

### 6.3.6   Aspects of Realization

**Sign Change of the Switching Function**

Let $D \subset \mathbb{R}^{n+1}$, $f : D \times \Omega \to \mathbb{R}^n$ be the right hand side of the differential equation with $\Omega := \{-1, 1\}^{n_q}$. From the assumptions follows that $f \in C^l(D \times \Omega, \mathbb{R}^n)$, i.e. $f$ is sufficiently smooth for fixed sign $s \in \Omega$, see Assumption 6.2.2. During the course of the integration $s = $const. has to be guaranteed, i.e. $f$ must be evaluated for the

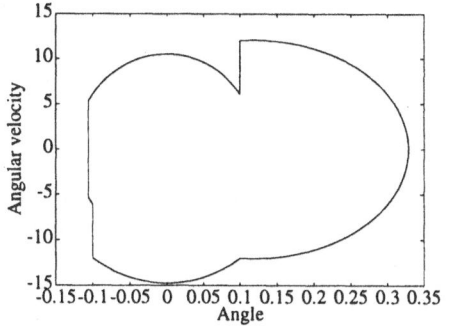

Figure 6.6: Phase diagram

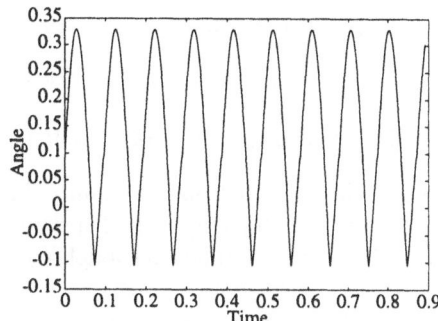

Figure 6.7: Rotation angle of the woodpecker

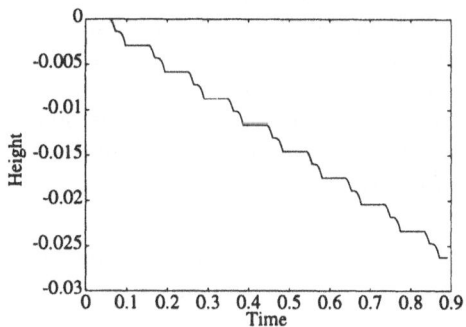

Figure 6.8: $z$-coordinate of the woodpecker

same "branch". After every integration step it is tested if this was feasible, i.e. if $q$ has had a sign change. If this is the case the root is localized and the sign-vector $s$ is changed. Note that for this reason the subroutine computing the right hand side should never take decisions on the basis of sign $q$. It should only depend on $s$. This variable must be a "read-only variable" in the right hand side and is changed only after localizing a discontinuity.

**Recomputation of the Jacobian**
Discontinuities may not only appear in the right hand side function and in the solution but also in the system's Jacobian. They must be taken into consideration if Newton iteration is used to solve nonlinear systems in the course of an implicit discretization method or in steady state computation. Different impacts of a discontinuity must be handled:

**a:** Changes in the equations with no effect on the Jacobian,

**b:** Changes in the values of the Jacobian elements,

**c:** Changes in the number of equations and variables.

Each of these discontinuities require different actions with respect to convergence control (**a**), recomputation and decomposition of the Jacobian (**b**) or even reinitialization of the simulation model (**c**).

### Choice of Switching Functions

Switching functions for one particular problem can have several forms, e.g. multiplication by a constant does not alter the switching point. However, the choice of the switching function influences the root-finding process. Note that the switching algorithm cannot find switching points which are double roots of the switching function. Threefold roots can be found by the secant rule, but Newton methods are not able to find this type of roots.
Switching functions should be as simple as possible, e.g. linear. It is advisable to use different switching functions for different discontinuities instead of multiplying them to obtain only one function.

## 6.3.7   Other Methods and Codes for Discontinuous Differential Equations

In [Man78, CR78] Newton's method is used to localize the switching point. Similar approaches can be found in [EF75, CS74, CS76, Hal76]. All these methods need many evaluations of the right hand side because they do not use a continuous representation of the solution.
In [Ell81, BS81] the use of a continuous representation is proposed. In [BS81] an Adams method is used, in [Ell81] a continuous representation is obtained by a third order Hermite polynomial for a discretization with a Runge–Kutta pair of order 3/4. Enright et al. [EJNT86a] use a Runge–Kutta method of order $p$ and a corresponding local interpolation for the localization. The switching point is computed by bisection.
In [Car78, CM78] a continuous representation of the switching functions itself is obtained by introducing additional differential equations for the switching functions and the use of a continuous representation of the solution (Nordsieck). If many switching functions are present this is not advisable because the additional effort for the integration becomes high. The continuous representation of $q$ as obtained above from inserting a continuous representation of $y$ into $q$ requires much less effort.
There are some articles proposing methods for localizing discontinuities by monitoring the error estimation, which grows significantly if a discontinuity is crossed

[GO84, Mod88, EJNT86b]. In [GO84, Mod88] a step size for the crossing step is
determined which holds the local error below the required tolerance. This demands
the knowledge of the order of the discontinuity which is often not known.

## 6.3.8   Special Aspects of Differential-Algebraic Equations with Discontinuities

We consider a DAE of the form

$$\dot{y} \;=\; f(y, z, s)$$
$$0 \;=\; g(y, z, s).$$

with $s = \operatorname{sign} q$ and the switching function $q(y, z, t)$.
An example for a discontinuity in the algebraic equation is a mass gliding on a
non-smooth surface.
When using a continuous representation of the solution $P^y, P^z$, the values at the
switching point $t^*$ are in general not consistent with the algebraic equation, i.e. we
may have

$$q(P^y(t^*), P^z(t^*),) \;=\; 0$$
$$g(P^y(t^*), P^z(t^*), s) \;\neq\; 0.$$

For the index-1 case one might instead of solving

$$q(P^y(t), P^z(t), t) = 0$$

for $t$ solve the system

$$g(P^y(t), z) \;=\; 0 \qquad\qquad (6.3.2a)$$
$$q(P^y(t), z, t) \;=\; 0 \qquad\qquad (6.3.2b)$$

for $(\hat{z}, \hat{t})$, [Tho93]. Then $\hat{t}$ is a switching point. Using $P^y(\hat{t})$ as value for $y$ and $\hat{z}$
as value for $z$ is consistent with the algebraic equations.
The applicability of this techniques in the general case cannot be always assured:

Consider the following constraint for a higher index system $g = g(y, s)$ and take
the switching function $q(y, z, t) = t - t_1$, where $t_1$ is given. Then with approach
(6.3.2) one has to solve

$$0 \;=\; g(P^y(t), s)$$
$$0 \;=\; t - t_1$$

for the unknowns $z, t$. The corresponding Jacobian

$$J = \begin{pmatrix} 0 & g_y \frac{dP^y}{dt} \\ 0 & 1 \end{pmatrix}$$

is singular, such that the Inverse Function Theorem 3.2.1 cannot be applied. Alternatively, one may think of taking the end point of the integration step as an additional variable. Assume that in $[t_n, t_{n+1}]$ $q$ changes its sign. For simplicity we describe this method for the implicit Euler discretization and the corresponding linear representation of $y$. This leads to the system

$$
\begin{aligned}
y &= y_n + (t - t_n)f(y, z, s) \\
0 &= g(y, z, s) \\
0 &= q(y, z, t)
\end{aligned}
$$

which is now solved for $y, z, t$, where $t$ becomes then the switching point.

In order to keep computational costs small (this is an $(n_y + n_z + 1) \times (n_y + n_z + 1)$ system in contrast to the 1-dimensional root finding which have to be carried out in the ODE case using interpolants), a good starting value for $\hat{y}, \hat{z}, \hat{t}$ can be obtained using techniques discussed above for the ODE case.

**Remark 6.3.2** *Newton's method involves the computation of the derivatives $\frac{d}{dt}q(t, x(t))$. For an ODE they can be computed numerically by a forward difference using Eq. (6.3.1). This is done e.g. in [Bock87, Bul71]. Note that in the DAE case this foregoing cannot be used, because in general the derivatives of the algebraic variables are not known and therefore cannot be substituted by the right hand side of the differential equation. Thus, methods involving derivatives of the switching function should not be taken into consideration in this context.*

# 6.4   Coulomb Friction: Difficulties with Switching

Friction phenomena are often modeled by a law dated back to Coulomb. The friction force is assumed to be proportional to the normal force $N$ on the surface between the two bodies in the contact point. The proportionality constant is the coefficient of friction $\mu$ which depends on the material properties of the surfaces and may depend on the velocity.

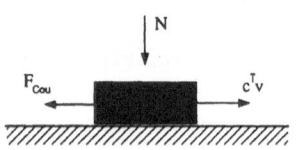

In Coulomb's original model $\mu > 0$ has been assumed to be constant:

$$
|F_{\text{Cou}}| = \mu|N|.
$$

The friction force is directed tangential to the friction surface and is opposite to the direction of motion of the body:

$$
F_{\text{Cou}} = -\mu|N|c\,\text{sign}\,(c^T\dot{p}).
$$

$c$ is a unit vector parallel to the friction surface, $c^T \dot{p}$ describes the relative tangential velocity. Later we will consider also models with a non constant friction coefficient. In this case we assume $\mu$ being continuous.

Often this Coulomb friction is called *dry friction*.

## 6.4.1   Introductory Example

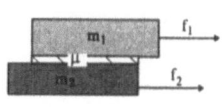

Let us consider the example of a simple two mass oscillator with dry friction between the bodies:

$$m_1 \ddot{p}_1 = f_1 - \mu |N| \text{sign}(\dot{p}_1 - \dot{p}_2) \tag{6.4.1a}$$
$$m_2 \ddot{p}_2 = f_2 + \mu |N| \text{sign}(\dot{p}_1 - \dot{p}_2), \tag{6.4.1b}$$

where $p_1$ describes the position of the first and $p_2$ the position of the second body. $f_1, f_2$ are applied forces. $\mu$ is the coefficient of friction. We use $m_1 = m_2 = 1$, $f_1 = \sin t$, $f_2 = 0$, $p_1(0) = p_2(0) = 1$, $\dot{p}_1(0) = \dot{p}_2(0) = 0$, $|N| = 1$ and use MATLAB's integrator ODE45 with a tolerance of $10^{-4}$ for the numerical integration of this system. The solutions obtained for different values of $\mu$ are shown in Figs. 6.9, 6.11 and 6.13. The relative velocity between the bodies is given in Figs. 6.10, 6.12 and 6.14.

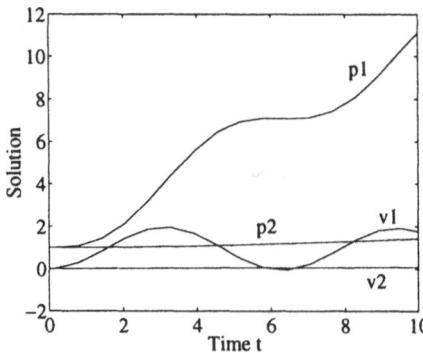

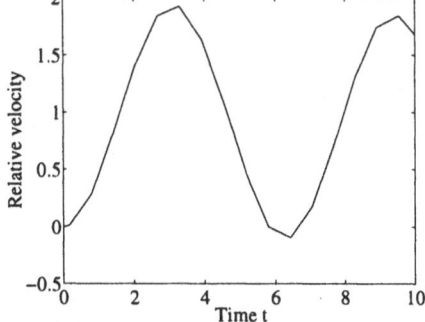

Figure 6.9: Solution for $\mu = 0.01$      Figure 6.10: Relative velocity for $\mu = 0.01$

For small $\mu$ the friction between the bodies is too low to get the second body into motion. For $\mu = 0.4$ the second body moves and then there are time intervals where both bodies move together as can be seen from Fig. 6.12 and Fig. 6.14. For $\mu = 1$ the friction is so large that except in a small initial time interval both bodies move together all the time.

The required number of steps is

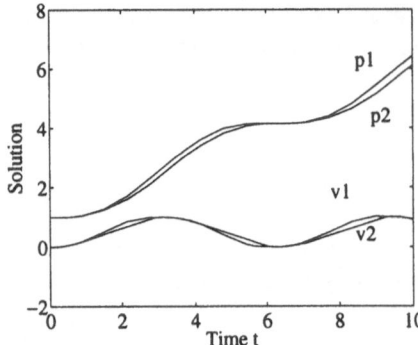

Figure 6.11: Solution for $\mu = 0.4$

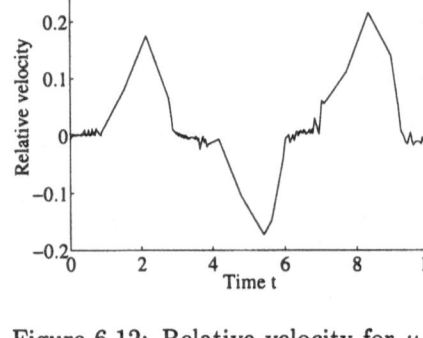

Figure 6.12: Relative velocity for $\mu = 0.4$

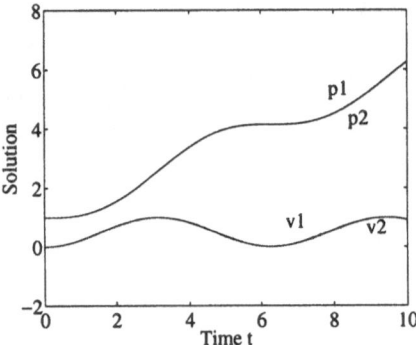

Figure 6.13: Solution for $\mu = 1$.

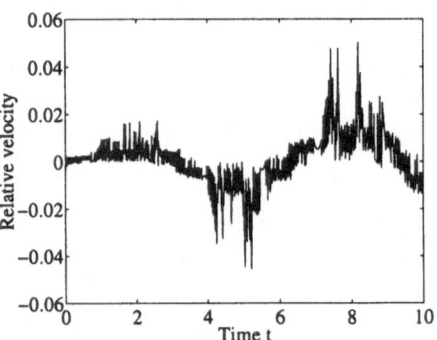

Figure 6.14: Relative velocity for $\mu = 1$.

| $\mu$ | 0.01 | 0.4 | 1.0 |
|-----|------|-----|------|
| NST | 19 | 214 | 1488 |

What is the reason for this high effort for large values of the friction coefficient? The answer can be obtained from Figs. 6.12 and 6.14. For $\mu = 0.4$ the relative velocity is oscillating around zero in some time intervals and for $\mu = 1$ even over the entire time interval. This can be seen also directly when considering method for (6.4.1). After subtracting both equations we get

$$
\begin{aligned}
v_{\text{rel},n+1} &= \dot{p}_{1,n+1} - \dot{p}_{2,n+1} \\
&= \dot{p}_{1,n} - \dot{p}_{2,n} + h\left(\sin t_n - \mu \operatorname{sign}\left(\dot{p}_{1,n} - \dot{p}_{2,n}\right)\right) \\
&= v_{\text{rel},n} + h\sin t_n - 2h\mu \operatorname{sign}\left(v_{\text{rel},n}\right)
\end{aligned}
$$

If $\mu$ is large enough the third term on the right hand side is dominating. As this term has always the opposite sign as $v_{\text{rel}}$ the velocity oscillates around zero. Physically, this oscillation does not occur: for large $\mu$ both bodies move together and the relative velocity is zero. For this case our model cannot be applied, since sign(0) is not defined and cannot be defined in such a way that $v_{\text{rel}}$ remains zero. A way out is to redefine the model for $v_{\text{rel}} = 0$. If the applied force is less than the friction force $\left| \frac{f_1}{m_1} - \frac{f_2}{m_2} \right| < \mu \left( \frac{1}{m_1} + \frac{1}{m_2} \right)$ we may set:

$$(m_1 + m_2)\ddot{p}_1 = (m_1 + m_2)\ddot{p}_1 = f_1 + f_2.$$

Using this formulation of the right hand side together with MATLAB's ODE45 integration method we obtain the relative velocities shown in Figs. 6.15 and 6.16. The oscillation disappeared. The number of integration steps is reduced drastically:

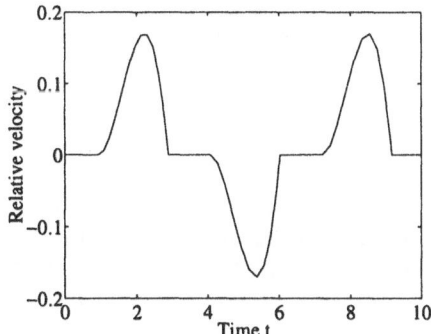

Figure 6.15: Relative velocity for $\mu =$ 0.4 (Refined model)

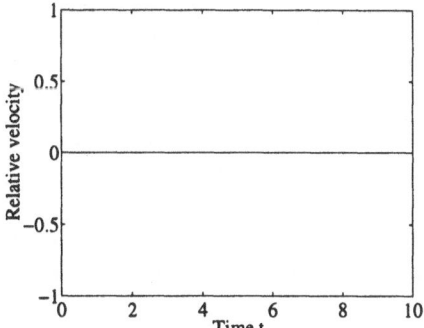

Figure 6.16: Relative velocity for $\mu =$ 1. (Refined model)

| $\mu$ | 0.01 | 0.4 | 1.0 |
|-----|------|-----|-----|
| NST | 19 | 56 | 23 |

In the sequel, we show how an adequate model formulation can be obtained for the case of dry friction. First we describe the mathematical background and then discuss the application. We then apply this concept to the truck driving over different road coverings.

## 6.4.2 Coulomb Friction: Mathematical Background

In the preceding section we used a simple model containing Coulomb friction to demonstrate the problems when solving the equations of motion. We gave a refined model for this example. Now, we want to generalize this foregoing, i.e. we want to construct refined models for the general case of systems with jump discontinuities in the right hand side. For that purpose we have to understand what is going on in these cases.

In the sequel we describe the problem of inconsistent switching and briefly introduce Filippov's approach. On this basis we construct an automatic treatment of this situation by a *generalized three-valued switching logic*.

The problem of inconsistent switching is described for ordinary differential equations of the form

$$\dot{x} = f(t, x, s) = \begin{cases} f^+(t, x) & \text{for } s = +1 \\ f^-(t, x) & \text{for } s = -1 < 0 \end{cases}$$

with $q : I\!R^{n+1} \to I\!R$ and $s = \text{sign } q$.

In the example given at the beginning of this section we have

$$q(t, x) \;=\; \dot{p}_1 - \dot{p}_2$$

$$f^+ = \left( \begin{array}{c} \frac{1}{m_1}(f_1 - \mu) \\ \frac{1}{m_2}(f_2 + \mu) \end{array} \right) \qquad f^- = \left( \begin{array}{c} \frac{1}{m_1}(f_1 + \mu) \\ \frac{1}{m_2}(f_2 - \mu) \end{array} \right).$$

Let $Dq^+, Dq^-$ denote the auxiliary switching functions

$$\begin{aligned} Dq^+ &:= q_x f^+ + q_t \\ Dq^- &:= q_x f^- + q_t. \end{aligned}$$

$Dq^+, Dq^-$ can be interpreted as directional derivatives of the switching functions in the direction of $f^+$ and $f^-$. The question if the solution can cross the discontinuity can be answered by the sign of $Dq^+$ and $Dq^-$.

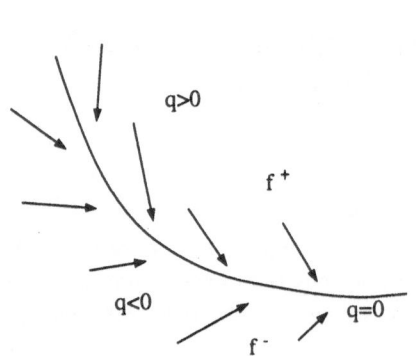

Figure 6.17: Direction field

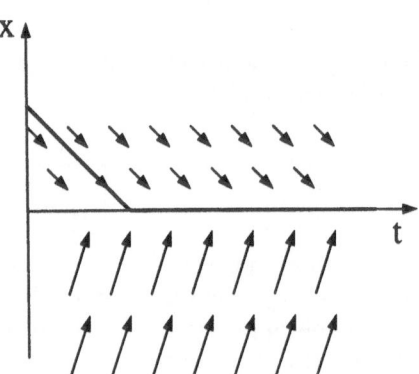

Figure 6.18: Direction field and solution

Assume that $q(\hat{t}, x(\hat{t})) = 0$ holds: If

$$Dq^+ < 0 \text{ and } Dq^- > 0,$$

both directions fields point to the manifold $q = 0$ and the solution cannot leave the manifold. Thus there is a point beyond which no classical solution exists (e.g. [Fil64, Man73, Tau76, Bock87]). The solution must stick in the manifold.

The expression "classical solution" is used to characterize functions which fulfill the differential equation almost everywhere. This corresponds to the point of view from the last sections, where the differential equation is fulfilled everywhere except at the switching points. If the differential equation is discontinuous along a manifold $q = 0$ it has only a classical solution if the solution crosses the discontinuity. If the direction field of the differential equation allows an approximation to the manifold from both directions, i.e. a crossing of the manifold is not possible, a solution in the classical sense does not exist. In the context of switching functions the fact that the manifold cannot be crossed is called *inconsistent switching*. It is not consistent in the sense that a change of the right hand side does not lead to a sign change in the switching function.

**Example 6.4.1**

$$\dot{x} = 1 - 2\operatorname{sign} x = \begin{cases} -1 & x > 0 \\ 3 & x < 0 \end{cases}, \qquad x(0) = x_0 > 0.$$

*We take $q(t, x) = x$. The solution of this system is $x(t) = x_0 - t$ for $t < x_0$. At $t = x_0$ we have $x(t) = 0$. For $Dq^+, Dq^-$ holds: $Dq^+ = -1$, $Dq^- = 3$. Since the direction field points towards the manifold $\mathcal{M} = \{(x,t)|x = 0\}$ (cf. Fig. 6.18) the solution cannot leave the manifold, therefore we should have $x(t) = 0$ for $t \geq x_0$. But this function does not fulfill the differential equation.*

To overcome this difficulty, Filippov [Fil64] extended the definition of a solution for ordinary differential equations by allowing the right hand side to be any value of a *set valued right hand side* which consists of all convex combinations of $f^+$, $f^-$:

$$\dot{x} = \varepsilon f^+ + (1 - \varepsilon)f^-, \quad \varepsilon \in [0, 1]. \tag{6.4.2}$$

By this definition the solution can be continued beyond these critical points. In Example 6.4.1 $\varepsilon = \frac{3}{4}$ gives the desired result.

Now, an element of the convex set has to be chosen, i.e. $\varepsilon$ has to be determined for the general case. Because the solution cannot leave the manifold, it can be determined from

$$q(t, x(t)) = 0. \tag{6.4.3}$$

as proposed in [Bock87, Eich91]. Differentiation leads to

$$q_t + q_x \dot{x} = 0. \tag{6.4.4}$$

Inserting (6.4.2) into (6.4.4) results in

$$\varepsilon = -\frac{Dq^-}{Dq^+ - Dq^-}.$$

The differential equation (6.4.2) then takes the form

$$\dot{x} = \frac{Dq^+ f^- - Dq^- f^+}{Dq^+ - Dq^-} \tag{6.4.5}$$

as long as the *conditions of inconsistency*

$$Dq^+ < 0 \quad \text{und} \quad Dq^- > 0$$

are fulfilled.

The solution can leave the manifold if $\varepsilon = 0$ or $\varepsilon = 1$. This corresponds to a sign change of one of the auxiliary switching functions $Dq^+$ or $Dq^-$, respectively.

Equation (6.4.3) together with the differential equation (6.4.2) can be interpreted as a system of differential-algebraic equations of index 2 for $(x, \varepsilon)$:

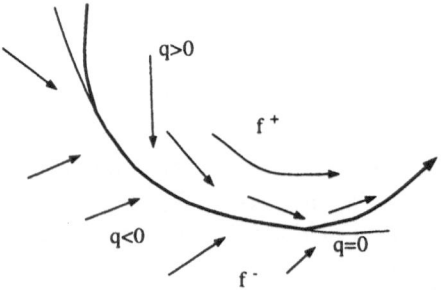

$$\begin{aligned} \dot{x} &= \varepsilon f^+(t, x) + (1 - \varepsilon) f^-(t, x) \\ 0 &= q(t, x). \end{aligned}$$

(6.4.4) together with Eq. (6.4.2) is an index-1 system which can be solved for $\varepsilon$. Inserting the solution $\varepsilon$ into (6.4.2), results in (6.4.5).

### 6.4.2.1 Treatment of Inconsistent Switching by a Generalized Three-valued Switching Logic

The treatment of inconsistent switching makes it necessary to extend the classical two-valued switching logic ($s = -1$, $s = 1$) to a three-valued logic, i.e. $s = -1$, $s = 1$, $s = 0$. We set $s = 0$ if $q(t) = 0$ and $Dq^+ < 0, Dq^- > 0$. This switching logic is given in Fig. 6.19.

In the different phases the following differential equations are used

$$\dot{x} = \begin{cases} f(t, x, +1) =: f^+ & \text{for } s = 1 \\ f(t, x, -1) =: f^- & \text{for } s = -1 \\ \dfrac{Dq^+ f^- - Dq^- f^+}{Dq^+ - Dq^-} & \text{for } s = 0 \end{cases}$$

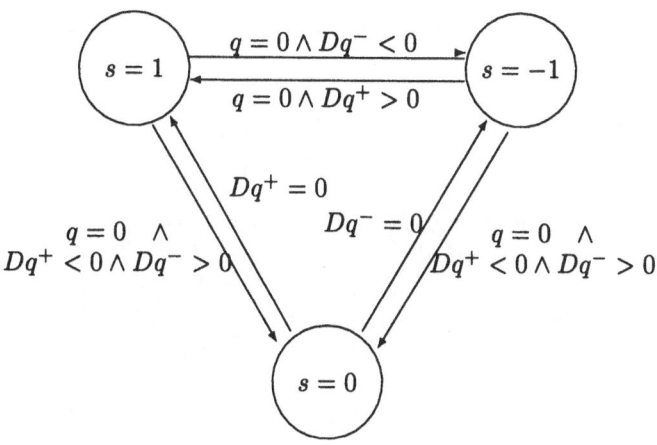

Figure 6.19: Three valued switching logic

## Dry Friction, Coulomb-friction

Before we are going to extend the introductory example to the case where the friction surface is modeled by the algebraic equation $g(p) = 0$, we want to demonstrate what happens when the jump discontinuity in Coulomb's model is smoothed.

**Example 6.4.2** *We consider the truck with Coulomb friction between the wheels and the road. The Coulomb coefficient is assumed to be $\mu = 0.9$ (dry road). We are only interested in the motion of the wheels and thus neglect the other parts.*

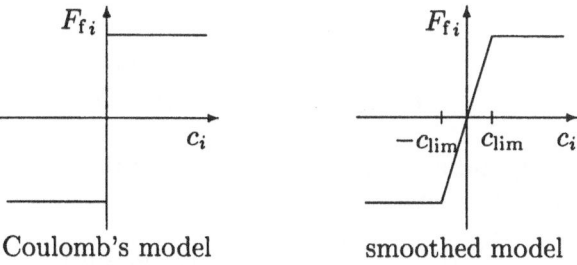

Figure 6.20: Different friction models

*The equations of motion are then given as*

$$m\dot{v} = F_{\mathrm{fr}} + F_{\mathrm{ff}}$$
$$l_r \dot{\omega}_r = M_{\mathrm{ar}} - F_{\mathrm{fr}}$$
$$l_f \dot{\omega}_f = M_{\mathrm{af}} - F_{\mathrm{ff}}$$

*with the friction forces $F_{\mathrm{fr/f}} = \mu \cdot \frac{mg}{2} \mathrm{sign}\, c_i$, $c_{r/f} = 1 - \frac{\omega_{r/t}}{v}r$ and the driving forces $M_{\mathrm{ar/f}} = d_{r/f} \max(\sin t, 0)$ with $d_f = 7 \cdot 10^3\, N$ and $d_r = 15 \cdot 10^3\, N$ and the wheel radius $r = 0.5\, m$. The constants are given as $l_{r/f} = 10\, kg\, m/s$, $m = 18 \cdot 10^3\, kg$. We investigate the computational effort when using MATLAB's integrator ODE45 with a required accuracy of $10^{-3}$, see Tab. 6.2. Using the refined model for Coulomb friction is the best we can do. Alternatively, when using a model with a smoothed discontinuity the effort grows significantly, see Fig. 6.21 and Tab. 6.2.*

|     |     | $c_{\mathrm{lim}}$ |     |     | refined |
|-----|-----|-----|-----|-----|---------|
|     | 0.4 | 0.2 | 0.1 | 0.05 | model   |
| NST | 1549 | 3098 | 6194 | 12311 | 72 |

Table 6.2: Number of integration steps for different parameters of the smoothed model

We now want to extend our introductory example to the case where the friction surface is modeled by the algebraic equation $g(p) = 0$. Then the normal force is the constraint force, i.e. $N = G^T\lambda$, $G = \frac{\partial g}{\partial p}$. Thus we obtain as equations of motion

$$M\ddot{p} = f_{\mathrm{a}}(p, \dot{p}) - G(p)^T \lambda - \mu |N| c(p) \,\mathrm{sign}\,(c(p)^T \dot{p})$$
$$0 = g(p).$$

The switching function is

$$q(p, \dot{p}) = c(p)^T \dot{p}$$

and consequently,

$$Dq^+ = c^T M^{-1}(f_{\mathrm{a}} - G^T\lambda - \mu |N| c) + \dot{c}^T \dot{p} \qquad (6.4.6a)$$
$$Dq^- = c^T M^{-1}(f_{\mathrm{a}} - G^T\lambda + \mu |N| c) + \dot{c}^T \dot{p}, \qquad (6.4.6b)$$

with $\dot{c}(p) = \frac{d}{dt} c(p(t))$. Inconsistent switching occurs if the forces in the direction of $c$ are smaller than the maximal friction force:

$$Dq^+ \cdot Dq^- < 0 \Leftrightarrow |c^T M^{-1}(f_{\mathrm{a}} - G^T\lambda) + \dot{c}^T \dot{p}| < \mu |N| c^T M^{-1} c.$$

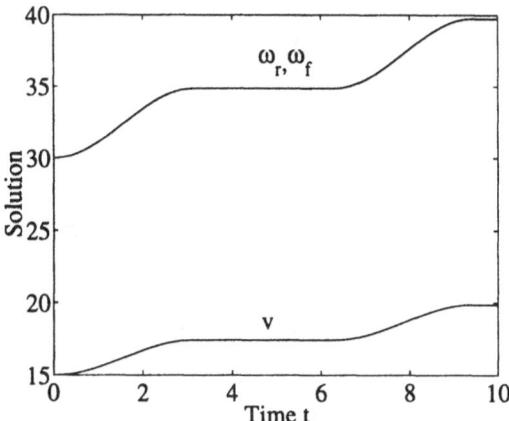

Figure 6.21: Solution of the truck model driving over a dry road

Filippov's concept results in

$$M\ddot{p} = f_a - G^T\lambda + (1 - 2\varepsilon)\mu|N|c \tag{6.4.7a}$$
$$0 = g(p) \tag{6.4.7b}$$
$$0 = c(p)^T\dot{p}. \tag{6.4.7c}$$

with the index-2-variable $\varepsilon$.
If stiction $\dot{q} = c^T\ddot{p} + \dot{c}^T\dot{p} \overset{!}{=} 0$ occurs the differential-algebraic equation

$$M\ddot{p} = f_a - G^T\lambda - \frac{c^T M^{-1}(f_a - G^T\lambda) + \dot{c}^T\dot{p}}{c^T M^{-1}c}c \tag{6.4.8a}$$
$$0 = g(p) \tag{6.4.8b}$$

is obtained.

**Example 6.4.3 (Mass sliding down a parabola)** *Let us consider the example
of a unit mass point sliding in a parabola* $g(p_1, p_2) = p_2 - p_1^2 = 0$ *and* $G(p_1, p_2) = (-2p_1, 1)$.
*Thus,* $c = \frac{1}{\sqrt{1+4p_1^2}}G(p_1, p_2)^T$. *Then the equations of motion are given as*

$$\begin{pmatrix} \ddot{p}_1 \\ \ddot{p}_2 \end{pmatrix} = \begin{pmatrix} 0 \\ -g_{gr} \end{pmatrix} - \begin{pmatrix} -2p_1 \\ 1 \end{pmatrix}\lambda$$
$$- \mu|\lambda|\begin{pmatrix} 1 \\ 2p_1 \end{pmatrix}\text{sign}\,(\dot{p}_1 + 2p_1\dot{p}_2)$$
$$0 = p_2 - p_1^2$$

*Eliminating $\lambda$ using the corresponding index-1 equation*

$$0 = \ddot{p}_2 - 2p_1\ddot{p}_1 - 2\dot{p}_1^2 = -g - \lambda \cdot 2\mu|\lambda|p_1\text{sign}\,q - 2p_1\left(2p_1\lambda - \mu|\lambda|\text{sign}\,q\right) - 2\dot{p}_1^2$$

*leads to*

$$\lambda = -\frac{g_{\text{gr}} + 2\dot{p}_1^2}{1 + 4p_1^2}$$

$$\ddot{p}_1 = (2p_1 + \mu\text{sign}\,\dot{p}_1)\left(-\frac{g_{\text{gr}} + 2\dot{p}_1^2}{1 + 4p_1^2}\right)$$

$$p_2 = p_1^2$$

*and we have $q = \dot{p}_1$ due to $\dot{p}_1 + 2p_1\dot{p}_2 = \dot{p}_1 + 2p_1(2p_1\dot{p}_1) = (1 + 4p_1^2)\dot{p}_1$. Stiction thus occurs for*

$$|p_1| < \frac{1}{2}\mu.$$

*The differential equation for this case is*

$$\ddot{p}_1 = 0.$$

*Solutions for the cases $\mu = 0.1$ (ice), $\mu = 0.5$, $\mu = 0.9$ are given in Figs. 6.22–6.24. In Tab. 6.3 the required number of integration for the refined formulation for the stiction phase versus the formulation involving the "sign" in the right hand side is given.*

| model | refined | original |
|-------|---------|----------|
| $\mu = 0.9$ | 16 | 1739 |
| $\mu = 0.5$ | 30 | 1172 |
| $\mu = 0.1$ | 109 | 433 |

Table 6.3: Number of integration steps for Ex. 6.4.3

Alternatively, the equations of motion for stiction can be obtained by adding the algebraic equation $q = c^T\dot{p} = 0$ and the Lagrange parameter $\lambda_S$

$$M\ddot{p} = f_a - G^T\lambda + c\lambda_S \qquad (6.4.9a)$$
$$0 = g(p) \qquad (6.4.9b)$$
$$0 = c^T\dot{p}. \qquad (6.4.9c)$$

Differentiation of (6.4.9c) yields

$$\lambda_S = -\frac{c^T M^{-1}(f_a - G^T\lambda) + \dot{c}^T}{c^T M^{-1}c},$$

such that (6.4.8) is obtained by elimination of $\lambda_S$.

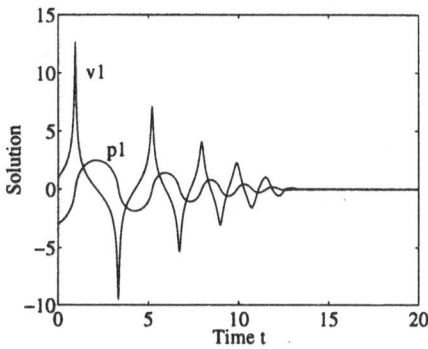

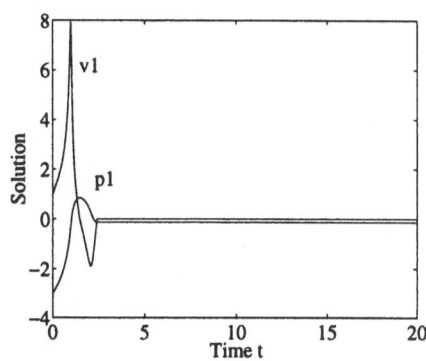

Figure 6.22: Solution for $\mu = 0.1$        Figure 6.23: Solution for $\mu = 0.5$

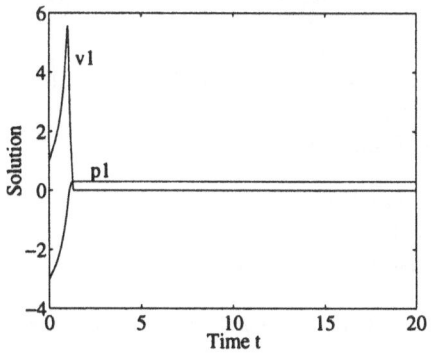

Figure 6.24: Solution for $\mu = 0.9$

**Example 6.4.4 (Wheel suspension with Coulomb friction)** *We consider the model of a wheel suspension subject to Coulomb friction given in Fig. 6.25. The equations of motion are given as*

$$m_A \ddot{z}_A = -C_A(z_A - z_R) - \rho_A(\dot{z}_a - \dot{z}_R) - R_C \operatorname{sign}(\dot{z}_A - \dot{z}_R)$$
$$m_R \ddot{z}_R = C_A(z_A - z_R) + \rho_A(\dot{z}_A - \dot{z}_R) + R_C \operatorname{sign}(\dot{z}_A - \dot{z}_R) + C_R(z_S - z_R).$$

*$z_A$ describes the vertical motion of the car and $z_r$ the vertical motion of the wheel. $z_S$ is used to describe the street profile. We want to simulate the car driving over a ramp*

$$z_S(t) = \begin{cases} 0.1\,t & for\ t < 0.1 \\ 0.01 & for\ t \geq 0.1\,. \end{cases}$$

*The data are given as $m_A = 10^3$ kg, $m_R = 10^2$ kg, $C_A = 4\pi^2 \cdot 10^3$ kg/s$^2$, $C_R = 4\pi^2 \cdot 10^4$ kg/s$^2$, $R_C = 400\,N$, $\rho_A = 1.2\pi^2 \cdot 10^3$ kg/s.*

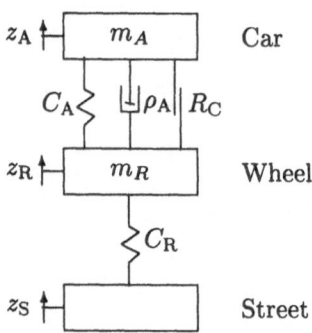

Figure 6.25: Model of a wheel suspension

*Discontinuities are present due to the ramp and due to the Coulomb friction term,*
*sign $(\dot{z}_A - \dot{z}_R)$, in the right hand side.*
*The following cases are discussed*

- *Case 1 (Fig. 6.26 and 6.27)*
  *Exact treatment of the model:*
  *In the stiction phase we use the model*

$$
\begin{aligned}
(m_A + m_R)\ddot{z}_A &= C_R(z_S - z_R) \\
\ddot{z}_R &= \ddot{z}_A \\
\dot{z}_R &= \dot{z}_A
\end{aligned}
$$

- *Case 2 (Fig. 6.28 - 6.29)*
  *Smoothing the Coulomb model:*
  *In the "stiction phase" (i.e. for $|\dot{z}_A - \dot{z}_R| < c_{\lim}$) we use*

$$
\begin{aligned}
m_A\ddot{z}_A &= -C_A(z_A - z_R) - \rho_A(\dot{z}_a - \dot{z}_R) - \frac{R_C}{c_{\lim}}(\dot{z}_A - \dot{z}_R) \\
m_R\ddot{z}_R &= C_A(z_A - z_R) + \rho_A(\dot{z}_A - \dot{z}_R) + \frac{R_C}{c_{\lim}}(\dot{z}_A - \dot{z}_R) + C_R(z_S - z_R).
\end{aligned}
$$

*In order to describe the discontinuities at the end points of the ramp we have*
*in this case additionally to take the switching functions*

$$
q_1 = \dot{z}_A - \dot{z}_R - \frac{1}{c_{\lim}} \qquad q_2 = \dot{z}_R - \dot{z}_A - \frac{1}{c_{\lim}}.
$$

*into consideration.*

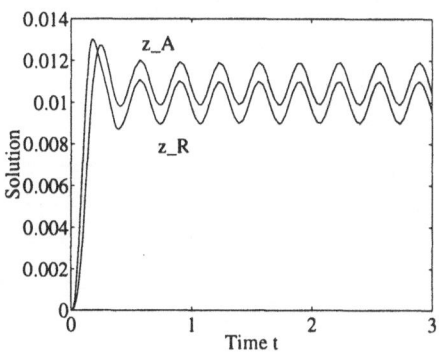

Figure 6.26: Case 1: Solution with discontinuous right hand side

Figure 6.27: Case 1: Relative velocity $\dot{z}_A - \dot{z}_R$ using the exact discontinuous model

|  | $c_{\lim}$ | NFE | IJC | IMC | NRED | NST | CPU |
|---|---|---|---|---|---|---|---|
| **Case 1:** | - | 741 | 28 | 6 | 1 | 364 | 1 |
| **Case 2:** | $10^{-3}$ | 817 | 36 | 6 | 1 | 386 | 1.06 |
|  | $10^{-4}$ | 881 | 66 | 22 | 10 | 400 | 1.15 |
|  | $10^{-5}$ | 849 | 71 | 31 | 19 | 399 | 1.15 |
|  | $10^{-6}$ | 935 | 106 | 53 | 25 | 445 | 1.31 |
|  | $10^{-7}$ | 1221 | 187 | 107 | 43 | 658 | 1.93 |

Table 6.4: Effort for the solution of the wheel suspension. (NFE: Number of function evaluation, IJC: Number of Jacobian decompositions, IMC: Number of Jacobian evaluations, NRED: number of rejected steps, NST: number of integration steps, CPU: normalized CPU time)

Table 6.4 gives the effort for both cases using an accuracy of $RTOL=10^{-6}$, $ATOL=10^{-8}$.

We see that using the smoothing approach

- the solution is damped for large $c_{\lim}$ (cf. Fig. 6.28), and we get qualitatively different solutions

- for small $c_{\lim}$ the system becomes stiff and its solution requires a high effort. This even holds for stiff integrators as BDF: this is due to the fact that the system is nearly always in the transient of a stiff equation.

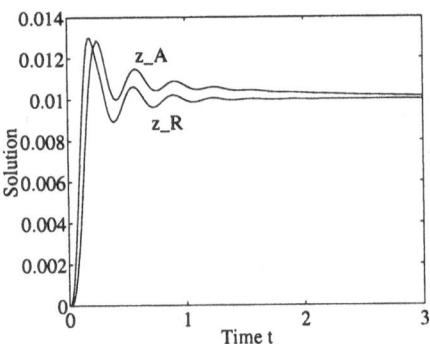

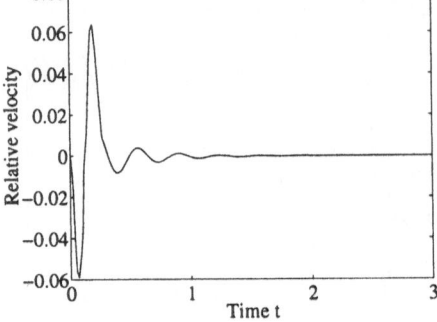

Figure 6.28: Case 2: Solution with smoothed right hand side with $c_{\lim} = 10^{-2}$

Figure 6.29: Case 2: Relative velocity $\dot{z}_A - \dot{z}_R$ using a smoothed right hand side with $c_{\lim} = 10^{-2}$

When using a non-stiff integration method as MATLAB's Runge–Kutta method ODE45, much more effort is necessary to obtain a solution as can be seen from the table beside.

| $c_{\lim}$ | $NST$ |
|-----------|-------|
| $10^{-2}$ | $444$ |
| $10^{-3}$ | $3297$ |

## 6.5   Other Special Classes of Discontinuities

Some classes of discontinuities occur so frequently that it is worth while to develop special techniques for their treatment. This accelerates the integration makes multibody simulation codes more user friendly.

### 6.5.1   Time Events

Time events can be handled very efficiently by stopping and reinitializing the integration routine at the corresponding time point. No localization of the switching point is necessary.

### 6.5.2   Structure Varying Contact Problems with Friction

In this section we consider contact problems, where contact may occur or disappear. In each contact point we assume friction due to Coulomb's law as discussed in the last section. Due to these phenomena the number of degrees of freedom varies with time. First, we discuss the equations of motion for these problems and then consider the question how to decide which of the contacts is active and if in the point of contact stiction or sliding is present.

### 6.5.2.1 Equations of Motion

In Sec. 5.5 we showed how the equations of motion for contact problems can be derived. In this section we describe structure variant systems based on the work of [GP92].

The distance $g_N$ must be greater or equal to zero and is given by

$$g_N = -n_0^T d = n_1^T d \geq 0$$

where $n_i$, $i = 0, 1$ are the normals, cf. Fig. 5.10. For contact we have

$$g_N = 0. \tag{6.5.1}$$

Note that $d$ and $n_i$ depend on the actual state of the system described by the coordinates $p$.

Formulating the corresponding constraint on velocity level yields

$$\dot{g}_N = n_1^T \dot{d} + \underbrace{\dot{n}_1^T d}_{=0} = n_1^T \dot{d} = 0 \tag{6.5.2}$$

where the last equality holds because of $\|n_1\| = 1$ and $d \parallel n_1$. Thus, the normal velocity is the orthogonal projection of the change of the distance vector. $\dot{d}$ is the difference of the velocities of the possible contact points: $\dot{d} = v_1 - v_0$, where due to the linearity of the constraints on velocity level in $\dot{p}$, $v_i$ can be written as linear combination of $\dot{p}$

$$v_i = J_i(t, p)\dot{p} + j_i(t, p), \quad i = 0, 1.$$

Now, (6.5.2) can be rewritten as

$$\begin{aligned}
\dot{g}_N &= n_1^T \underbrace{[(J_1\dot{p} - J_0\dot{p}) + j_1 - j_0]}_{=\dot{d}} = \underbrace{n_1^T (J_1 - J_0)}_{=:G_N}\dot{p} + \underbrace{n_1^T (j_1 - j_0)}_{=:z_N} \\
&= G_N\dot{p} + z_N(t, p).
\end{aligned}$$

In case of contact we need the velocity in tangential direction for the determination of the friction force (see Sec. 6.4). Its value is the projection of the change of the distance vector on the tangent $\tau_1$

$$\dot{g}_T = \tau_1^T \dot{d} = \underbrace{\tau_1^T (J_1 - J_0)}_{=:G_T}\dot{p} + \underbrace{\tau_1^T (j_1 - j_0)}_{z_T} = G_T\dot{p} + z_T(t, p).$$

$\dot{g}_T$ is the sliding velocity. Thus for

$$\dot{g}_T = 0 \tag{6.5.3}$$

we have stiction. In case of $\dot{g}_T \neq 0$ the friction force is given by

$$F_{\text{Cou}} = \tau \|F_N\| \, (-\mu(\dot{g}_T)\text{sign}\,\dot{g}_T)$$

with $F_N = G_N^T \lambda_N$.

Now, we set up the total set of equations. Let the system without the structure variant equations be described as usual by

$$
\begin{align}
M(p)\ddot{p} &= f(t,p,\dot{p}) - G^T(p)\lambda && \text{(6.5.4a)} \\
g(p) &= 0. && \text{(6.5.4b)}
\end{align}
$$

For an active constraint we have to add condition (6.5.1) and the corresponding constraint force. If in this point stiction occurs we have to add condition (6.5.3) and the corresponding constraint force, if sliding occurs we have to add the corresponding friction force $F_{\text{Cou}}$ due to Coulomb's law. Assuming $n_N$ active contacts, where the $n_T$ first points show stiction we end up with the following system of equations

$$
\begin{align}
M(p)\ddot{p} &= f(t,p,\dot{p}) - G^T(p)\lambda && \text{(6.5.5a)} \\
&\quad - \sum_{i=1}^{n_N} G_{N,i}^T(p)\lambda_{N,i} - \sum_{i=1}^{n_T} G_{T,i}^T(p)\lambda_{T,i} && \text{(6.5.5b)} \\
&\quad - \sum_{i=n_T+1}^{n_N} t_i \|G_{N,i}^T \lambda_{N,i}\| \, (-\mu_i(\dot{g}_{T,i})\text{sign}\,\dot{g}_{T,i}) \\
g(p) &= 0 && \text{(6.5.5c)} \\
g_{N,i}(p) &= 0, \quad i = 1,\ldots,n_N && \text{(6.5.5d)} \\
\dot{g}_{T,i}(p) &= 0, \quad i = 1,\ldots,n_T && \text{(6.5.5e)}
\end{align}
$$

### 6.5.2.2  Determination of the New System State

In the case where contact in different points may occur or disappear one has to decide in which of these points contact will be active in the future.

**Necessary Conditions for Contact**

The necessary conditions for a contact point to become active are

$$g_N = 0, \qquad \dot{g}_N = 0. \tag{6.5.6}$$

These can be used as switching function when the contact point is not active. If the contact breaks the normal force becomes zero

$$\lambda_N = 0. \tag{6.5.7}$$

This can be used as switching function to monitor the break of a contact.

**Sufficient Conditions for Contact**
During the contact we have $g_N = 0$ and the normal force $\lambda_N$ must be greater than zero

$$\lambda_N \geq 0. \tag{6.5.8}$$

If the normal force becomes zero and if the relative acceleration is positive

$$\ddot{g}_N > 0 \tag{6.5.9}$$

the contact breaks. Thus sufficient conditions for the contact together with the first instance after the contact break are given by

$$\lambda_N \geq 0, \qquad \ddot{g}_N \geq 0, \qquad \lambda_N \ddot{g}_N = 0. \tag{6.5.10}$$

Since $\ddot{g}_N$ is a linear function of $\lambda_N$ these conditions form a *linear complementarity problem*[2], which can be solved by using Simplex-like algorithms as the one given in [Sey93, GP92]. This has to be done at any time-point where the necessary conditions for structure changing (6.5.6), (6.5.7) become active.

**Necessary Conditions for Stiction**
The necessary condition for stiction in a contact point is

$$\dot{g}_T = 0. \tag{6.5.11}$$

This can be used as switching function to monitor the start of a stiction phase. A necessary condition for the stiction phase say in contact point number $N_T + 1$ is that the actual tangential forces in the point of contact are less than the maximal transferable force

$$\|G_{T,n_T+1}^T M^{-1} F\| < \|G_{T,n_T+1}^T M^{-1} G_{T,n_T+1}^T G_{N,n_T+1} \lambda_{N,n_T+1} \mu_{n_T+1}\| \tag{6.5.12}$$

with

$$
\begin{aligned}
F \;=\; & f(t,p,\dot{p}) - G^T(p)\lambda - \sum_{i=1}^{n_N} G_{N,i}^T \lambda_{N,i} - \sum_{i=1}^{n_T} G_{T,i}^T \lambda_{T,i} \\
& - \sum_{i=n_T+2}^{n_N} \tau_i \|G_{N,i}^T \lambda_{N,i}\| \left(-\mu_i(\dot{g}_{T,i}) \operatorname{sign} \dot{g}_{T,i}\right).
\end{aligned}
$$

This can be deduced also by using $q = \dot{g}_{T,n_T+1}$ as switching function and the concept of inconsistent switching as given in the preceding sections. Thus a necessary condition for a stiction phase to begin is Eq. (6.5.11). A necessary condition for a stiction phase to end is Eq. (6.5.12).

---

[2] Note that it is not linear but bilinear.

**Sufficient Conditions for Stiction**

Similarly, sufficient conditions for a stiction phase and the instance before/after stiction can be formulated. These conditions can again be reformulated as linear complementarity problem by splitting the values into their positive and negative part to get rid of the absolute values [GP92].

### 6.5.3   Computer-controlled Systems

In order to achieve a desired behavior a given mechanical system is controlled by adding a special control device. Such a device processes a signal obtained from the system's output and feeds the resulting signal back to the system's input. Mathematically, such a control is described by a control law, which might be among other possibilities

- in the *continuous case* a set of additional differential equations and additional state variables $s(t)$, or

- in the *discrete case* a set of difference equations and states $s^i$.

In this section we briefly discuss aspects of the numerical treatment of the later system class, often denoted as *computer-controlled systems*, [ÅW90].

The system equations of a computer-controlled system consist of a coupled set of differential and difference equations

$$\dot{x} \quad = \quad f(x, u, t) \tag{6.5.13a}$$

$$s^{i+1} \quad = \quad r(x(t^i), s^i, t^i) \tag{6.5.13b}$$

$$u(t) \quad := \quad s^{i+1} \quad \text{for} \quad t^i \leq t < t^{i+1} \tag{6.5.13c}$$

with sample points $t^{i+1} := t^i + \Delta t$ and initial values $x(0) = x_0, s^0 := s_0$.

Sometimes the controller state $s$ can admit only discrete values, e.g. $s^i \in \{-1, 0, 1\}$. These systems are obviously discontinuous, but as the discontinuities occur at explicitly given *time events*, the sampling points, their numerical treatment is much easier than the general discontinuous case. The only task is to synchronize the integration step size $h$ with the sampling period $\Delta t$ in such a way that the sampling points are hit, so that the integration can be stopped and restarted with the new input $u$, according to (6.5.13c).

The state of the controller in general does not change for every $t^i$. The conditions under which this is the case can be formulated as roots of switching functions $q$. A restart is only necessary if

$$q(t^i, x(t^i)) \cdot q(t^{i+1}, x(t^{i+1})) < 0,$$

otherwise the integration can be continued.

We will study this class of discontinuities by means of the example of an Anti Blocating System.

**Example 6.5.1 (Anti Blocating System)** *Our truck is now provided with an Anti-Blocating system acting independently on the front and the rear wheel. We consider the reduced truck model, like in Example 6.4.2 and introduce additionally an anti-blocating system, which controls the creepage of the wheels on roads with different surface properties.*
*Thus we have as continuous variables*
 $v$      *forward speed,*
 $w_{r/f}$    *angular velocity of the rear and front wheel, respectively,*
 $P_{r/f}$    *pressure in front and rear brake cylinder, respectively.*
*We assume three different forces acting on each wheel:*

1. *Acceleration forces due to the motor.*
 *This is an external input representing the driver's action. We assume an acceleration torque $M_a$ given as*

$$M_{ar/f} = d_{r/f} \cdot \max\left(0, \sin 10\, t\right)$$

 *with r/f denoting the rear and front wheel, respectively.*

2. *Friction forces between wheel and road.*
 *Here we use Coulomb's law:*

$$F_{Fr/f} = \mu \cdot F_{Nr/f} \cdot \operatorname{sign} c_{r/f}$$

 *with the creepage $c_{r/f}$ defined as*

$$c_{r/f} = 1 - \frac{\omega_{r/f}}{v} r$$

 *and the wheel radius $r = 0.5\, m$. Thus we have for the creepage*

 - *$c_{r/f} < 0 \Leftrightarrow \omega_{r/f} \cdot r > v$, i.e. the wheel is spinning. This may be the case in the acceleration phase.*
 - *$c_{r/f} = 0 \Leftrightarrow \omega_{r/f} \cdot r = v$, i.e. the wheel is perfectly rolling.*
 - *$c_{r/f} > 0 \Leftrightarrow \omega_{r/f} \cdot r < v$, i.e. the wheel is blocating. This may be the case in the braking phase. The goal of an anti-blocating system is to avoid wheel blocating.*

3. *Forces due to the brake.*
 *The torques acting on each wheel due to the brake are given as*

$$M_{br/f} = c_{Br/f} \cdot P_{r/f} \cdot \operatorname{sign} \omega_{r/f}$$

*The pressure $P_i$ in each brake cylinder is governed by*

$$\dot{P}_{r/f} = \begin{cases} c_{inc}(P - P_{r/f}) & \text{if } s_{r/f} = 1 \\ -c_{dec}P & \text{if } s_{r/f} = -1 \\ 0 & \text{if } s_{r/f} = 0. \end{cases}$$

$P$ is the pressure in the main braking cylinder representing the driver's action. It is assumed to be

$$P(t) = -\min(0, \sin 10\, t)$$

i.e. braking and acceleration phases alternate.
The valve position is determined every $\Delta t$ seconds. It can assume three different states:

$$s_{r/f} := \begin{cases} 0 & \text{the valve is closed} \\ 1 & \text{the valve is open to increase the pressure} \\ -1 & \text{the valve is open to decrease the pressure} \end{cases}$$

The valve position depends on the acceleration and on the creepage between road and wheel. To define this dependency, we introduce two integer variables:

- angular acceleration $a_i = \dot{\omega}_i$
  A logical variable $a_i^{\log}$ is computed based on the acceleration threshold $c^a$

$$a_i^{\log} = \begin{cases} 1 & \text{if } 0 \le a_i(t_j) \\ 0 & \text{if } c^a < a_i(t_j) < 0 \\ -1 & \text{if } a_i(t_j) \le c^a \end{cases} , \qquad i = r, f$$

- creepage $c_i = 1 - \dfrac{\omega_i r}{v}$
  Again, a logical variable is introduced

$$c_i^{\log} = \begin{cases} 0 & \text{if } c_i \le c_1^c \\ 1 & \text{if } c_1^c < c_i < c_2^c \\ 2 & \text{if } c_2^c \le c_i \end{cases} , \qquad i = r, f.$$

with upper and lower creepage thresholds $c_j^c, j = 1, 2$.

From these logical variables the valve position $s_i$ is determined as

|       |                  | $a_i^{\log}$ | | |
|-------|------------------|:------------:|:------------:|:------------:|
|       |                  | -1 | 0 | 1 |
|       | $c_i^{\log}$     | high | medium | acceleration |
|       |                  | slow down | slow down | |
| 0 | low creepage | 0 | 1 | 1 |
| 1 | medium creepage | -1 | 0 | 0 |
| 2 | high creepage | -1 | -1 | 0 |

*Summarizing we end up with the following equations of motion*

$$m\dot{v} = F_{\mathrm{Fr}} + F_{\mathrm{Ff}} \tag{6.5.14a}$$

$$l_{\mathrm{f}}\dot{\omega}_{\mathrm{f}} = -F_{\mathrm{Ff}} + M_{\mathrm{af}} - M_{\mathrm{bf}} \tag{6.5.14b}$$

$$l_{\mathrm{r}}\dot{\omega}_{\mathrm{r}} = -F_{\mathrm{Fr}} + M_{\mathrm{ar}} - M_{\mathrm{br}} \tag{6.5.14c}$$

$$\dot{P}_{\mathrm{f}} = \begin{cases} c_{\mathrm{inc}}(P - P_{\mathrm{f}}) & \text{if } s_{\mathrm{f}} = 1 \\ -c_{\mathrm{dec}}P & \text{if } s_{\mathrm{f}} = -1 \\ 0 & \text{if } s_{\mathrm{f}} = 0. \end{cases} \tag{6.5.14d}$$

$$\dot{P}_{\mathrm{r}} = \begin{cases} c_{\mathrm{inc}}(P - P_{\mathrm{r}}) & \text{if } s_{\mathrm{r}} = 1 \\ -c_{\mathrm{dec}}P & \text{if } s_{\mathrm{r}} = -1 \\ 0 & \text{if } s_{\mathrm{r}} = 0. \end{cases} \tag{6.5.14e}$$

*The constants are given in Table 6.5.*

| | |
|---|---|
| total mass $m$ | $m = 18000\,\mathrm{kg}$ |
| moments of inertia $l_i$ | $l_{\mathrm{r}} = l_{\mathrm{f}} = 10\,\mathrm{kgm}^2$ |
| acceleration constants $d_i$ | $d_{\mathrm{r}} = 45 \cdot 10^3\,\mathrm{N}$ , $d_{\mathrm{f}} = 32 \cdot 10^3\,\mathrm{N}$ |
| braking constants $c_{\mathrm{B}i}$ | $c_{\mathrm{Br}} = 1.5 \cdot 10^4\,\mathrm{m}^2$, $c_{\mathrm{Bf}} = 10^4\,\mathrm{m}^2$ |
| creepage thresholds $c_i^{\mathrm{c}}$ | $c_1^{\mathrm{c}} = 0.05, c_2^{\mathrm{c}} = 0.1$ |
| acceleration threshold | $c^{\mathrm{a}} = -0.5\,\mathrm{m/s}^2$ |
| time interval $\Delta t$ | $\Delta t = 0.01\,\mathrm{s}$ |
| normal forces $F_{\mathrm{N}i}$ | $F_{\mathrm{Nr}} = -1.2195 \cdot 10^6\,\mathrm{N}$ , $F_{\mathrm{Nf}} = -4.7435 \cdot 10^4\,\mathrm{N}$ |

Table 6.5: Data for the truck with ABS

*In Figs. 6.30 and 6.31 simulation results for different cases are presented[3].*

## 6.5.4 Hysteresis

Hysteresis is characterized by the fact that the right hand side $f$ does not only depend on the solution values at the actual time but also on the "history" of the motion, i.e. on solution values at previous time points.

Hysteresis occurs, e.g., in the modeling of elasto-plastic behavior, e.g. in crash-test simulation.

We introduce a hysteresis parameter $\sigma$ in order to describe these phenomena:

$$f = f(t, x, \sigma), \quad \sigma = \sigma(t_1, x(t_1), \ldots, t_n, x(t_n)).$$

$\sigma$ summarizes all information from previous time points which are necessary to compute the right hand side at the actual time. Fig. 6.32 shows a hysteresis curve

---

[3]Note, this example is for demonstration purposes only. This is no state of the art ABS, the brakes do not work as they should.

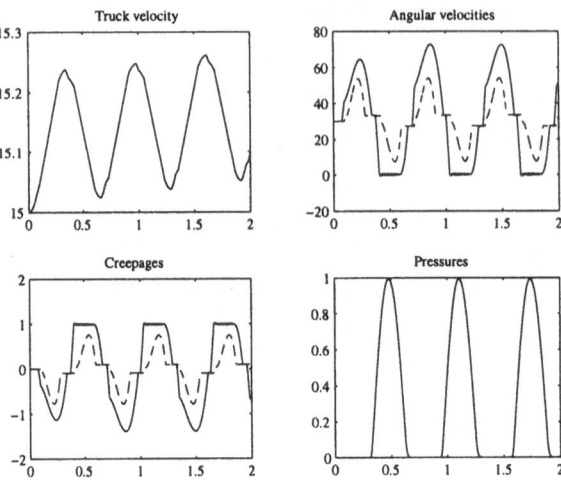

Figure 6.30: Truck without ABS, $\mu = 0.1$ (Icy road)

where the hysteresis is given by two sets of functions. The functions of each set can be described by the parameter $\sigma_i$. The choice of the set is also stored in $\sigma$. A change between the sets occurs for $q_{change} = \dot{h}_{hyst} = 0$.

The realization in a program can be done with the help of an additional memo-vector, which corresponds to $\sigma$. In the right hand side it must be "read-only". It is allowed to be changed only by the switching algorithm after a root of of $q_{change}$ has been localized.

This foregoing overcomes the difficulties occurring if the hysteresis parameter is changed in the right side when steps are rejected by the integrator.

## 6.5.5   Approximations by Piecewise Smooth Functions

Physical laws often cannot be described in closed form, or their representation in explicit form is very expensive. Frequently, the only information available is a set of measurement data. In these cases the laws are often represented as piecewise smooth interpolating functions (linear, splines, ... ) or simple tables.

The so-called *characteristic lines* can be defined by using the variables given in Fig. 6.33:

$$
k(w_{\text{thresh}}(t, x(t))) = \begin{cases} k_0(w_{\text{thresh}}) & \text{for } w_{\text{thresh}} \leq s_0 \\ k_1(w_{\text{thresh}}) & \text{for } s_1 < w_{\text{thresh}} \leq s_1 \\ \quad \vdots \\ k_{n_k+1}(w_{\text{thresh}}) & \text{for } s_{n_k} \leq w_{\text{thresh}}. \end{cases}
\tag{6.5.15}
$$

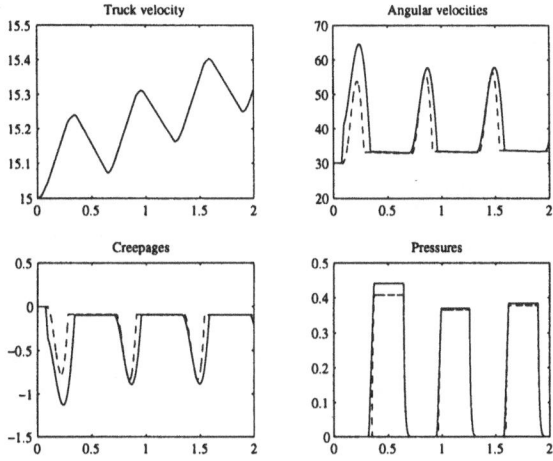

Figure 6.31: Truck with ABS, $\mu = 0.1$ (Icy road)

The characteristic line $k$ is given as function of $w_{\text{thresh}}(x(t), t)$, which itself depends on the state variables $x(t)$ and time.

Sometimes damping elements are described by such characteristic lines. In this case the damping force $f_D$ depends on the relative velocity $v_{\text{rel}}$. This yields

$$w_{\text{thresh}} = v_{\text{rel}}, \qquad k(w_{\text{thresh}}) = f_D.$$

If $w_{\text{thresh}}$ crosses one of the *threshold values* $s_i$ this leads to a discontinuity. The treatment with standard switching algorithms involves the computation of all $n_k + 1$ switching functions

$$q_i(t, x(t)) = w_{\text{thresh}}(t, x(t)) - s_i, \qquad i = 0, \ldots, n_k + 1.$$

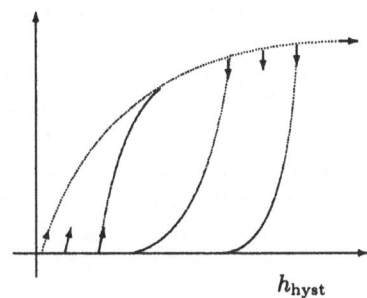

Figure 6.32: Hysteresis

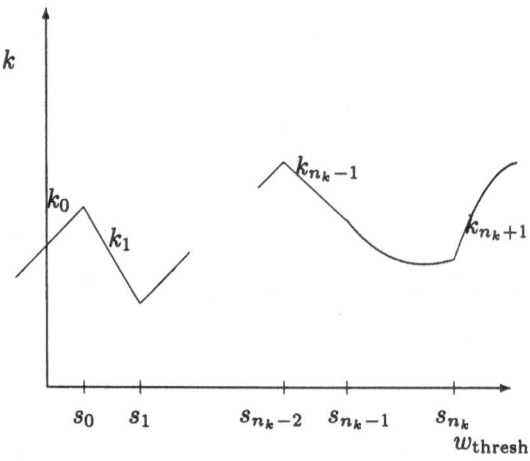

Figure 6.33: Characteristic line

Since for $s_j \leq w_{\text{thresh}} \leq s_{j+1}$ only the switching functions $q_j, q_{j+1}$ may encounter a sign change, it is only necessary to monitor their sign. Thus, for each characteristic line only two switching functions are necessary. For multidimensional characteristic fields the gain is much higher.

Furthermore, the computation of $k$ is less expensive and easier than its determination on the basis of the sign of general switching functions: The right hand side has direct access to the number of the interval $j$ avoiding the computation of this number. The recomputation is only necessary if a change in one of the neighboring intervals occurs. The gain obtained with this trick can be seen from the following example:

**Example 6.5.2 (Container on a ship)** *Ship container are stored not only in the body of a ship but also on the deck. They are brought into the body by large hatchways in the deck. On the hatchway covers additional containers are stored and are fixed on the hatchways.*

*Due to the motion of the sea the body of the ship is deformed whereas the hatchway covers behave much less elastic. This results in relative motions between the parts and thus to stresses resulting in damages at the hatchway covers and bearings. The stresses rely on the friction force between covers and bearings. This is the reason to study different materials reducing the stress.*

*In addition to a characteristic line describing the friction coefficient we have to deal with Coulomb friction in points of contact between container and the deck of the ship.*

*The coefficients of friction have been obtained from measurements and are given by the following table*

| $v$ [mm/s] | 0. | 0.05 | 0.1 | 0.2 | 0.3 | 0.5 | 1. | 2. | 3. | 5. | 10. |
|---|---|---|---|---|---|---|---|---|---|---|---|
| $\mu$ | | 0.3 | 0.3 | 0.3 | 0.38 | 0.34 | 0.33 | 0.34 | 0.32 | 0.32 | 0.32 | 0.34 |

The intermediate values are obtained by linear interpolation.
Fig. 6.34 shows the multibody model. The model is based on work presented in
[Leh87].

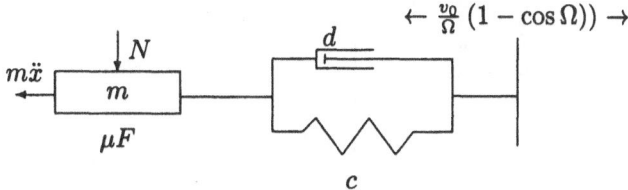

Figure 6.34: Model for a slipping container

As equations of motion we obtain

$$m\ddot{x} = f_K - 2\mu(|\dot{x}|)\mathrm{sign}\,\dot{x}N$$
$$f_K = d(v_0 \sin \Omega t - \dot{x}) + c\left(\frac{v_0}{\Omega}(1 - \cos \Omega t) - x\right),$$

where $\frac{v_0}{\Omega}$ is the maximal amplitude of the ship's rolling motion.
$\mu(|\dot{x}|)\mathrm{sign}\,\dot{x}N$ is the Coulomb friction term, $f_K$ is used to model the sinusoidal
excitation by the rolling. The constants are given as $N = 18.000\ N$, $v_0 = 4\ mm/s$
and $\Omega = \frac{\pi}{6}\ Hz$, $c = 2000\ kg/s^2$, $d = 0\ kg/s$.
The model contains the following difficulties:

- The friction coefficient $\mu(|\dot{x}|)$ has to be evaluated by table-look-up.

- Because of the term $\mathrm{sign}\,\dot{x}$ we have Coulomb friction, the problem cannot be
treated with the classical two-valued switching logic .

Fig. 6.35–6.37 show the solution for position coordinates, forces and switching functions. From the figures showing the switching function the stiction intervals can be
obtained.
The gain obtained from taking the special structure into consideration of the characteristic lines can be seen from Tab. 6.6. The computing time has been reduced
by a factor of 60 %.

As concluding example we give the example of a percussion drill:

**Example 6.5.3 (Percussion drill)** In this example we investigate the dynamic
behavior of a percussion drill using the model given in [Lei88]. We simulate the
motion of the striker (index S), the piston (index P), the intermediate mass (index
M), and the casing (index C). The working principle is as follows:

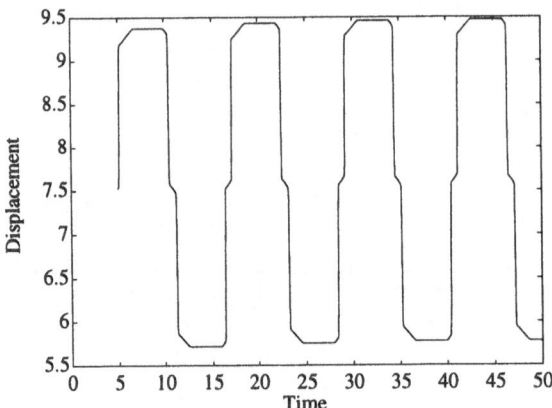

Figure 6.35: Solution $x$

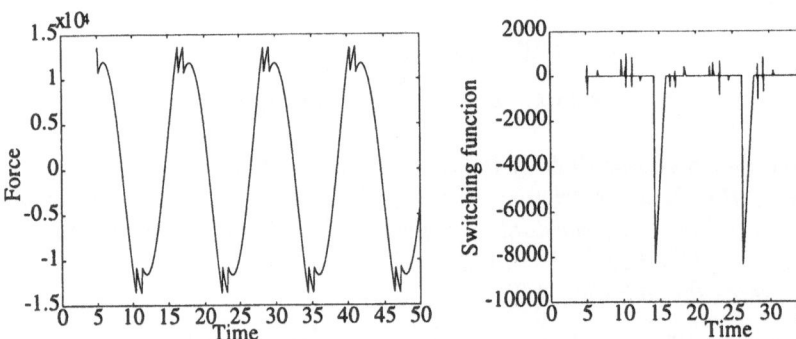

Figure 6.36: Force $f_K$          Figure 6.37: Switching function $q = \dot{x}$

The piston is driven periodically by the slider crank mechanism (intermediate mass). This piston acts on the striker through an air bolster. The striker gives its energy to the striking mechanism. The motion is described by the following differential equations

$$
\begin{pmatrix}
m_{\mathrm{S}} & 0 & 0 \\
0 & m_{\mathrm{M}} + m_{\mathrm{P}} & 0 \\
0 & 0 & m_{\mathrm{C}}
\end{pmatrix}
\begin{pmatrix}
\ddot{p}_{\mathrm{S}} \\
\ddot{p}_{\mathrm{M}} \\
\ddot{p}_{\mathrm{C}}
\end{pmatrix}
=
\begin{pmatrix}
-f_{\mathrm{A}} - f_{\mathrm{fric}} \\
m_{\mathrm{P}} \ddot{S}(t) + c_{\mathrm{M}}(p_{\mathrm{C}} - p_{\mathrm{M}}) + f_{\mathrm{A}} \\
-f_{\mathrm{H}} - c_{\mathrm{M}}(p_{\mathrm{C}} - p_{\mathrm{M}}) + f_{I} + f_{fric}
\end{pmatrix}
$$

The motion of the piston is given by

$$
p_{\mathrm{P}}(t) = p_{\mathrm{M}}(t) - S(t).
$$

and the following notation is used

| TOL | $10^{-3}$ | $10^{-4}$ | $10^{-5}$ | $10^{-6}$ | $10^{-7}$ | $10^{-8}$ |
|---|---|---|---|---|---|---|
| CPU with | 1.0 | 1.3 | 1.6 | 2.2 | 2.8 | 3.4 |
| CPU without | 2.4 | 3.2 | 4.2 | 5.5 | 6.8 | 8.4 |

Table 6.6: Comparison of the normalized CPU-time for computations with and without exploiting the special structure of the characteristic lines

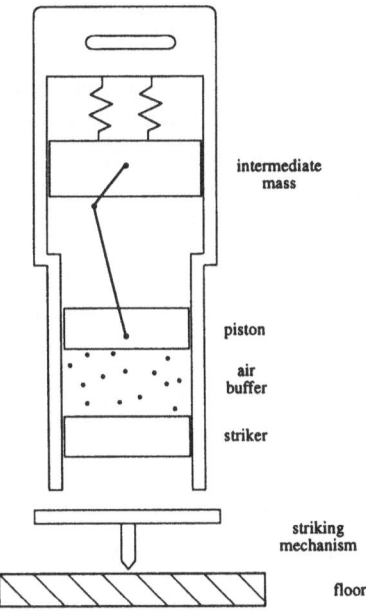

Figure 6.38: Model of a percussion drill

$S(t)$  describes the distance between intermediate mass and piston. The relationship between $S(t)$ and the crank angle is

$$S(t) = r \cos \omega t + q\sqrt{1 - \left(\frac{r}{q} \sin \omega t\right)^2}$$

where $r$ is the radius of the crank and $q$ the length of the connecting rod.

$f_{\text{fric}}$     describes the Coulomb friction between striker and the striker guide.

$$f_{\text{fric}} = \begin{cases} 0 & \text{for } t \leq 0.05 \ [s] \\ Ap_0 \text{sign} (\dot{p}_S - \dot{p}_C) & \text{for } t > 0.05 \ s. \end{cases}$$

$f_A$     describes the force of the air bolster between piston and striker:

$$f_A = Ap_0 \left( 1 - \frac{l_{\text{nom}}}{p_M(t) - p_S(t) - S(t)} \right)^\kappa$$

where $l_{\text{nom}}$ is the length of the bolster, when the pressure of the bolster is equal to the pressure of the atmosphere. $A$ is the area, $A = \pi \frac{d_l^2}{4}$. $\kappa$ is the adiabatic exponent.

$f_H$     describes the force of the hand.

$f_I$     describes the force resulting from the casing impact:

$$f_I = \begin{cases} 0 & \text{for } p_C \geq 0 \ m \\ -cp_C - d\dot{p}_C & \text{for } p_C < 0 \ m. \end{cases}$$

$c, d$ are the damping constants for the casing.

$c_M$     is the spring constant of the spring between intermediate mass and casing.

The impact of the striker on the floor is modeled by

$$\dot{p}_S^+ = -e\dot{p}_S^-.$$

The impact number $e$ is chosen to be $e = 0.17$. The remaining data is given in Table 6.8. Thus we have the following types of discontinuities in this example:

| Geometry data | Masses | Stiffness and damping coefficients | | Forces | |
|---|---|---|---|---|---|
| $l_{\text{nom}} = 0.0365 \, m$ | $m_S = 0.6 \, kg$ | $c =$ | $10^7 \, N/m$ | $f_H =$ | $200 \, N$ |
| $r = 0.022 \, m$ | $m_P = 0.335 \, kg$ | $c_M =$ | $10^4 \, N/m$ | $\mu =$ | $0.2$ |
| $q = 0.076 \, m$ | $m_M = 2.5 \, kg$ | $d = 8156.283 \, N/m$ | | $p_0 =$ | $10^5 \, N/m^2$ |
| $d_l = 0.044 \, m$ | $m_C = 8 \, kg$ | | | $\kappa =$ | $1.4$ |
| | | | | $\omega = \frac{\pi}{30} \cdot 1340 \, \text{min}^{-1}$ | |

Table 6.8: Data of the percussion drill

- impact: $q_1(t) = g_N = p_C$ (striker on the floor), $q_2(t) = p_S$ (casing)

- *Coulomb friction:* $q_3(t) = \dot{p}_S - \dot{p}_C$

- *general state dependent switching functions (see impact)*

- *time dependent switching functions:* $q_4(t) = t - 0.05$

*Fig. 6.39 gives the result of the simulation. Beginning from the top, the graphic shows the motion of the casing $0.2 + p_H$ (dash-dotted line), the motion of the intermediate mass $0.1346 + p_M$ (dotted line), the motion of the piston $0.1346 + p_M - S$ (dashed line), and the motion of the striker $p_S$ (solid line).*

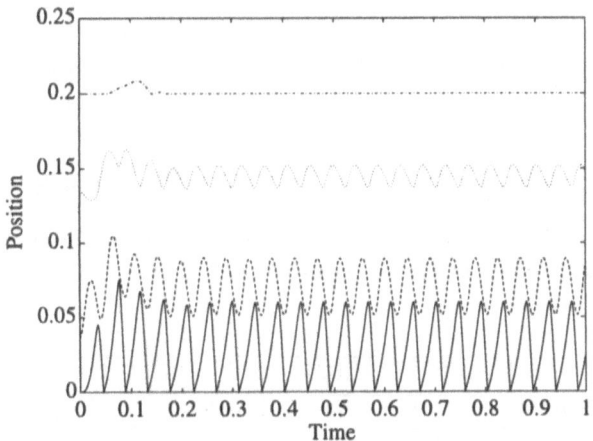

Figure 6.39: Solution for position coordinates

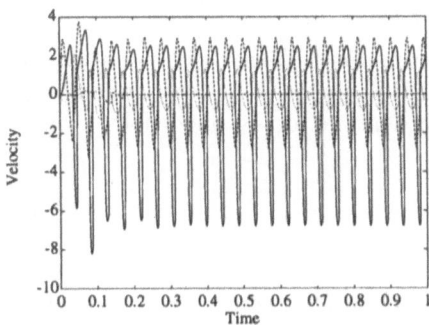

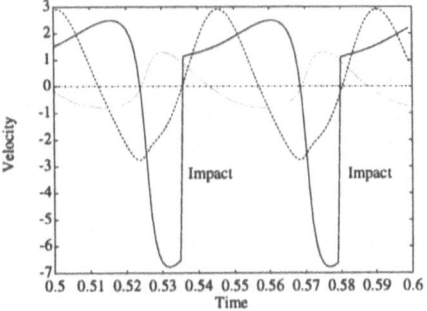

Figure 6.40: Solution for velocity coordinates

Figure 6.41: Solution for velocity coordinates, zoom

*Tables 6.9, 6.10 show the results effort and accuracy for two different foregoings*

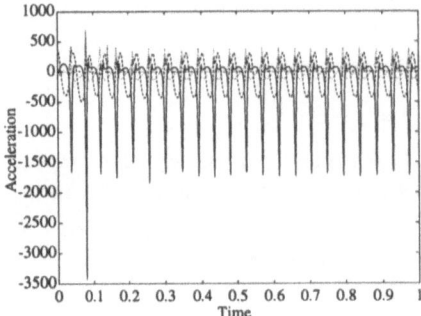

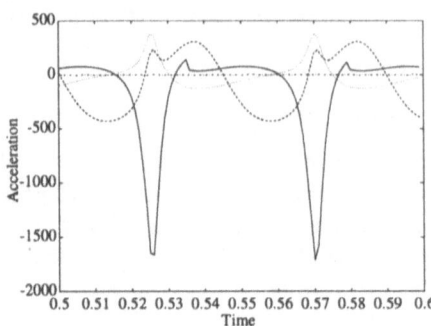

Figure 6.42: Solution for acceleration coordinates

Figure 6.43: Solution for acceleration coordinates, zoom

- *All discontinuities are localized using the switching algorithm.*

- *The discontinuities are not localized explicitly but through IF statements in the right hand side of the equations of motion. It is only the impact which must be localized*[4].

| TOL | NFE | NRED | NST | ERR/TOL | CPU |
|---|---|---|---|---|---|
| $10^{-4}$ | 4741 | 24 | 2236 | $1.9 \cdot 10^3$ | 1 |
| $10^{-5}$ | 6071 | 41 | 2932 | $1.8 \cdot 10^3$ | 1.17 |
| $10^{-6}$ | 8344 | 49 | 4146 | $1.5 \cdot 10^3$ | 1.57 |
| $10^{-7}$ | 11032 | 34 | 5564 | $1.7 \cdot 10^3$ | 1.98 |
| $10^{-8}$ | 14793 | 12 | 7517 | — | 59.84 |
| $10^{-9}$ *5 | 20282 | 2 | 10347 | — | 78.68 |

Table 6.9: Effort when localizing all discontinuities. (TOL: required tolerance, NFE: number of function evaluations, NRED: number of rejected steps, NST: number of integration steps, ERR/TOL: obtained accuracy divided by required accuracy, CPU: normalized CPU time)

*Interpretation of results:*

- *The results when localizing switching points are more accurate then those without localization when requiring the same accuracy.*

- *The effort with localization of switching points is in most cases less. This is in contradiction to the common opinion that the localization is a time consuming*

---

[4]Jumps in variables can only be handled by localizing the corresponding time points.

| TOL | NFE | NRED | NST | ERR/TOL | CPU |
|-----|-----|------|-----|---------|-----|
| $10^{-4}$ | 4568 | 171 | 1844 | $2.2{\cdot}10^3$ | 0.76 |
| $10^{-5}$ | 7042 | 292 | 2906 | $8.0{\cdot}10^3$ | 1.15 |
| $10^{-6}$ | 9914 | 440 | 4219 | $1.3{\cdot}10^4$ | 1.57 |
| $10^{-7}$ | 13806 | 638 | 5985 | $3.0{\cdot}10^4$ | 2.18 |
| $10^{-8}$ | 18551 | 786 | 8207 | $6.6{\cdot}10^4$ | 2.88 |
| $10^{-9}*$ | 25021 | 992 | 11313 | — | 91.56 |

Table 6.10: Effort without localizing all discontinuities

*task because the integration has to be restarted at every discontinuity although in this case about 100 switching points are localized.*

- *The numerical results are more reliable when using switching functions: the ratio between required and obtained accuracy ERR/TOL is nearly constant for the case of localizing the switching points.*

- *The number of rejected steps is high when not localizing the switching points.*

## 6.5.6  Implementation of Switching Algorithms in Multibody Codes

In order to get a user-friendly code the switching function should be generated automatically. With each discontinuous force element, e.g., the multibody program therefore has not only to setup the equations of the element but in addition the corresponding switching function as well as an indicator giving the type of the discontinuity. The standard integrator has to be substituted by one which is able to locate roots of switching functions, as shown in Fig. 6.3. The evaluation of switching functions as well as the root finding process should take the type of the discontinuity into consideration as shown above.

# 7  Parameter Identification Problems

In the preceding chapters we started from a mathematical description of a mechanical system and used it to predict its behavior.

This chapter is devoted to the inverse task: We use now knowledge about the system's behavior in order to determine unknown parameters in the mathematical description.

A typical example for *parameter identification* (PI) in vehicle dynamics is the identification of unknown parameters in a tire model: One starts by setting up a mathematical model with a couple of unknown parameters. Then the tire itself or in combination with other components of a vehicle is investigated under various conditions. Measurements are taken in order to determine the unknown parameters.

In most cases the unknown parameters cannot be measured directly, they might even have no direct physical interpretation. The selection of the quantities to be measured and the way how the measurements are set up are important engineering questions with a strong impact on the quality of the parameter identification.

As the measurements are usually randomly perturbed by measurement noise, the unknown parameters are determined in such a way that they describe the system with the largest likelihood (*maximum likelihood approach*).

We will state in this chapter the mathematical task of parameter identification and discuss the corresponding numerical methods. Techniques from various branches of numerical mathematics are required, e.g. numerical solution of differential equations, numerically solving nonlinear problems especially large-scale constrained nonlinear *least squares problem*. Thus, some of the methods discussed in the previous chapters will reappear here. We will see how parameter identification problems can be treated efficiently by *boundary value problem (BVP)* methods and extend the discussion of solution techniques for initial value problems (IVPs) to those for BVPs.

Furthermore, this chapter serves as an example for *optimization problems* in differential-algebraic equations. Another example for this class of problems is optimal control.

In the first part of this chapter we will give a mathematical formulation of parameter identification problems for unconstrained differential equations. In the second part we discuss essential aspects of the numerical treatment of these systems. In the last part we will discuss the application and/or extension of these method to differential-algebraic equations and discontinuous systems.

For an application of parameter identification in the context of railway dynamics see [Gru95]. There, the complete working path from setting up measurements to special numerical methods for parameter identification for constrained multibody systems is described, see also Sec. 7.3.1.

## 7.1   Problem Formulation

We will start by discussing the smooth ODE case. DAEs and discontinuous system dynamics will be considered in Secs. 7.3.1, 7.3.2, respectively.

Other than in the simulation task, where the behavior of a mechanical system could be studied even before it has been build, the PI problem requires a set of measurements on the "real life" mechanical system in order to improve the mathematical model and the knowledge about the parameters therein.

Let us be given a mathematical model for a multibody system with the states $x$ together with a model for the measurement (output function) $y$. Let us assume that in this model there are $n_\theta$ unknown model parameters $\theta \in I\!\!R^{n_\theta}$

$$\dot{x} = f(t, x, \theta) \qquad\qquad (7.1.1a)$$
$$y = b(x, \theta). \qquad\qquad (7.1.1b)$$

We assume that $f$ and $b$ are sufficiently smooth functions.

The solutions $x, y$ depend on the parameter $\theta$ and the *initial value* $x(t_0) =: s_0$. Often, the initial value is an unknown system parameter too, which has to be identified. Thus the complete set of quantities to be identified is

$$s := (\theta^T, s_0^T)^T \in I\!\!R^{n_s}$$

with $n_s = n_\theta + n_x$. In order to emphasis the dependency of $x$ on parameters and initial values we use one of the notations $x(t; \theta, s_0)$, $x(t; s)$ or $x(t)$ according to the context. The same holds for $y$.

The $n_y$ measurements at $n_t$ time points $t_i$ are denoted by

$$\eta_{ij} := \eta_j(t_i), \quad i = 1, \dots, n_t, \ j = 1, \dots, n_y.$$

The measurements are in general polluted by some random measurement error $\varepsilon_{ij}$. We make the following assumptions.

## Assumption 7.1.1

1. *There is a set of parameters and initial values* $\hat{s}$, *s.t.*

$$\eta_{ij} = y_j(t_i, \hat{s}) + \varepsilon_{ij}$$

   *with* measurement errors $\varepsilon_{ij}$.

2. *The measurement errors are normal distributed with zero mean and standard deviation* $\sigma_{ij}$.

3. *The individual measurement errors are stochastically independent, i.e.* $E(\varepsilon_{ij}\varepsilon_{kl}) = 0$. *Especially, the error in* $\eta_j(t_i)$ *is uncorrelated with the error made at another time point* $\eta_j(t_k)$.

By these assumptions, errors in setting up the mathematical model for the mechanical system are excluded. It is assumed that there is at least one set of parameters $\hat{\theta}$ which lets the mathematical result coincide with the measurement if no measurement error is made.

We now have to discuss how an optimal set of parameters should look like. The intuitive question, "What are the parameters which are correct with the highest probability?" is not meaningful, since there is only one correct set of parameters, and it is the data set which is statistical in nature. Thus, we have to ask the other way round: "What are the parameters such that this data set could have occurred with highest probability?".

We precise this question in order to define an objective function.

The joint probability density function of the measurements is

$$p(\eta) = p(\eta, \hat{s}) = (2\pi)^{-n_t n_y/2} \prod_{i,j} \sigma_{ij}^{-1} \exp\left(-\frac{(\eta_{ij} - y_j(t_i; \hat{s}))^2}{2\sigma_{ij}^2}\right).$$

This is a symmetric function centered around its maximum at $\hat{\eta}$, with $\hat{\eta}_{ij} := y_j(t_i; \hat{s})$.

From a given set of measurements $\eta^*$ one attempts to determine $s^*$ in such a way, that

$$p(\eta^*, s^*) = \max_s p(\eta^*, s). \tag{7.1.2}$$

In practice $\hat{\eta}$ and $\hat{s}$ are not known. If $s^*$ were the correct parameters and initial values, then it would be most likely to obtain measurements in a neighborhood of $\eta^*$, i.e. for any $\delta > 0$ we have $P(\eta \in \mathcal{U}(\eta^*, \delta)) = \max_\xi P(\eta \in \mathcal{U}(\xi, \delta))$, with $P$ being the joint probability function defined by the density function $p(., s^*)$. This property motivates the name *maximum likelihood estimation*.

As $P(\eta \in \mathcal{U}(\eta^*, \delta)) = 2\delta p(\eta^*, s) + \mathcal{O}(\delta^2)$ the maximum is obtained by maximizing $p(\eta^*, s)$. Taking logarithms results in the condition

$$\sum_{i,j} \frac{(\eta^*_{ij} - y_j(t_i; s))^2}{2\sigma^2_{ij}} = \min_s . \qquad (7.1.3)$$

We define

$$r_{1,ij}(x(t_i; s), s) := \frac{\eta^*_{ij} - y_j(t_i; s)}{\sigma_{ij}} = \frac{\eta^*_{ij} - b_j(t_i, x(t_i; s), \theta)}{\sigma_{ij}}$$

and assemble these components in the vector

$$r_1 = r_1(x(t_1; s), \ldots, x(t_{n_i}; s), s) \in \mathbb{R}^{n_1} \quad \text{with} \quad n_1 := n_t n_y.$$

By (7.1.3) the objective is to minimize the function

$$\|r_1(x(t_1; s), \ldots, x(t_{n_i}; s), s)\|_2.$$

A priori information on certain solution points ("*exact measurements*", interior point conditions and end point information) is formulated as an *equality constraint*

$$r_2(x(t_0; s), x(t_1; s), \ldots, x(t_e; s), s) = 0 \in \mathbb{R}^{n_2} \qquad (7.1.4)$$

while *bounds* on certain components of $s$ or of the solution $x$ at specific points $t_i$ give raise to *inequality constraints*

$$r_3(x(t_0; s), x(t_1; s), \ldots, x(t_e; s), s) \geq 0 \in \mathbb{R}^{n_3} . \qquad (7.1.5)$$

Summarizing, the parameter identification problem can now be formulated as the following nonlinear constrained optimization problem

**Problem 7.1.2** *Determine* $s \in \mathbb{R}^{n_s}$, $x : [t_0, t_e] \to \mathbb{R}^{n_x}$ *which solve*

$$\|r_1(x(t_1; s), \ldots, x(t_{n_i}; s), s)\|_2 = \min$$

*subject to the constraints*

$$
\begin{aligned}
r_2(x(t_0; s), \ldots, x(t_e; s), s) &= 0 \\
r_3(x(t_0; s), \ldots, x(t_e; s), s) &\geq 0 \\
\dot{x}(t; s) &= f(t, x(t; s), \theta) \qquad (7.1.6)
\end{aligned}
$$

**Example 7.1.3 (Truck)** *As a model problem we consider in this chapter the model of the unconstrained truck in its variant with nonlinear pneumatic springs, cf. Sec. A.2. A typical parameter, which in practice often is unknown and which needs to be identified is the constant adiabatic coefficient* $\kappa$.

We restrict ourselves to the case of equality constraints. Inequality constraints can be handled by the same techniques combined with an *active set strategy* [GMW81].

# 7.2 Numerical Solution of Parameter Identification Problems

## 7.2.1 Elimination of the ODE: Integration

The objective function in Problem 7.1.2 depends on $n_s$ unknown parameters $s$ and on an unknown function $x$ evaluated at $n_t$ different points. The ODE (7.1.6) as constraint provides information about the function $x$ at infinitely many time points. Thus, such a constrained optimization problem can be viewed as a problem with an infinite dimensional constraint. In a first step the problem is reduced to a finite dimensional problem by eliminating these constraints.

If we could express $x(t_i; s)$ as a function of the initial values and parameters $s = (s_0, \theta)$, then Problem 7.1.2 would reduce to the finite dimensional problem:

**Problem 7.2.1** *Determine a parameter vector $s \in \mathbb{R}^{n_s}$ satisfying*

$$\|F_1(s)\| = \min \qquad (7.2.1a)$$
$$F_2(s) = 0 \qquad (7.2.1b)$$

*with $F_1(s) := r_1(x(t_1; s), \dots, x(t_{n_t}; s), s)$ and $F_2(s) := r_2(x(t_0; s), \dots, x(t_e; s), s)$.*

Unfortunately, in most cases $x$ cannot be expressed as a function of $s$ directly. In these cases $x$ is obtained by numerically integrating the ODE

$$\dot{x} = f(t, x, s) \quad \text{with} \quad x(t_0; s) = s_0. \qquad (7.2.2)$$

Thus the ODE constraint (7.1.6) can only be eliminated for a given parameter value. Solving the entire problem requires an iterative procedure, which suggests at every iteration step a value $s^{(k)}$ for the next step as a parametric input of the ODE (7.2.2).

This corresponds to the following basic algorithm:

1. Choose initial estimates $s^{(0)}$ for the unknown parameters and initial values. Set $k := 0$.

2. Solve (7.2.2).

3. Evaluate the objective function $\|F_1(s^{(k)})\|$ and the constraint $F_2(s^{(k)})$.

4. Determine a new estimate $s^{(k+1)}$ by some optimization procedure.

5. Set $k := k + 1$.

6. If a solution of Problem 7.2.1 has been obtained with sufficient accuracy then stop the process else continue with Step 2.

This parameterization of the solutions of the initial value problems by the initial values and parameters is in the context of boundary value problems the well-known *single shooting* technique [SB93].

The constrained nonlinear least squares system 7.2.1 can be solved iteratively by a *Gauß–Newton method* (Step 4). This will be discussed in the next section.

Gauß–Newton methods as well as other modern optimization methods require derivatives of the objective function and constraints with respect to the unknowns, i.e. the computation of so-called *sensitivity matrices*. Their evaluation will be discussed in Sec. 7.2.3.

Often it is difficult to apply single shooting. For too bad initial guesses $s_0^{(0)}$ a solution of the initial value problem may not exist. A way out of this dilemma is to use multiple shooting techniques instead. This will be the topic of Sec. 7.2.5.

## 7.2.2  Iterative Solution of Constrained Nonlinear Least Squares Problems by Gauß–Newton Methods

Problem 7.2.1 is a constrained nonlinear least squares problem. This can be solved iteratively by solving a sequence of constrained linear least squares problems. The Gauß–Newton method generates a sequence of iterates $s^{(k)}$ by

$$
\begin{aligned}
\|F_1^k + J_1^k \Delta s^{(k)}\|_2 &= \min && (7.2.3a)\\
F_2^k + J_2^k \Delta s^{(k)} &= 0 && (7.2.3b)\\
s^{(k+1)} &= s^{(k)} + \Delta s^{(k)} && (7.2.3c)
\end{aligned}
$$

with

$$
F_i^k := F_i(s^{(k)}), \qquad J_i^k := \frac{\mathrm{d}F_i}{\mathrm{d}s}(s^{(k)}).
$$

After introducing Lagrange multipliers $\mu$ and taking squares in (7.2.3a) we obtain the Lagrangian

$$
L_{GN}(\Delta s^{(k)}, \mu^{(k)}) = \|F_1^k + J_1^k \Delta s^{(k)}\|_2^2 + (F_2^k + J_2^k \Delta s^{(k)})^T \mu^{(k)}.
$$

By applying the *Kuhn-Tucker conditions* $\nabla_{(\Delta s^{(k)}, \mu^{(k)})} L_{GN}(\Delta s^{(k)}, \mu^{(k)}) = 0$, this sequence can be written as a sequence of linear equations of the form

$$
\begin{pmatrix} 2J_1^T J_1 & J_2^T \\ J_2 & 0 \end{pmatrix}^{(k)} \begin{pmatrix} \Delta s^{(k)} \\ \mu^{(k)} \end{pmatrix} + \begin{pmatrix} 2J_1^T F_1 \\ F_2 \end{pmatrix}^{(k)} = 0. \qquad (7.2.4)
$$

The numerical solution of these linear problems has been discussed in Sec. 2.3.

## Relationship to Other Methods

Applying the *Kuhn-Tucker conditions* directly to Problem 7.2.1 with the Lagrangian

$$L(s,\mu) = \|F_1(s)\|_2^2 + \mu^T F_2(s)$$

results in the system of nonlinear equations

$$\nabla_{(s,\mu)} L(s,\mu) = \begin{pmatrix} 2J_1^T(s)F_1(s) + \mu^T J_2(s) \\ F_2(s) \end{pmatrix} = 0.$$

Applying to this Newton's iteration requires the solution of the sequence of linear equations

$$\begin{pmatrix} 2J_1^T J_1 + 2\frac{\mathrm{d}}{\mathrm{d}s}(J_1^T)F_1 + \frac{\mathrm{d}}{\mathrm{d}s}(J_2^T \mu^{(k)}) & J_2^T \\ J_2 & 0 \end{pmatrix}^{(k)} \begin{pmatrix} \Delta s^{(k)} \\ \mu^{(k)} \end{pmatrix} = -\begin{pmatrix} 2J_1^T F_1 \\ F_2 \end{pmatrix}^{(k)}.$$

$$(7.2.5)$$

By comparing (7.2.4) with (7.2.5) the convergence properties of the Gauß–Newton method can be related to those of Newton's method. In (7.2.4) the second derivatives $\frac{\mathrm{d}}{\mathrm{d}s}(J_1^T)F_1$ and $\frac{\mathrm{d}}{\mathrm{d}s}(J_2^T \mu^{(k)})$ are missing. As a consequence of Assumption 7.1.1 it can be expected that $F_1(s^*)$ and $\mu^*$ are small if the measurement errors $\varepsilon_{ij}$ are small. By this, at least in the neighborhood of the solution, the Gauß–Newton method has a similar convergence behavior as Newton's method.

This can also be seen when considering Theorem 3.4.1 for the case of Gauß–Newton methods:

$$\text{Let } F = \begin{pmatrix} F_1 \\ F_2 \end{pmatrix}, J = \begin{pmatrix} J_1 \\ J_2 \end{pmatrix}, B := \begin{pmatrix} J_1 \\ J_2 \end{pmatrix}^{\text{CLSQ+}}.$$

- The curvature condition (3.4.2) for $\omega$ is fulfilled if $\|B(y)\|$ is bounded and $J$ Lipschitz continuous.

- From Sec. 2.3 follows that $B := \begin{pmatrix} J_1 \\ J_2 \end{pmatrix}^{\text{CLSQ+}}$ is an (1,2,4) inverse. We use property (2), $BJB = B$, of the pseudo-inverse for the following transformation of the residual condition (3.4.3):

$$\begin{aligned} \|B(\tilde{s})R(s)\| &= \|B(\tilde{s})(I - J(s)B(s))F(s)\| \\ &= \|(B(\tilde{s}) - B(s))(I - J(s)B(s))F(s)\| \\ &= \|(B(\tilde{s}) - B(s))R(s)\| \overset{!}{\leq} \kappa\|\tilde{s} - s\|, \quad \kappa \overset{!}{<} 1 \end{aligned}$$

This requirement can be fulfilled if $B$ is Lipschitz continuous and the residual $R(s) = (I - JB)F$ not too large.

$\kappa$ reflects the effect of omitting of the second derivatives $\frac{d}{ds}(J_1^T)F_1$ and $\frac{d}{ds}(J_2^T\mu^{(k)})$ of Newton's method. Thus, Gauß–Newton methods fail to converge due to the neglection of second derivatives if Assumption 7.1.1 is not valid or if the measurement errors are too large.

It can be shown that if $\kappa > 1$, there are always measurement errors such that the solution runs away unboundedly [Bock87].

To improve the convergence behavior this method can be combined with additional techniques such as line search, cf. Sec. 3.7.

In practice, faced with a convergence failure of the identification program, one has several alternatives. One may try

- to use higher accuracies in the numerical routines;

- to get better initial guesses;

- to rescale the problem;

- to get better measurements;

- to revise the model. This means one has to review simplifications and neglections in the model.

There is often no receipt to decide which of these actions has to be taken. Analyzing the reasons for failure of the method is the most difficult problem in parameter identification.

## 7.2.3    Evaluation of Functions and Jacobians

Performing iteration (7.2.3) requires the evaluation of $F_i$ and $J_i = (\frac{\partial F_i}{\partial s_0}, \frac{\partial F_i}{\partial \theta})$ for $i = 1, 2$. From the definition of $F_1$ it can be seen that the solutions $x(t; s^{(k)})$ of the ODE (7.1.6) have to be determined and evaluated at the $n_t$ time points $t_i$. Additionally, the derivatives of these solutions with respect to the parameters $s = (s_0^T, \theta^T)^T$ are required. These derivatives are

$$\frac{\partial F_i}{\partial s_0} = \frac{\partial r_i}{\partial x(t_1)}W^0(t_1) + \cdots + \frac{\partial r_i}{\partial x(t_{n_t})}W^0(t_{n_t}) \tag{7.2.6a}$$

$$\frac{\partial F_i}{\partial \theta} = \frac{\partial r_i}{\partial x(t_1)}W^\theta(t_1) + \cdots + \frac{\partial r_i}{\partial x(t_{n_t})}W^\theta(t_{n_t}) + \frac{\partial r_i}{\partial \theta} \tag{7.2.6b}$$

with the *sensitivity matrices*

$$W^0(t_i) := \frac{\partial x(t_i; s_0, \theta)}{\partial s_0} \text{ and } W^\theta(t_i) := \frac{\partial x(t_i; s_0, \theta)}{\partial \theta} \quad i = 1, \ldots, n_t.$$

The computation of sensitivity matrices is the most time consuming part of the algorithm and the way how these matrices are computed has a large impact on the overall performance of the method.

**Variational Differential Equations.**
One way to compute sensitivity matrices is to integrate the *variational differential equations.*
Differentiating the integral form of the ODE (7.1.6)

$$x(t; s_0, \theta) = s_0 + \int_{t_0}^{t} f(\tau, x(\tau; s_0, \theta), \theta) d\tau$$

with respect to the initial value $s_0$ and the other parameters $\theta$ gives

$$\frac{\partial x}{\partial s_0}(t; s_0, \theta) = I + \int_{t_0}^{t} \frac{\partial f}{\partial x}(\tau, x(\tau; s_0, \theta), \theta) \frac{\partial x}{\partial s_0}(\tau; s_0, \theta) d\tau$$

$$\frac{\partial x}{\partial \theta}(t; s_0, \theta) = \int_{t_0}^{t} \left( \frac{\partial f}{\partial x}(\tau, x(\tau; s_0, \theta), \theta) \frac{\partial x}{\partial \theta}(\tau; s_0, \theta) + \frac{\partial f}{\partial \theta}(\tau, x(\tau; s_0, \theta), \theta) \right) d\tau.$$

By rewriting these equations as ODEs we obtain the *variational differential equations* (VDE)

$$\dot{W}^0 = \frac{\partial f}{\partial x}(t, x(t; s_0, \theta), \theta) W^0 \tag{7.2.7a}$$

$$\dot{W}^\theta = \frac{\partial f}{\partial x}(t, x(t; s_0, \theta), \theta) W^\theta + \frac{\partial f}{\partial \theta}(t, x(t; s_0, \theta), \theta) \tag{7.2.7b}$$

with initial values $W^0(t_0) = I$ and $W^\theta(t_0) = 0$.
The variational differential equations are two matrix differential equations corresponding to a system of $n_x(n_x + n_\theta)$ differential equations. The right hand side of these equations depends on $\frac{\partial f}{\partial x}, \frac{\partial f}{\partial \theta}$ evaluated along the trajectory $x(t; s)$. If these quantities can be provided analytically by means of special library elements in a multibody formalism these differential equations can be solved by standard numerical integration methods. These are normally the same as the one used for integrating (7.1.6). Sometimes it seems to be advantageous to use integration schemes which use many right hand side evaluations at the same time point, like the classical Runge–Kutta–Fehlberg fourth order method. This may save expensive evaluations of the derivatives $\frac{\partial f}{\partial x}, \frac{\partial f}{\partial \theta}$, [BHK94].

**Numerical Differentiation of the Discretization Scheme**
If the derivatives $\frac{\partial f}{\partial x}, \frac{\partial f}{\partial \theta}$ are not given analytically the sensitivity matrices can be computed alternatively by numerical differentiation of the discretization scheme. For motivating this foregoing we consider Fig. 7.1:
When computing the *sensitivity matrices* from the variational differential equation the differential equation (7.1.6) is differentiated first and then the resulting equations are discretized for numerical integration. For a discretization in terms of the explicit Euler method this corresponds to the upper right path in Fig. 7.1.

However, if the variational differential equation is not known explicitly the derivatives are calculated the other way round: First we apply a discretization scheme to the ODE (7.1.6), and then we differentiate the resulting equation (lower left path in the diagram).

We discuss this procedure in terms of the explicit Euler method as discretization scheme

$$x_{n+1}(s_0, \theta) = x_n(s_0, \theta) + h f(t_n, x_n(s_0, \theta), \theta)$$

where $n$ denotes the step number and $x_n(s_0, \theta)$ denotes the corresponding numerical solution obtained from starting values $x(t_0) = s_0$.

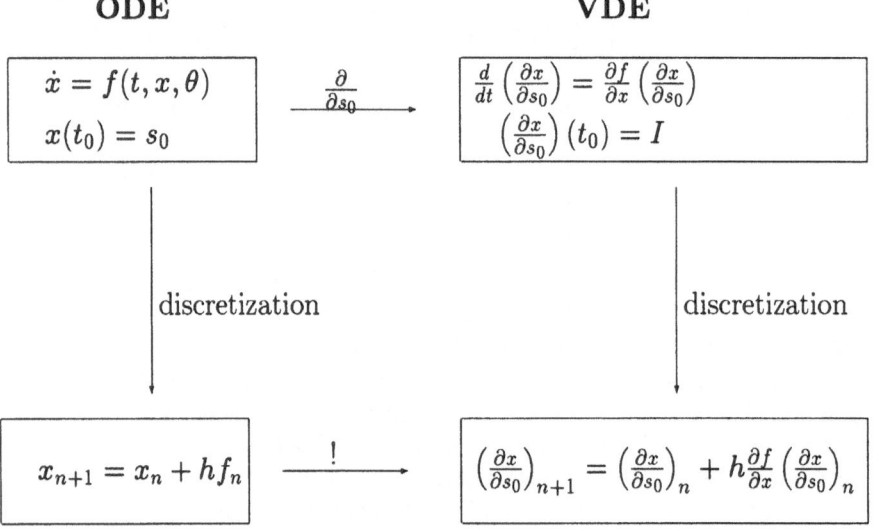

Figure 7.1: Commutativity relationship for the computation of sensitivity matrices

Differentiating with respect to $s_0$ results in

$$\frac{\partial x_{n+1}}{\partial s_0} = \frac{\partial x_n}{\partial s_0} + h \frac{\partial f}{\partial x}(t_n, x_n(s_0, \theta), \theta) \frac{\partial x_n}{\partial s_0} + \frac{\partial h}{\partial s_0} f(t_n, x_n(s_0, \theta), \theta).$$

The interesting term in this expression is the derivative of the step size $h$. If no special care is taken, this derivative may not be defined due to fact that the integration procedure with its IF's, ... is not a smooth mapping. Even if the numerical integration is implemented in a smooth way the diagram in Fig. 7.1 is only commutative if the step size is independent on the actual initial values, i.e. $\frac{\partial h}{\partial s_0} = 0$.

This requirement provokes often the use of fixed step size integrators. As the differentiation has to be replaced in practice by computing forward differences, the use of integration codes with "frozen" step size sequences is often taken into consideration as an alternative to fixed step size codes. There, a step size sequence used for initial values $s_0$ is re-used when varying initial values.

In order to compute a forward difference approximation to the derivatives we have to compute $x_{n+1}(s_0, \theta)$ and $x_{n+1}(\tilde{s}_0, \theta)$ by solving the differential equation numerically for different initial values $s_0, \tilde{s}_0$:

$$
\begin{aligned}
x_{n+1}(s_0, \theta) &= x_n(s_0, \theta) + h(s_0)f(t_n, x_n(s_0, \theta), \theta), & x_0(t_0) = s_0 \\
x_{n+1}(\tilde{s}_0, \theta) &= x_n(\tilde{s}_0, \theta) + h(\tilde{s}_0)(f(t_n, x_n(\tilde{s}_0, \theta), \theta), & x_0(\tilde{s}_0) = \tilde{s}_0 = s_0 + \varepsilon \cdot e_j
\end{aligned}
$$

and then to approximate

$$
(W_{n+1}^0)_{:,j} \approx \frac{x_{n+1}(\tilde{s}_0, \theta) - x_{n+1}(s_0, \theta)}{\varepsilon},
$$

cf. Sec. 3.5.

If the computations for $s_0, \tilde{s}_0$ use the same step size $h = h(s_0) = h(\tilde{s}_0)$ we get

$$
\begin{aligned}
\frac{x_{n+1}(\tilde{s}_0, \theta) - x_{n+1}(s_0, \theta)}{\varepsilon} &= \frac{x_n(\tilde{s}_0, \theta) - x_n(s_0, \theta)}{\varepsilon} \\
&\quad + \frac{h}{\varepsilon}\left(f(t_n, x_n(\tilde{s}_0, \theta), \theta) - f(t_n, x_n(s_0, \theta)), \theta)\right),
\end{aligned}
$$

which is an approximation of the discretized variational differential equation

$$
\frac{\partial x_{n+1}}{\partial s_0} = \frac{\partial x_n}{\partial s_0} + h\frac{\partial f}{\partial x}(t_n, x_n(s_0, \theta), \theta)\frac{\partial x_n}{\partial s_0} + \mathcal{O}(\varepsilon).
$$

Note that this is a good approximation only in the case when both solutions don't differ to much.

The same approach applies also for other discretization schemes. There, also the order of the method and the number of Newton iterations in every step must be "frozen" in the same way as the step size above.

Freezing these parameters can be implemented in an easy way by simultaneously integrating the system for all initial values required for computing the forward differences. Furthermore, especially when using implicit integration schemes, a lot of extra work can be saved by simultaneous integration.

## 7.2.4   Summary of the Algorithm

1. Choose initial guesses $s_0^{(0)}, \theta^{(0)}$

2. Solve the constrained optimization problem

$$\|F_1(s)\|_2 = \min_s$$
$$F_2(s) = 0$$

with $s = (s_0, \theta)$

- by applying Gauß–Newton iterations,

$$s^{(k+1)} = s^{(k)} + \alpha^{(k)} \Delta s^{(k)}$$
$$\|F_1^k + J_1^k \Delta s^{(k)}\|_2 = \min$$
$$F_2^k + J_2^k \Delta s^{(k)} = 0$$
$$s^{(k+1)} = s^{(k)} + \alpha^{(k)} \Delta s^{(k)}.$$

- This requires the computation of sensitivity matrices which are needed to get $J_1, J_2$. For this end $n_x + n_\theta + 1$ initial value problems have to be solved by using identical stepsize and order sequences.

## 7.2.5   The Boundary Value Approach

At this point let us review the approach in order to see a potential shortcoming. The PI problem has as input data among others the measurements and as output data the initial values $s_0$ and unknown system parameters $\theta$. For solving this task an initial value problem (IVP) must be integrated, which has as input values initial values and system parameters and as output approximations to the measurements. Thus, the PI problem and the IVP can be viewed as inverse to each other. Consequently, both problems have also opposite stability properties. This can be easily seen, when considering initial values. A stiff IVP is extremely stable, while identifying the initial values of such an ODE is an ill-conditioned task. On the other hand, identifying initial values of an unstable or chaotic ODE is a well-posed problem, while integrating this ODE is ill-posed [BBE93].

Furthermore, at least for bad initial guesses of initial values or parameters the initial value problem may even have no solution at all, while the original PI problem is stable and well conditioned.

**Example 7.2.2** *This can be seen, when solving the truck example for different values of the adiabatic coefficient $\kappa$. When $\kappa$ exceeds a value between 1.542 and 1.545, the relative translation $\rho_{10}$ changes its sign. This leads to the physically unrealistic situation shown in Fig. 7.2. Although technically irrelevant, these values of $\kappa$ can occur within the optimization loop as intermediate results.*

To overcome this difficulty one attempts to feed into the solution process already during integration information from measurements[Bock87]. This makes multiple

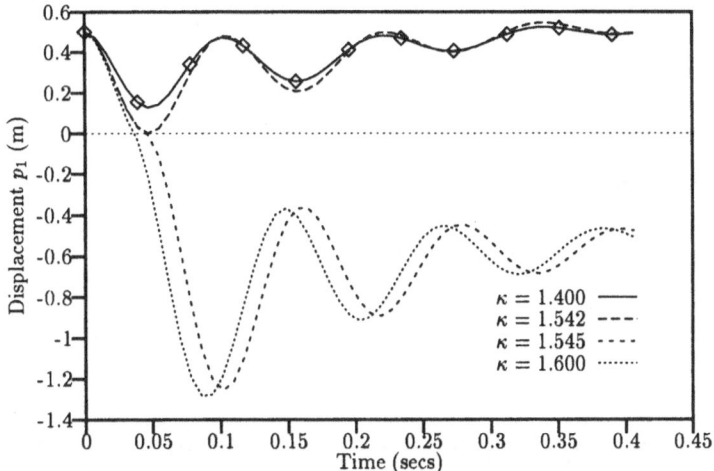

Figure 7.2: Influence of small changes in $\kappa$ on $p_1$

shooting techniques attractive. This method class is often applied to boundary value problems.

One chooses a mesh covering the time interval under consideration

$$t_0 = \tau_0 < \cdots < \tau_m = t_e$$

in such a way that it includes a subset of the measurement points and points with inner point conditions. Additionally, we choose as before guesses for the initial values, but now also "initial values" $s_j$ at the grid points $\tau_j$. These additional variables $s_j$ are estimates of the states $x(\tau_j)$. This is the point where information available from the measurements can be brought in. Then, we solve the $m$ independent initial value problems

$$\dot{x} = f(x, \theta), \quad x(\tau_j) = s_j, \quad t \in [\tau_j, \tau_{j+1}] \quad j = 0, \ldots, m-1 \tag{7.2.8}$$

separately in each subinterval.

The corresponding trajectory is in general discontinuous, see Fig. 7.3. For $x$ to be a solution of the overall initial value problem we have to require *continuity* in the nodes:

$$x(\tau_{j+1}; s_j, \theta) - s_{j+1} = 0, \quad j = 0, \ldots, m-1. \tag{7.2.9}$$

Thus we have $(m + 1) \cdot n_x$ equations, the $n_x$ boundary conditions and the $m \cdot n_x$ *continuity conditions* or *matching conditions* (7.2.9) for the $(m + 1) \cdot n_x$ variables $s_0, \ldots, s_m$.

Formal insertion of the discontinuous representation of the solution into Problem 7.1.2 yields a constrained least squares problem of the form

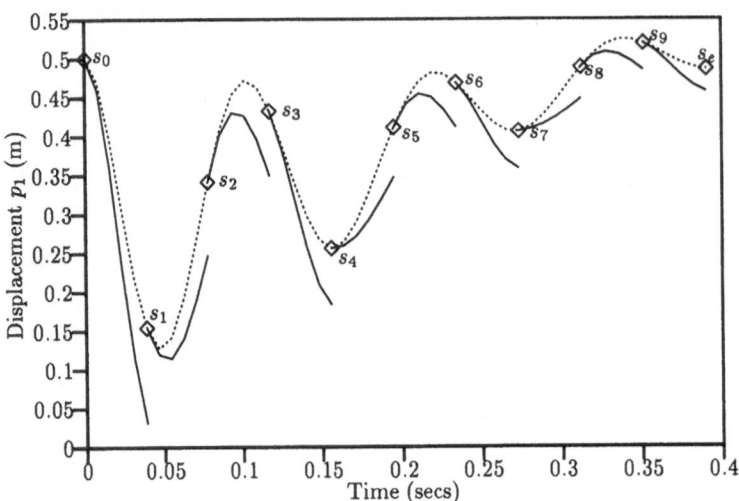

Figure 7.3: Multiple shooting technique demonstrated on the unconstrained truck

**Problem 7.2.3** *Determine $\theta \in I\!\!R^{n_\theta}$, $s_0, \ldots, s_m \in I\!\!R^{n_x}$ which minimize the objective function*

$$\|R_1(s_0, s_1, \ldots, s_m, \theta)\|_2 = \min$$

*subject to*

$$
\begin{aligned}
R_2(s_0, s_1, \ldots, s_m, \theta) &= 0 \\
x(\tau_{j+1}; s_j, \theta) - s_{j+1} &= 0 \qquad j = 0, \ldots, m-1
\end{aligned}
$$

*where $R_i$ corresponds to $r_i$ with $x$ parameterized by $s_j, \theta$:*

$$R_i(s_0, s_1, \ldots, s_m, \theta) := r_i(x(t_1, \theta), \ldots, x(t_e, \theta)).$$

The advantages of this approach compared to the simpler initial value problem approach is that information about the solution available from measurements can be brought in to improve the initial guesses by choosing $s_j^{(0)}$ appropriately. Now the dynamic variables are no longer eliminated in favor of the unknown parameters. This damps out the influence of poor initial guesses and improves the convergence behavior of the optimization procedure[Bock87].

**Gauß–Newton Method for Multiple Shooting**
Again, we have to deal with a problem of the form

$$
\begin{aligned}
\|F_1(s)\|_2^2 &= \min & \text{(7.2.10a)} \\
F_2(s) &= 0 & \text{(7.2.10b)}
\end{aligned}
$$

where now $s := (s_0^T, s_2^T, \ldots, s_m^T, \theta^T)^T \in \mathbb{R}^{n_s}$. $F_2 : \mathbb{R}^{n_s} \to \mathbb{R}^{n_2}$ summarizes the original equality constraints and matching conditions:

$$F_1(s) = R_1(s) \qquad\qquad\qquad \text{(measurement conditions)}$$

$$F_2(s) = \begin{pmatrix} R_2(s) \\ x(\tau_1; s_0, \theta) - s_1 \\ \vdots \\ x(\tau_m; s_{m-1}, \theta) - s_m \end{pmatrix} \qquad \text{(multipoint and matching conditions)}.$$

**Evaluation of Functions and Jacobians**
We have to evaluate and solve the linearized system (7.2.3)

$$\|F_1^k + J_1^k \Delta s^{(k)}\|_2 = \min \qquad (7.2.11a)$$
$$F_2^k + J_2^k \Delta s^{(k)} = 0 \qquad (7.2.11b)$$

with

$$\begin{pmatrix} J_1^k \\ J_2^k \end{pmatrix} \Delta s^{(k)} = \begin{pmatrix} D_1^0 & D_1^1 & \cdots & & & D_1^\theta \\ D_2^0 & D_2^1 & \cdots & & & D_2^\theta \\ W_1 & -I & 0 & & & W_1^\theta \\ 0 & W_2 & -I & & & W_2^\theta \\ & & & \ddots & \ddots & 0 & \vdots \\ & & & & W_m & -I & W_m^\theta \end{pmatrix} \begin{pmatrix} \Delta s_0 \\ \Delta s_1 \\ \vdots \\ \Delta s_m \\ \Delta\theta \end{pmatrix}$$

$$\begin{pmatrix} F_1^k \\ F_2^k \end{pmatrix} = \begin{pmatrix} R_1 \\ \hline R_2 \\ x(\tau_1; s_0, \theta) - s_1 \\ \vdots \\ x(\tau_m; s_{m-1}, \theta) - s_m \end{pmatrix}$$

and

$$D_i^j := \frac{\partial R_i}{\partial s_j}, \qquad D_i^\theta := \frac{\partial R_i}{\partial\theta} \qquad W_j = \frac{\partial x(\tau_j; s_{j-1}, \theta)}{\partial s_{j-1}}, \qquad W_j^\theta = \frac{\partial x(\tau_j; s_{j-1}, \theta)}{\partial\theta}.$$

Thus in every Gauß–Newton iteration step one has to evaluate

- solutions of the differential equations $x(t; s_j, \theta)$,

- their derivatives with respect to initial values and parameters $W_j, W_j^\theta$, $j = 0, \ldots, m$, i.e. the sensitivity matrices,

- the objective function $R_1$ and the multipoint conditions $R_2$

- their derivatives $D_i^j, D_i^\theta, \ i = 1, 2; \ j = 0, \dots, m$ with respect to initial values and parameters,

- the matching conditions $x(\tau_{j+1}; s_j, \theta) - s_{j+1} = 0$ with $j = 0, \dots, m - 1$.

Again, the computation of the sensitivity matrices is the most time consuming part of the algorithm. Aspects of this task have been discussed in Sec. 7.2.3.

**Solution of Linearized Systems**
System (7.2.11) is solved by a so-called *condensing algorithm* or *Block-Gauß–Newton method*, which exploits the particular structure. We successively eliminate the variables $\Delta s_m, \dots, \Delta s_1$ by solving (7.2.11) from bottom to top.
Solving the last equation for $\Delta s_m$ yields

$$\Delta s_m = W_m \Delta s_{m-1} + W_m^\theta \Delta\theta - h_{m-1}$$

with $h_i := x(t_i, s_{i-1}, \theta) - s_i$. Inserting this into the upper block of equations results in

$$\|D_1^0 \Delta s_0 + \dots + D_1^{m-2}\Delta s_{m-2} + E_1^{m-1}\Delta s_{m-1} + C_1^{m-1}\Delta\theta - u_1^{m-1}\|_2 \ = \ \min$$
$$D_2^0 \Delta s_0 + \dots + D_2^{m-2}\Delta s_{m-2} + E_2^{m-1}\Delta s_{m-1} + C_2^{m-1}\Delta\theta - u_2^{m-1} \ = \ 0$$

with $E_i^{m-1} := \left(D_i^{m-1} + D_i^m W_m\right)$, $C_i^{m-1} := \left(D_i^\theta + D_i^m W_m^\theta\right)$, and $u_i^{m-1} := (R_i + D_i^m h_{m-1})$.
In the next step we eliminate $\Delta s_{m-1}$ by using the last but one equation and so on. This results in a recursion, which is initialized by

$$u_i^m := R_i, \quad C_i^m := D_i^\theta, \quad E_i^m := D_i^m, \qquad i = 1, 2.$$

Then the following quantities are computed for $j = m, \dots, 1$ and $i = 1, 2$:

$$u_i^{j-1} := u_i^j + E_i^j h_{j-1}, \quad C_i^{j-1} := C_i^j + E_i^j W_j^\theta, \quad E_i^{j-1} := D_i^{j-1} + E_i^j W_j.$$

The recursion finally leads to the *condensed system*

$$\|u_1^0 + E_1^0 \Delta s_0 + C_1^0 \Delta\theta\|_2 \ = \ \min \qquad\qquad (7.2.12a)$$
$$u_2^0 + E_2^0 \Delta s_0 + C_2^0 \Delta\theta \ = \ 0 \qquad\qquad (7.2.12b)$$

which is of much smaller dimension (equal to the dimension of the initial value problem approach).
This linear system can be solved using the least squares method presented in Sec. 2.3.

## 7.3 Extension to Constrained and Discontinuous Multibody Systems

### 7.3.1 Differential Algebraic Equations

The extension of the approach to constrained multibody systems and differential-algebraic equations affects the formulation of the multiple shooting method and the computation of sensitivity matrices. The former requires a more sophisticated treatment because variations of initial values and parameters may no longer be consistent with the algebraic equations. The latter can be done efficiently by exploiting the fact that the number of degrees of freedom of the system is reduced due to the presence of constraints.

#### 7.3.1.1 Shooting Methods for Index 1 DAEs

We first consider semi-explicit index 1 DAEs of the form

$$\dot{y} = f_D(y, \lambda, \theta)$$
$$0 = f_C(y, \lambda, \theta), \qquad \frac{\partial f_C}{\partial \lambda} \text{ regular.}$$

As discussed in Sec. 5.1.2 a solution exists only for consistent initial values. However, if the initial values at the shooting nodes $s_j^y, s_j^\lambda$ are degrees of freedom for the Gauß–Newton method, consistency cannot be guaranteed during the iteration process.

To overcome this problem it has been proposed in [BES87] to solve in each subinterval $[\tau_j, \tau_{j+1}]$ the $m-1$ *consistent extended initial value problems*

$$\dot{y} = f_D(y, \lambda, \theta) \tag{7.3.1a}$$
$$0 = f_C(y, \lambda, \theta) - f_C(s_j^y, s_j^\lambda, \theta) \tag{7.3.1b}$$

with the initial values

$$y(\tau_j) = s_j^y, \qquad \lambda(\tau_j) = s_j^\lambda.$$

In order to ensure a continuous solution it has to be required additionally to the matching conditions (7.2.9) for $y$ that the solution of the boundary value problem fulfills the original algebraic equations. Thus the pointwise constraints

$$r_\lambda(s_j^y, s_j^\lambda, \theta) := f_C(s_j^y, s_j^\lambda, \theta) = 0$$

are added to the matching conditions.

This allows the treatment of problems where the differential and algebraic variables are not consistent with the algebraic equations during the iteration, as it is often the case in PI problems for the initial guesses. However, the solution of the overall problem satisfies the original DAEs.

The other possibility would be to solve $f_C(s_j^y, s_j^\lambda, \theta) = 0$ for $s_j^\lambda$ as function of $s_j^y$ and thus $s_j^\lambda$ is no longer a degree of freedom for the Gauß Newton method. The former type of procedure is often called *infeasible path* method as for the iterates the constraints are not enforced. This is in contrast to the latter type which belongs to *feasible path* methods. Feasible path methods are known to converge in general slower than infeasible path methods.

Additionally, the infeasible path method avoids the elimination of the algebraic variables $\lambda$ in favor of $y, \theta$. The information available from measurements for both types of variables can be brought in. Although $f_C(s_j^y, s_j^\lambda, \theta) \neq 0$ during the solution process the solution of the DAE remains a well-posed problem because the consistent extension of the IVP always leads to consistent initial values for the DAE (7.3.1).

### 7.3.1.2 Shooting Methods for Higher Index DAEs

Constrained multibody systems are described by index-3 DAEs. Index reduction by differentiating the constraints transforms these into an index-1 problem, for which we just described an adequate formulation for applying shooting techniques. Unfortunately, index-1 problems suffer from the so-called drift-off effect, i.e. the index-2 (velocity) and index-3 (position) constraints will be no longer met in the presence of numerical errors. Furthermore, the residual in these constraints increases with time. In Sec. 5 we described several projection techniques to overcome this problem. After applying modifications to the index-2 and index-3 constraints to cover the situation of inconsistent iterates for the initial values $s_0^y, s_0^\lambda$ like in the previous section, coordinate projection can also applied in this context. We briefly describe the main ideas of this approach and refer the reader to [EMS93] for details.

**Sensitivity Matrices for Higher Index DAEs**

We describe the foregoing by considering the first subinterval and drop the parameters $\theta$ for ease of notation.

As in Sec. 5.1.4 we eliminate $\lambda$ in Eq. (7.3.1a) by solving Eq. (7.3.1b) for $\lambda$ and obtain an explicit ODE of the form[1]

$$\dot{y} = f(y; s_0), \quad y(t_0) = s_0^y \quad \text{with} \quad y = \begin{pmatrix} p \\ v \end{pmatrix}. \tag{7.3.2}$$

Note that $f$ depends by construction on the residual $r_\lambda$ and therefore on $s_0$. Corresponding to the integral invariants of the index reduced system in Sec. 5.1.4, the solution of Eq. (7.3.2) has the following property

$$\varphi(t, y, s_0) := \bar{\varphi}(t, y, s_0) - \bar{\varphi}(t_0, s_0, s_0) = 0 \tag{7.3.3}$$

---

[1] In practice, the equivalent index-1 system is integrated.

with

$$\bar{\varphi}(t, y, s_0) = \begin{pmatrix} G(p)v - r_\lambda(s_0^p, s_0^v, s_0^\lambda)(t - t_0) \\ g(p) - r_v(s_0^p, s_0^v)(t - t_0) - \frac{1}{2}r_\lambda(s_0^p, s_0^v, s_0^\lambda)(t - t_0)^2 \end{pmatrix},$$

$r_v(s_0^p, s_0^v) := G(s_0^p)s_0^v.$

Since Eq. (7.3.3) holds for arbitrary initial values we get by total differentiation with respect to the initial value $s_0$

$$\varphi_y(t, y(t; s_0), s_0)\frac{\partial y}{\partial s_0} + \varphi_{s_0}(t, y(t; s_0), s_0) = 0. \qquad (7.3.4)$$

This gives us a property of the sensitivity matrices $W = \frac{\partial y}{\partial s_0}$ which will be used later in this section.

In order to ensure that the Gauß-Newton iteration converges towards consistent values $s_0$ we have to extend the constraining function $F_2$ in (7.2.1b) by additionally requiring

$$\phi(s_0) = \begin{pmatrix} r_\lambda(s_0^p, s_0^v, s_0^\lambda) \\ r_v(s_0^p, s_0^v) \\ r_p(s_0^p) \end{pmatrix} = 0, \qquad (7.3.5)$$

where we set $r_p(s_0^p) := g(s_0^p)$.

For performing the Gauß-Newton iteration we have to integrate Eq. (7.3.2) and to compute the sensitivity matrix with respect to the initial values. In both subtasks we make use of the solution properties by applying coordinate projection.

Let us assume, that the numerical solution of Eq. (7.3.2) at $t = t_n$ is $\tilde{y}_n$, then we project this solution in order to enforce Eq. (7.3.3), cf. Sec. 5.3.3:

$$\|y_n - \tilde{y}_n\|_2 = \min_{y_n} \qquad (7.3.6a)$$

$$\varphi(t_n, y_n, s_0) = 0. \qquad (7.3.6b)$$

The solution of this system is computed by iteratively solving linear constraint least squares problems as described in Sec. 5.3.1.

In general it is sufficient for avoiding the drift-off effect to perform only a single step of this iteration which then reads

$$y_n := \tilde{y}_n - H(\tilde{y}_n)^+ \varphi(t_n, \tilde{y}_n, s_0) \qquad (7.3.7)$$

with $H(y) := \frac{\partial}{\partial y}\varphi(t_n, y, s_0)$ and $H^+$ denoting the Moore–Penrose inverse of $H$.

When computing the sensitivity matrix with respect to $s_0$ we get by the chain rule

$$W_n := \frac{\partial y_n}{\partial s_0} = \frac{\partial y_n}{\partial \tilde{y}_n}\frac{\partial \tilde{y}_n}{\partial s_0} - H(\tilde{y}_n)^+ \varphi_{s_0}(t_n, \tilde{y}_n, s_0).$$

From (7.3.7) follows

$$\frac{\partial y_n}{\partial \tilde{y}_n} = I - \frac{\partial H(\tilde{y}_n)^+}{\partial \tilde{y}_n}\varphi(t_n, \tilde{y}_n, s_0) - H(\tilde{y}_n)^+ H(\tilde{y}_n). \tag{7.3.8}$$

If $\varphi$ is sufficiently smooth and $(HH^T)^{-1}$ is bounded, then $\frac{\partial H(\tilde{y}_n)^+}{\partial \tilde{y}_n}$ exists and is bounded, too. Since $\varphi(t_n, \tilde{y}_n, s_0)$ can be assumed to be small during the integration process we can approximate

$$\frac{\partial y_n}{\partial \tilde{y}_n} \approx I - H(\tilde{y}_n)^+ H(\tilde{y}_n) =: P(\tilde{y}_n) \tag{7.3.9}$$

where $P$ is an orthogonal projector onto the nullspace of $H(\tilde{y}_n)$.
Thus we obtain

$$W_n = \frac{\partial y_n}{\partial s_0} = P \underbrace{\frac{\partial \tilde{y}_n}{\partial s_0} - H(\tilde{y}_n)^+ \varphi_{s_0}(t_n, \tilde{y}_n, s_0)}_{=: \widetilde{W}_n}.$$

Note that $W_n$ fulfills equation (7.3.4).
Furthermore, the matrices used to compute this projection are the same as for the nominal trajectory and thus the effort for this additional projection is very small. Thus we end up with the following procedure

$$\begin{array}{lllll}
\cdots \rightarrow y_{n-1} & \overset{\text{Discret.}}{\rightarrow} & \tilde{y}_n & \overset{\text{Proj.}}{\rightarrow} & y_n = \tilde{y}_n - H^+\varphi(\tilde{y}_n) & \overset{\text{Discret.}}{\rightarrow} \\
\cdots \rightarrow W_{n-1} & \overset{\text{Discret.}}{\rightarrow} & \widetilde{W}_n = \frac{\partial \tilde{y}_n}{\partial s_0} & \overset{\text{Proj.}}{\rightarrow} & W_n = P(\tilde{y}_n)\widetilde{W}_n - H^+\varphi_{s_0} & \overset{\text{Discret.}}{\rightarrow}
\end{array}$$

In the case that more iterations of the projection step are necessary the same differentiation procedure as above can be applied.


**Reducing the Number of Partial Derivatives: Index-1 Case**
The effort for the computation of sensitivity matrices can be reduced significantly observing that only directional derivatives in the directions of the degrees of freedom of the system are necessary. This leads to ideas used earlier also in the context of multistage least squares systems see [BES87].
We will demonstrate the approach for a typical subinterval $[\tau_j, \tau_{j+1}]$ and consider first index-1 DAEs.
We have to solve at the multiple shooting nodes the matching conditions and the consistency conditions for the algebraic equations

$$\begin{aligned}
y(\tau_{j+1}; s_j^y, s_j^\lambda) - s_{j+1}^y &= 0 \\
f_C(s_j^y, s_j^\lambda) &= 0.
\end{aligned}$$

When applying a Gauß–Newton method these constraints of the optimization problem have to be linearized in every iteration step. This leads to

$$W_j^y \Delta s_j^y + W_j^\lambda \Delta s_j^\lambda - \Delta s_{j+1}^y = -r_y \qquad (7.3.10a)$$

$$f_{C_y} \Delta s_j^y + f_{C_\lambda} \Delta s_j^\lambda = -r_\lambda \qquad (7.3.10b)$$

with $r_y := y\left(\tau_{j+1}; s_j^y, s_j^\lambda\right) - s_{j+1}^y$ and $r_\lambda := f_C(s_j^y, s_j^\lambda)$. (We omitted iteration indices for notational simplicity).

The matrices $f_{C_y} = \frac{\partial f_C}{\partial y}$ and $f_{C_\lambda} = \frac{\partial f_C}{\partial \lambda}$ can be calculated easily by finite differences of $f_C$, whereas the computation of $W_j^y, W_j^\lambda$ is much more expensive since it involves the solution of a DAE system of $(n_y + n_\lambda)^2$ equations. We now show how this effort can be reduced to solving only $n_y^2$ equations.

Because of the index-1 assumption (7.3.10b) can be solved for $\Delta s_j^\lambda$:

$$\Delta s_j^\lambda = -f_{C_\lambda}^{-1} f_{C_y} \Delta s_j^y - f_{C_\lambda}^{-1} r_\lambda.$$

Inserting this into (7.3.10a) leads to

$$\begin{pmatrix} W_j^y & W_j^\lambda \end{pmatrix} \begin{pmatrix} I \\ -f_{C_\lambda}^{-1} f_{C_y} \end{pmatrix} \Delta s_j^y - \Delta s_{j+1}^y = -r_y + W_j^\lambda f_{C_\lambda}^{-1} r_\lambda.$$

The term $\begin{pmatrix} W_j^y & W_j^\lambda \end{pmatrix} \begin{pmatrix} I \\ -f_{C_\lambda}^{-1} f_{C_y} \end{pmatrix}$ is the directional derivative with respect to

initial values of $y(\tau_{j+1}; s_j^y, s_j^\lambda)$ in the direction of the $n_y$ columns of $\begin{pmatrix} I \\ -f_{C_\lambda}^{-1} f_{C_y} \end{pmatrix}$.

For the $k$-th column we obtain by using finite differences

$$\left( \begin{pmatrix} W_j^y & W_j^\lambda \end{pmatrix} \begin{pmatrix} I \\ -f_{C_\lambda}^{-1} f_{C_y} \end{pmatrix} \right)_{:,k} =$$

$$\frac{y(\tau_{j+1}; s_j^y + \varepsilon e_k, s_j^\lambda - \varepsilon f_{C_\lambda}^{-1} f_{C_y} e_k) - y(\tau_j; s_j^y, s_j^\lambda)}{\varepsilon} + \mathcal{O}(\varepsilon)$$

Thus, only this smaller number of directional derivatives (plus an additional directional derivative in the direction of $f_{C_\lambda}^{-1} r_\lambda$ to evaluate the right hand side) is necessary for using this approach of simultaneous generation and solution of the linearized system.

### Reducing the Number of Partial Derivatives: Higher Index Case

Here we have to solve at every node the matching conditions and the consistency conditions (cf. Eq. (7.3.5))

$$y(\tau_{j+1}; s_j) - s_{j+1}^y = 0$$

$$\phi(s_j) = 0.$$

In every step of the Gauß–Newton procedure these equations are linearized

$$W_j \Delta s_j - \Delta s_{j+1}^y = -(y(\tau_{j+1}; s_j) - s_{j+1}^y) =: -h_{j+1} \qquad (7.3.11a)$$
$$B_j \Delta s_j = -\phi(s_j) \qquad (7.3.11b)$$

with $B_j := \frac{\partial \phi}{\partial s}(s_j)$. Let the $n_y \times (n_y - n_\phi)$ matrix $V_j$ denote the nullspace of $B_j$. Then, the general solution of (7.3.11b) has the form

$$\Delta s_j = V_j \eta_j - B_j^+ \phi_j,$$

where $\eta_j$ denotes the remaining degrees of freedom. Inserting this into (7.3.11a) leads to

$$W_j V_j \eta_j - \Delta s_{j+1}^y = -h_{j+1} + B_j^+ \phi_j.$$

Let $Z_j(t) := W_j(t) V_j$ be the directional derivative of $W_j$ in the directions spanned by the columns of $V_j$, i.e. the free directions of motion. These directional derivatives can be calculated by only $n_y - n_\phi$ variations of the initial values in the directions of $V_j$. Analogous to the index-1 case described above, it is not necessary to compute the matrices $W_j$. The number of directional derivatives equals the number of degrees of freedom actually present in the system. There are originally $n_s = 2n_p + n_\lambda$ components of $s_j$ to vary. This number is reduced by $n_\phi = 3n_\lambda$ due to the technique presented above. Thus only $2(n_p - n_\lambda)$ degrees of freedom remain. This is the actual number of physical degrees present in the system.

## 7.3.2   Discontinuous Systems

As seen in Sec. 6 in mechanical systems one often has to deal with discontinuities. For the computation of sensitivity matrices this requires some extra effort.
We consider the system

$$\dot{x} = f(t, x, \theta, \operatorname{sign} q(t, x, \theta)). \qquad (7.3.12)$$

$q$ denotes the switching functions permitting the formulation of problems with discontinuities, see Sec. 6.
We assume that $f$ is sufficiently smooth as long as $\operatorname{sign} q = \text{const}$. Additionally, the state variables are allowed to have jump discontinuities when a sign change of $q$ occurs:

$$x(t^+) = x(t^-) + z(t^-, x(t^-), \theta) \quad \text{if} \quad q_j(t^-, x(t^-), \theta) = 0.$$

The differentiability of the solution with respect to initial values and parameters in the presence of discontinuities can also be guaranteed under mild assumptions as has been shown in [Bock87]. These assumptions include the differentiability of the

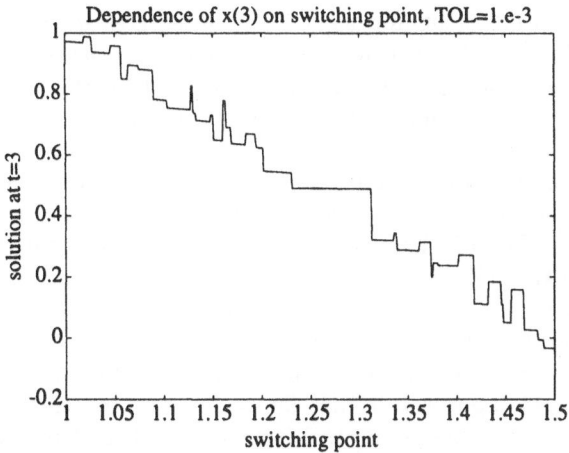

Figure 7.4: Dependence of numerical solution on parameter for TOL$=10^{-3}$

switching and jump functions. Additionally, one has to require that the switching structure does not change in the neighborhood of the solution $x$, especially that $\hat{t}$ is a simple zero of $q(t, x(t), \theta)$.

The explicit location of discontinuities is essential for the computation of sensitivity matrices with a prescribed and guaranteed accuracy. This is demonstrated by the example:

Consider the system

$$\dot{x} = \operatorname{sign}(t - \theta), \qquad x(0) = 0.$$

The exact solution of this system is

$$x(t, \theta) = \begin{cases} -t & \text{if } t < \theta \\ t - 2\theta & \text{if } t \geq \theta. \end{cases}$$

Let us consider the solution at $t = 3$

$$x(3, \theta) = 3 - 2\theta \qquad \text{if } \theta \leq 3.$$

Numerical integration with MATLAB's integration routine ODE45 without explicit location of the discontinuity for various values of $\theta$ leads to the results presented in Fig. 7.4.

We are looking for derivatives of the solution with respect to the parameter $\theta$. Numerically calculating derivatives means to use the slope of a straight line connecting two points of the graph. In this example this may lead to numerical derivatives even with a wrong sign.

The situation looks better at first sight if one integrates with a higher accuracy $10^{-5}$ instead of $10^{-3}$ as above. However, focusing gives a similar picture.

Again, the non-smoothness of the solution is caused by the integration routine which is a non-smooth mapping of the initial value to the value at a specified point, because e.g. error tests, ... require IF statements lead to discontinuities. This is much more important in the presence of discontinuities because most integrators with step size and order selection strategies have severe difficulties to pass a discontinuity (cf. Ex. 6.1.1) and use many heuristics and security IF-statements to manage this critical situation.

Thus, computing the derivatives from these integration results by forward differences can lead to totally misleading derivatives.

The explicit localization of the discontinuities leads to more accurate derivatives.

The drawback of this approach is that for each perturbed trajectory one has to localize a different switching point resulting in an inefficient behavior of the integrator. In addition, this concept does not fully fit into the concept of differentiation of the discretization scheme, see Sec. 7.2.7.

To overcome this difficulty in [Bock87] a correction is made at the switching point which takes into consideration the dependence of the switching point on the initial values and parameters. All perturbed trajectories are forced to switch at the same point and the sensitivity matrices are corrected by a term depending on the different switching points.

# A  The Truck Model

## A.1  Data of the Truck Model

Every body has a body fixed reference frame attached to its center of mass (CM). Initially the axes of these frames are parallel to the corresponding axis of the inertial frame. The location of the centers of mass is given with respect to the inertial frame, i.e. in absolute coordinates.

| CM1 | CM2 | CM3 | CM4 | CM5 |
|-----|-----|-----|-----|-----|
| (-2.06,0.5) | (0.0,2.0) | (2.44,0.5) | (2.74,2.90) | (-1.46,2.90) |

Table A.1: Center of mass location of the truck model's initial configuration (unconstrained and constrained case)

Springs, dampers and constraints are located between attachment points on the adjacent bodies. The coordinates of the attachment points are given relative to the respective body fixed frame.

| | Attachment Point 1 | Attachment Point 2 |
|---|---|---|
| Intercon. 10 | $(a_{12}, 0.0)_0 = (-2.06, 0.0)_0$ | $(0.0, 0.0)_1$ |
| Intercon. 12 | $(0.0, 0.0)_1$ | $(a_{12}, h1)_2 = (-2.06, -1.34)_2$ |
| Intercon. 20 | $(a_{23}, 0.0)_0 = (2.44, 0.0)_0$ | $(0.0, 0.0)_3$ |
| Intercon. 23 | $(0.0, 0.0)_3$ | $(a_{23}, h_1)_2 = (2.44, -1.34)_2$ |
| Intercon. 24 | $(a_{24}, h_2)_2 = (1.94, 0.0)_2$ | $(b_{24}, h_3)_4 = (-0.8, 0.0)_4$ |
| Intercon. 42 | $(a_{42}, h_2)_2 = (3.64, 0.0)_2$ | $(b_{42}, h_3)_4 = (0.9, 0.0)_4$ |
| Intercon. 25 | $(a_{25}, h_2)_2 = (0.98, 0.0)_2$ | $(c_{25}, h_3)_5 = (2.44, 0.0)_5$ |
| Intercon. 1d | $(a_{52}, h_2)_2 = (-3.07, 0.0)_2$ | $(c_{d1}, h_3)_5 = (-1.91, 0.0)_5$ |
| Intercon. 2d | $(a_{52}, h_2)_2 = (-3.07, 0.0)_2$ | $(c_{d2}, h_3)_5 = (-1.31, 0.0)_5$ |
| Joint | $(a_{c1}, a_{c2})_2 = (-3.07, 0.15)_2$ | $(c_{c1}, c_{c2})_5 = (-1.61, -0.75)_5$ |

Table A.2: Coordinates of the attachment points (subscripts indicate the coordinate system)

| Masses ($m_i$) and inertia ($l_i$): | |
|---|---|
| $m_1 = 1450\,\mathrm{kg}$ | |
| $m_2 = 3335\,\mathrm{kg}$ | $l_2 = 14313\,\mathrm{kg\,m^2}$ |
| $m_3 = 600\,\mathrm{kg}$ | |
| $m_4 = 1100\,\mathrm{kg}$ | $l_4 = 948\,\mathrm{kg\,m^2}$ |
| $m_5 = 11515\,\mathrm{kg}$ | $l_5 = 33000\,\mathrm{kg\,m^2}$ |

Table A.3: Masses and Inertias of the Truck model (unconstrained)

| stiffness ($k_{ij}$) | damping ($d_{ij}$) |
|---|---|
| $k_{10} = 44.0\,10^5\,\mathrm{N/m}$ | $d_{10} = 600\,\mathrm{Ns/m}$ |
| $k_{12} = 8.247\,10^5\,\mathrm{N/m}$ | $d_{12} = 21593\,\mathrm{Ns/m}$ |
| $k_{30} = 22.0\,10^5\,\mathrm{N/m}$ | $d_{30} = 300\,\mathrm{Ns/m}$ |
| $k_{23} = 2.711\,10^5\,\mathrm{N/m}$ | $d_{12} = 38537\,\mathrm{Ns/m}$ |
| $k_{24} = 1.357\,10^5\,\mathrm{N/m}$ | $d_{24} = 12218\,\mathrm{Ns/m}$ |
| $k_{42} = 1.357\,10^5\,\mathrm{N/m}$ | $d_{42} = 12218\,\mathrm{Ns/m}$ |
| $k_{25} = 9.0\,10^5\,\mathrm{N/m}$ | $d_{25} = 38500\,\mathrm{Ns/m}$ |
| $k_{d1} = 7.7\,10^5\,\mathrm{N/m}$ | $d_{d1} = 33013\,\mathrm{Ns/m}$ |
| $k_{d2} = 7.7\,10^5\,\mathrm{N/m}$ | $d_{d2} = 33013\,\mathrm{Ns/m}$ |

Table A.4: Stiffness and damping coefficients of the truck model (unconstrained)

## A.2 Model of a Nonlinear Pneumatic Spring

The equations of the truck model given in Sec. 1 are nonlinear due to nonlinear kinematics. The force laws are linear in the relative displacements and velocities. For some computational experiments in this book we need also nonlinear force elements. For this end we replace the sprig/damper element between the wheels and the chassis of the unconstrained truck by a nonlinear pneumatic spring, which corresponds to a component also found in "real-life" trucks, see Fig. A.2. The schematic principle of the pneumatic spring is sketched in Fig. A.1. We assume *adiabatic transition of states* in the air

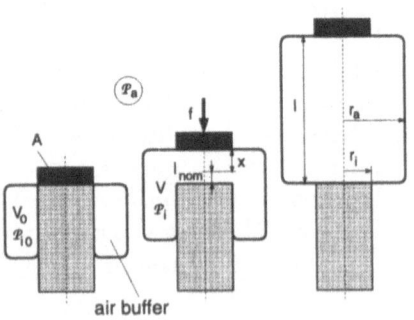

Figure A.1: Model of a pneumatic spring

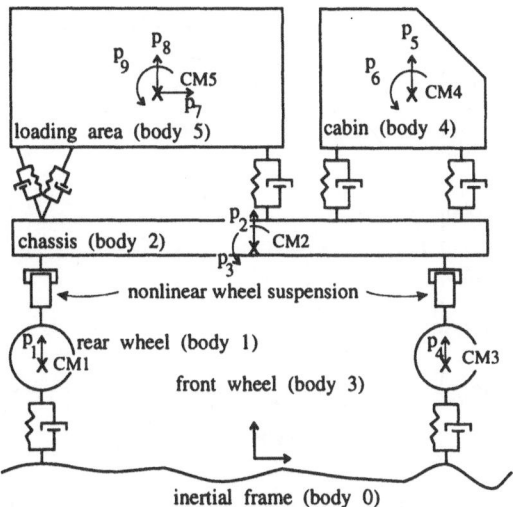

Figure A.2: Truck with nonlinear wheel suspensions

buffer. This means, there is no heat exchange between the air buffer and the environment. This assumption is reasonable for fast transitions. With $\mathcal{P}_i$ denoting the pressure in the air buffer and $V$ its volume, the adiabatic transition of states is characterized by

$$\mathcal{P}_i(x)V(x)^\kappa = \mathcal{P}_{i0}V_0^\kappa = \text{const.} \qquad (A.2.1)$$

where $\kappa = 1.4$ is the constant adiabatic coefficient. Furthermore, the surface area of the air buffer is supposed to be constant. Thus, the change of the buffer volume is described by

$$V_0 = \pi\,\frac{l}{2}\,(r_a^2 - r_i^2) \qquad (A.2.2)$$

$$V(x) = V_0 + \pi\,\frac{x + l_{\text{nom}}}{2}\,(r_a^2 + r_i^2) \qquad (A.2.3)$$

where $x$ denotes the displacement with respect to a nominal position $l_{\text{nom}}$. The force law is given by

$$f(x) = \left(\mathcal{P}_i(x) - \mathcal{P}_a\right)A \qquad (A.2.4)$$

with the constant pressure $\mathcal{P}_a$ of the atmosphere and the (constant) surface area $A$.

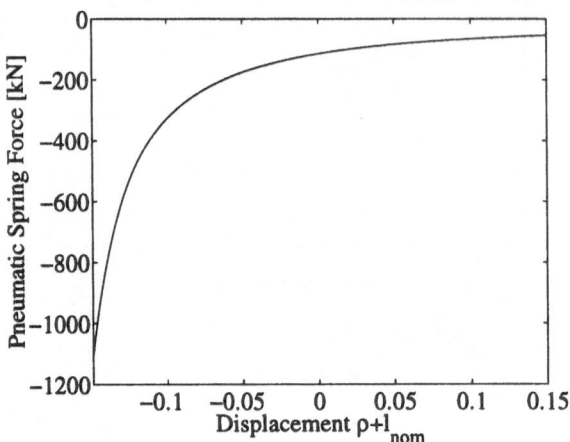

Figure A.3: Nonlinear force law (nominal force = -115 kN)

Combining (A.2.1), (A.2.2), (A.2.3), $F$ can be expressed as:

$$f(x) = \left( \frac{c}{V(x)^{\kappa}} - \mathcal{P}_{\mathrm{a}} \right) A = \left( \frac{\mathcal{P}_{i0}}{\left( 1 + \frac{(x + l_{\mathrm{nom}})}{l} \frac{(r_{\mathrm{a}}^2 + r_{\mathrm{i}}^2)}{(r_{\mathrm{a}}^2 - r_{\mathrm{i}}^2)} \right)^{\kappa}} - \mathcal{P}_{\mathrm{a}} \right) A \qquad \text{(A.2.5)}$$

where $\mathcal{P}_{i0}$ is the internal pressure for the volume $V_0$. $\mathcal{P}_{i0}$ can be replaced by the expression for the nominal force

$$F_{eq} = \left( \mathcal{P}_{\mathrm{i,eq}} - \mathcal{P}_{\mathrm{a}} \right) A = \left( \frac{\mathcal{P}_{i0}}{\left( 1 + \frac{l_{\mathrm{nom}}}{l} \frac{(r_{\mathrm{a}}^2 + r_{\mathrm{i}}^2)}{(r_{\mathrm{a}}^2 - r_{\mathrm{i}}^2)} \right)^{\kappa}} - \mathcal{P}_{\mathrm{a}} \right) A \qquad \text{(A.2.6)}$$

$$\mathcal{P}_{i,0} = \left( \frac{F_{eq}}{A} + \mathcal{P}_{\mathrm{a}} \right) \left( 1 + \frac{l_{\mathrm{nom}}}{l} \frac{(r_{\mathrm{a}}^2 + r_{\mathrm{i}}^2)}{(r_{\mathrm{a}}^2 - r_{\mathrm{i}}^2)} \right)^{\kappa}. \qquad \text{(A.2.7)}$$

With the abbreviation

$$s = \frac{1}{l} \cdot \frac{(r_{\mathrm{a}}^2 + r_{\mathrm{i}}^2)}{(r_{\mathrm{a}}^2 - r_{\mathrm{i}}^2)} \qquad \text{(A.2.8)}$$

and with $x = x_j$ with $j = 12, 23$ we finally get the nonlinear forces $f_{12}$ and $f_{23}$ in Example 1.3.1:

$$f_j(x_j) = \left( F_{eq,j} + \mathcal{P}_{\mathrm{a}} \cdot A \right) \cdot \left( \frac{1 + s \cdot x_{eq,j}}{1 + s \cdot (x_j + x_{eq,j})} \right)^{\kappa} - \mathcal{P}_{\mathrm{a}} \cdot A , \qquad j = 12, 23$$

$$\text{(A.2.9)}$$

Figure A.3 shows the force $f_{12}$ versus the displacement $x_{12}$.

| $\kappa = 1.4$ | $A = 0.0562\,\mathrm{m}^2$ | $l_{\mathrm{nom}} = 0.16\,\mathrm{m}$ | $P_a = 10^5\,\mathrm{N/m}^2$ | $s = 39.13228\,1/\mathrm{m}$ |
|---|---|---|---|---|

Table A.5: Additional data for the truck with nonlinear pneumatic springs

## Linearized Force Law

A linearization of (A.2.9) around the static equilibrium position $x_{eq,j}$ yields

$$\hat{f}_j(x_j) = F_{\mathrm{eq},j} - \kappa \cdot s \cdot \left( F_{\mathrm{eq},j} + \mathcal{P}_{\mathrm{a}} \cdot A \right) \cdot \left( 1 + s \cdot x_{\mathrm{eq},j} \right)^{-1} \cdot x_j, \qquad j = 12, 23.$$

$$(\mathrm{A.2.10})$$

For small displacements $x_j$, (A.2.10) represents a good approximation of the non-linear force law, see Figure A.3.

# A.3   Matlab m-files for the Unconstrained Truck

The following MATLAB m-file establishes the *right-hand side* function of the unconstrained truck model after dividing by the masses and torques.

```
function xdot=trurhs(t,x)
%  function xdot=trurhs(t,x)
%               Right hand side function for the
%               unconstrained truck model
%               Edda Eich-Soellner/ Claus Fuehrer   1997
%
global uu1
global uu2
m1=1450;m2=3335;m3=600;m4=1100;m5=11515;12=14313;14=948;15=33000;
%
a12=-2.06;a23=2.44;a24=1.94;a42=3.64;a25=0.98;a52=-3.07;
b42=0.9;b24=-0.8;c25=2.44;c1d=-1.91;c2d=-1.31;h1=-1.34;h2=0;h3=0;
%
k10=44d+5;k12=8.247E+5;k30=22d5;k23=2.711E+5;
k42=1.357d5;k24=k42;k25=9.0d5;
k1d=7.70E5;k2d=k1d;
d10=600;d12=21593;d30=300;d24=12218;d23=38537;
d42=12218;d25=38500;d1d=33013;d2d=d1d;
ggr=9.81;
%
% excitation (assumed truck speed 15 m/s
% 1 m takes tm secs)
```

```
v = 15.;
tm=1/15;
piv=pi*v;
ts= 2./15.;
td=(-a12+a23)*tm; % time delay between the wheels
%
[u1,u1d]=excite(t,ts+td,[piv,tm]);
[u2,u2d]=excite(t,ts,[piv,tm]);
%
p1=x( 1); p2=x( 2); p3=x( 3);
p4=x( 4); p5=x( 5); p6=x( 6);
p7=x( 7); p8=x( 8); p9=x( 9);
%
v1=x(10); v2=x(11); v3=x(12);
v4=x(13); v5=x(14); v6=x(15);
v7=x(16); v8=x(17); v9=x(18);
%
f10N =   -2.32914516200000e+6;
f12N =   -0.24687266200000e+6;
f30N =   -1.14743483800000e+6;
f23N =   -0.08492483800000e+6;
f24N =   -0.12784288235294e+6;
f42N =   -0.12720811764706e+6;
f25N =   -0.85490594111111e+6;
f2dN =   -0.76635491099974e+6;
f1dN =   -0.76635491099974e+6;
%
% relative vectors and their derivatives
%
[rho10,rho10d]=relat([0.,      u1,0.],[0.,u1d,0.],a12,0.,...
                      [a12,     p1,0.],[0.,v1, 0.],0., 0.);
[rho12,rho12d]=relat([a12,     p1,0.],[0.,v1, 0.],0., 0.,...
                      [0.,      p2,p3],[0.,v2, v3],a12,h1);
[rho23,rho23d]=relat([a23,     p4,0.],[0.,v4, 0.],0., 0.,...
                      [0.,      p2,p3],[0.,v2, v3],a23,h1);
[rho30,rho30d]=relat([0.,      u2,0.],[0.,u2d,0.],a23,0.,...
                      [a23,     p4,0.],[0.,v4, 0.],0., 0.);
[rho24,rho24d]=relat([0.,      p2,p3],[0.,v2, v3],a24,h2,...
                      [a24-b24,p5,p6],[0.,v5, v6],b24,h3);
[rho42,rho42d]=relat([0.,      p2,p3],[0.,v2, v3],a42,h2,...
                      [a24-b24,p5,p6],[0.,v5, v6],b42,h3);
[rho25,rho25d]=relat([0.,      p2,p3],[0.,v2, v3],a25,h2,...
```

```
                        [p7,     p8,p9],[v7,v8, v9],c25,h3);
[rhod2,rhod2d]=relat([0.,    p2,p3],[0.,v2, v3],a52,h2,...
                        [p7,     p8,p9],[v7,v8, v9],c2d,h3);
[rhod1,rhod1d]=relat([0.,    p2,p3],[0.,v2, v3],a52,h2,...
                        [p7,     p8,p9],[v7,v8, v9],c1d,h3);
%
[f101,f102,e10]=spridamp(k10,d10,rho10,rho10d,f10N);
[f121,f122,e12]=spridamp(k12,d12,rho12,rho12d,f12N);
[f301,f302,e30]=spridamp(k30,d30,rho30,rho30d,f30N);
[f231,f232,e23]=spridamp(k23,d23,rho23,rho23d,f23N);
[f241,f242,e24]=spridamp(k24,d24,rho24,rho24d,f24N);
[f421,f422,e42]=spridamp(k42,d42,rho42,rho42d,f42N);
[f251,f252,e25]=spridamp(k25,d25,rho25,rho25d,f25N);
[f2d1,f2d2,e2d]=spridamp(k2d,d2d,rhod2,rhod2d,f2dN);
[f1d1,f1d2,e1d]=spridamp(k1d,d1d,rhod1,rhod1d,f1dN);
%
xdot( 1) = v1; xdot( 2) = v2; xdot( 3) = v3;
xdot( 4) = v4; xdot( 5) = v5; xdot( 6) = v6;
xdot( 7) = v7; xdot( 8) = v8; xdot( 9) = v9;
%
co3=cos(p3);si3=sin(p3);co6=cos(p6);si6=sin(p6);
co9=cos(p9);si9=sin(p9);
xdot(10) = (-f102+f122)/m1 -ggr;
xdot(11) = (-f122-f232+f242+f422+f252+f2d2+f1d2)/m2-ggr;
xdot(12) = (...
            (a23*(-f232)+a12*(-f122)+h1*(-f231-f121))*co3-...
            (a23*(-f231)+a12*(-f121)+h1*(-f232-f122))*si3+...
            (a25*f252+ a52*(f2d2+f1d2)+h2*(f251+f2d1+f1d1))*co3-...
            (a25*f251+ a52*(f2d1+f1d1)+h2*(f252+f2d2+f1d2))*si3+...
            (a42*f422+  a24*f242  +h2*(f421+f241))*co3-...
            (a42*f421+  a24*f241  +h2*(f422+f242))*si3)/12;
xdot(13) = (-f302+f232)/m3-ggr;
xdot(14) = (-f422-f242)/m4-ggr;
xdot(15) = ((b42*(-f422)+b24*(-f242)+h3*(-f421-f241))*co6-...
            (b42*(-f421)+b24*(-f241)+h3*(-f422-f242))*si6)/14;
xdot(16) = (-f251-f2d1-f1d1)/m5;
xdot(17) = (-f252-f2d2-f1d2)/m5-ggr;
xdot(18) = ((c25*(-f252)+c2d*(-f2d2)+c1d*(-f1d2)
                        +h3*(-f251-f2d1-f1d1))*co9-...
            (c25*(-f251)+c2d*(-f2d1)+c1d*(-f1d1)...
                        +h3*(-f252-f2d2-f1d2))*si9)/15;
uu1=[uu1,u1];
```

```
uu2=[uu2,u2];
xdot=xdot';
```

As an example for a road-input function $u$ a smooth bump is taken:

```
function [u,ud]=excite(t,t0,param)
%   function [u,ud]=excite(t,t0,param)
%             excitation model (position and velocity)
%
%             Edda Eich-Soellner/ Claus Fuehrer   1997
%
if (t < t0)
   u=0.0d0; ud=0.0d0;
else
   u = 0.001*(t-t0)^6*exp(-(t-t0));  % a ca 12 cm hump
   ud= 0.006*(t-t0)^5*exp(-(t-t0))-0.001*(t-t0)^6*exp(-(t-t0));
end
```

The following MATLAB m-files are sub-procedures to establish the relative kinematics and the force laws of special force elements:

```
function [rho,rhod]=relat(fr,frd,fr1,fr2,to,tod,to1,to2)
%       function [rho,rhod]=relat(fr,frd,fr1,fr2,to,tod,to1,to2)
%          computes the relative vector and its dertivative between
%          the point (fr1,fr2) on body "from" described in the
%          body fixed coordinate system located in [fr(1),fr(2)] and
%          rotated with the angle fr(3).
%          the other point is located on body "to" and described
%          correspondingly.
%          Edda Eich-Soellner/Claus Fuehrer   1997
rotmto=[cos(to(3)),-sin(to(3));sin(to(3)),cos(to(3))];
rotmfr=[cos(fr(3)),-sin(fr(3));sin(fr(3)),cos(fr(3))];
rho =[to(1);to(2)]+rotmto*[to1;to2]-....
     ([fr(1);fr(2)]+rotmfr*[fr1;fr2]);
rhod=[tod(1);tod(2)]+...
     [-sin(to(3)),-cos(to(3));cos(to(3)),-sin(to(3))]
                                        *tod(3)*[to1;to2]...
     -([frd(1);frd(2)]+...
     [-sin(fr(3)),-cos(fr(3));cos(fr(3)),-sin(fr(3))]
                                        *frd(3)*[fr1;fr2]);
```

```
function [f1,f2,e]=spridamp(k,d,rho,rhod,fN)
%    [f1,f2,e]=spridamp(k,d,rho,rhod,fN)
```

```
%               forces of a planar spring and damper in
%               parallel, where
%               k,d   are the stiffness and damping ratio
%               rho   is the relative displacement vector and
%               rhod  is its time derivative
%               fN             is the nominal force
%               Edda Eich-Soellner / Claus Fuehrer  1997
%
length  = norm(rho);
lengthd = (rho(1)*rhod(1)+rho(2)*rhod(2))/length;
f=k*length+d*lengthd+fN;
e=[rho(1),rho(2)]/length;
f1=f*e(1); f2=f*e(2);
```

# Bibliography

[AFS96]   Arévalo C., Führer C., and Söderlind G. (1996) Stabilized multistep methods for index 2 Euler–Lagrange Equations. *BIT, Numerical Analysis* 36(1): 1–13.

[AFS97]   Arévalo C., Führer C., and Söderlind G. (1997) $\beta$-blocked multistep methods for Euler-Lagrange DAEs: Linear analysis. *Z. Angew. Math. Mech.* 77: 609–617.

[AG90]    Allgower E. L. and Georg K. (1990) *Numerical Continuation Methods.* Springer.

[AM97]    Arnold M. and Murua A. (1997) Non-stiff integrators for differential-algebraic systems of index 2. *Annals of Numerical Mathematics* , submitted for publication.

[AN96]    Arnold M. and Netter H. (1996) The approximation of contact conditions in the dynamical simulation of wheel-rail systems. Technical Report IB 515-96-08, German Aerospace Research Establishment, DLR Oberpfaffenhofen.

[And95]   Anderson et al. (eds.) E. (1995) *LAPACK users' guide.* SIAM.

[AP91]    Ascher U. M. and Petzold L. R. (1991) Projected implicit Runge-Kutta methods for differential-algebraic equations. *SIAM J. Numer. Anal.* 28: 1097–1120.

[AP93]    Ascher U. M. and Petzold L. R. (1993) Stability of computational methods for constrained dynamic systems. *SIAM J.Sci.Comp.* 14: 95–120.

[Are93]   Arévalo C. (1993) *Matching the Structure of DAEs and Multistep Methods.* PhD thesis, Dept. of Computer Science, Univ. Lund (Sweden).

[Arn78]   Arnold V. I. (1978) *Mathematical Methods of Classical Mechanics.* Springer.

[Arn95]    Arnold M. (1995) A perturbation analysis for the dynamical simulation of mechanical multibody systems. *Applied Numerical Mathematics* 18: 37–56.

[Arn98]    Arnold M. (1998) Half-explicit Runge–Kutta methods with explicit stages for differential-algebraic systems of index 2. *BIT, Numerical Analysis* , to appear.

[ÅW90]    Åström J. and Wittenmark B. (1990) *Computer-Controlled Systems, Theory and Design.* Prentice-Hall, Englewood Cliffs, N.J.

[Bar89]    Barrlund A. (1989) Constrained least squares methods, of implicit type, for the numerical solution to higher index linear variable coefficient differential algebraic systems. Technical Report UMINF-166.89, Institute of Information Processing, University of Umeå, Sweden.

[BBE93]    Baake M., Baake E., and Eich E. (1993) On the role of invariants for the parameter estimation problem in Hamiltonian systems. *Physics Letters A* 180.

[BCP89]    Brenan K. E., Campbell S. L., and Petzold L. R. (1989) *The Numerical Solution of Initial Value Problems in Ordinary Differential-Algebraic Equations.* North Holland Publishing Co.

[BES87]    Bock H. G., Eich E., and Schlöder J. P. (1987) Numerical solution of constrained least squares boundary value problems in differential-algebraic equations. In Strehmel (ed) *Proceedings NUMDIFF 87.* Teubner, Halle.

[BH93]    Brasey V. and Hairer E. (1993) Half-explicit Runge–Kutta methods for differential-algebraic systems of index 2. *SIAM J. Numer. Anal.* 30: 538–552.

[BHK94]    Buchauer O., Hiltmann P., and Kiehl M. (1994) Sensitivity analysis of initial-value problems with applications to shooting techniques. *Numer. Math.* 67: 151–159.

[BIAG73]    Ben-Israel Adi / Greville T. N. E. (1973) *Generalized Inverses: Theory and Applications.* John Wiley.

[BJO86]    Brandl H., Johanni R., and Otter M. (1986) An algorithm for the simulation of multibody systems with kinematic loops. In *Proc. IFAC/IFIP/IMACS International Symposium on The Theory of Robots, Vienna, Austria.*

[Bock87]    Bock H. G. (1987) *Randwertproblemmethoden zur Parameteridenti-fizierung in Systemen nichtlinearer Differentialgleichungen.* PhD thesis, Bonner Mathematische Schriften 183, Universität Bonn.

[Bock89]    Bock H. G. (1989) Course on "Optimization problems in ordinary differential equations" given at the University of Augsburg, Germany.

[Bra92]     Brasey V. (1992) A half-explicit method of order 5 for solving constrained mechanical systems. *Computing* 48: 191–201.

[Bro70]     Brockett R. W. (1970) *Finite Dimensional Linear Systems.* J. Wiley, New York.

[BS81]      Bock H. G. and Schlöder J. P. (1981) Numerical solution of retarded differential equations with state-dependent time-lages. *Z. Angew. Math. Mech.* page T 269.

[Bul71]     Bulirsch R. (1971) Die Mehrzielmethode zur numerischen Lösung von nichtlinearen Randwertproblemen und Aufgaben der optimalen Steuerung. Technical report, Carl-Cranz-Gesellschaft.

[But87]     Butcher J. C. (1987) *The Numerical Analysis of Ordinary Differential Equations: Runge–Kutta and General Linear Methods.* Wiley, Chichester.

[Cam80]     Campbell S. L. (1980) *Singular Systems of Differential Equations I.* Research Notes in Math.; 40. Pitman, Marshfield.

[Car78]     Carver M. B. (1978) Efficient integration over discontinuities in ordinary differential equation simulations. *Math. Comp. Simul.* 20(3): 190–196.

[CM78]      Carver M. B. and MacEwen S. R. (1978) Numerical analysis of a system described by implicitly-defined ordinary differential equations containing numerous discontinuities. *Applied Math. Modelling* 2: 280–286.

[CM79]      Campbell S. L. and Meyer C. D. (1979) *Generalized Inverses of Linear Transformations.* Pitman.

[CP76]      Cline R. E. and Plemmons R. J. (1976) $l_2$ - solutions to undertermined linear systems. *SIAM Review* 18(1): 92–106.

[CR78]      Cellier F. E. and Rufer D. F. (1978) Algorithms suited for the solution of initial value problems in engineering applications. *Math. Comp. Simul.* 20(3): 160–165.

[CS74]      Chartres B. A. and Stepleman R. S. (1974) Actual order of convergence
            of Runge-Kutta methods of differential equations with discontinuities.
            *SIAM J. Numer. Anal.* 11: 1193–1206.

[CS76]      Chartres B. A. and Stepleman R. S. (1976) Convergence of linear mul-
            tistep methods for differential equations with discontinuities. *Numer.
            Math.* 27: 1–10.

[Dav53]     Davidenko D. (1953) On approximate solution of systems of nonlinear
            equations. *Ukraine Mat. Z.* 5: 196–206. (in Russian).

[DH79]      Deuflhard P. and Heindl G. (1979) Affine invariant convergence theo-
            rems for Newton's method and extensions to related methods. *SIAM
            J. Numer. Anal.* 16: 501–516.

[DHZ87]     Deuflhard P., Hairer E., and Zugck J. (1987) One-step and extrapo-
            lation methods for differential-algebraic systems. *Numer. Math.* 51:
            501–516.

[DJ82]      Duffek W. and Jaschinski A. (1982) Efficient implementation of wheel-
            rail contact mechanics in dynamic curving. In Wickens A. (ed) *Proc.
            7th IAVSD-Symposium on the Dynamics of Vehicles on Roads and on
            Tracks*, pages 441 – 454. Swets & Zeitlinger, B.V. Lisse.

[DMBS79]    Dongarra J., Moler C., Bunch J., and Stewart G. (1979) *LINPACK
            User's Guide.* SIAM, Philadelphia.

[DP89]      Dormand J. and Prince P. (1989) Practical Runge–Kutta processes.
            *SIAM J. Sci. Stat. Comput.* 10(5): 977–989.

[DS80]      Deuflhard P. and Sautter W. (1980) On rank-deficient pseudo inverses.
            *Lin. Alg. Appl.* 29: 91–111.

[EF75]      Evans D. J. and Fatunla S. O. (1975) Accurate numerical determination
            of the intersection point of the solution of a differential equation with
            a given algebraic relation. *J. Inst. Math. Appl.* 16(3): 355–369.

[EFLR90]    Eich E., Führer C., Leimkuhler B., and Reich S. (1990) Stabilization
            and projection methods for multibody dynamics. Technical report,
            Helsinki University of Technology, Finland.

[Eich91]    Eich E. (1991) *Projizierende Mehrschrittverfahren zur numerischen
            Lösung der Bewegungsgleichungen technischer Mehrkörpersysteme mit
            Zwangsbedingungen und Unstetigkeiten.* PhD thesis, Institut für Math-
            ematik, Universität Augsburg. Also: VDI-Fortschrittsber. (18), 109,
            VDI-Verlag, Düsseldorf, 1992.

[Eich93]    Eich E. (1993) Convergence results for a coordinate projection method applied to constrained mechanical systems. *SIAM J. Numer. Anal.* 30(5): 1467–1482.

[EJNT86a]   Enright W. H., Jackson K. R., Nørsett S. P., and Thomson P. G. (1986) Effective solution of discontinuous IVPs using a Runge–Kutta formula pair with interpolants. *Appl. Math. Comp.* 27: 313–335.

[EJNT86b]   Enright W. H., Jackson K., Nørsett S. P., and Thomson P. G. (1986) Interpolants for Runge-Kutta formulas. *ACM Trans. Math. Softw.* 12(3): 193–218.

[Ell81]     Ellison D. (1981) Efficient automatic integration of ODEs with discontinuities. *Math. Comput. Simul.* 23(1): 12–20.

[EMS93]     Eich E., Mehlhorn R., and Sachs G. (1993) Stabilization of numerical solutions of boundary value problems exploiting invariants. Technical Report 16, Lehrstuhl für Flugmechanik und Flugregelung, TU München.

[Fea83]     Featherstone R. (Spring 1983) The calculation of robot dynamics using articulated body inertias. *The International Journal of Robotics Research* 2(1): 13–30.

[Fil64]     Filippov A. F. (1964) Differential equations with discontinuous right hand side. *AMS Transl.* 42: 199–231.

[Füh88]     Führer C. (1988) *Differential-algebraische Gleichungssysteme in mechanischen Mehrkörpersystemen.* PhD thesis, Mathematisches Institut, Technische Universität München.

[FW84]      Führer C. and Wallrapp O. (1984) A computer-oriented method for reducing linearized multibody equations by incorporating constraints. *Comp. Meth. Apl. Mech. Eng.* 46: 169 – 175.

[Gear71a]   Gear C. W. (1971) *Numerical Initial Value Problems in Ordinary Differential Equations.* Prentice-Hall.

[Gear71b]   Gear C W.. (1971) Simultaneous numerical solution of differential-algebraic equations. *IEEE Trans. Circuit Theory* CT-18(1): 89 – 95.

[Gear86]    Gear C. W. (1986) Maintaining solution invariants in the numerical solution of ODEs. *SIAM J. Sci. Stat. Comput.* 7(3): 734–743.

[GGL85]     Gear C. W., Gupta G., and Leimkuhler B. (1985) Automatic integration of the Euler-Lagrange equations with constraints. *J. Comp. Appl. Math.* 12 & 13: 77–90.

[GL83]      Golub G. and Loan van C. (1983) *Matrix Computations*. The John
            Hopkins University Press, Baltimore.

[GM86]      Griepentrog E. and März R. (1986) *Differential-Algebraic Equations
            and Their Numerical Treatment*. Teubner-Texte zur Mathematik No.
            88. BSB B.G. Teubner Verlagsgesellschaft, Leipzig.

[GMW81]     Gill P., Murray, and Wright M. (1981) *Practical Optimization*. Addison
            Wesley.

[GO84]      Gear C. and Osterby O. (1984) Solving ordinary differential equations
            with discontinuities. *ACM Trans. Math. Softw.* 10(1): 23–44.

[GP92]      Glocker C. and Pfeiffer F. (1992) Dynamical systems with unilateral
            contacts. *Nonlinear Dynamics* 3:245–259.

[Gri89]     Griewank A. (1989) On automatic differentiation. In *Mathematical
            Programming: Recent Development and Applications*, pages 83–108.
            Kluwer.

[Gru95]     Grupp F. (1995) *Parameteridentifizierung nichtlinearer mechanischer
            Deskriptorsysteme mit Anwendungen in der Rad-Schiene-Dynamik*.
            PhD thesis, Universitt Wuppertal. Also: VDI-Fortschrittsber. (8),
            550, VDI-Verlag, Düsseldorf, 1995.

[Hal76]     Halin H. J. (1976) Integration of ordinary differential equations con-
            taining discontinuities. In *Procedings of the Summer Computer Sim-
            ulation Conference 1976, La Jolla*, pages 46–53. SCI Press, La Jolla,
            California.

[Haug89]    Haug E. (1989) *Computer-Aided Kinematics and Dynamics of Mechan-
            ical Systems*, volume I. Allyn and Bacon, Boston.

[HIM97]     Hanke M., Izquierdo Macana E., and März R. (1997) On asymptotics in
            case of linear index-2 differential-algebraic equations. Technical Report
            97-3, Humboldt-Universität Berlin, Institut für Mathematik, D-10088
            Berlin.

[HLR89]     Hairer E., Lubich C., and Roche M. (1989) *The Numerical Solution
            of Differential-Algebraic Equations by Runge-Kutta Methods*. Lecture
            Notes in Mathematics Vol. 1409. Springer, Heidelberg.

[HNW87]     Hairer E., Nørsett S., and Wanner G. (1987) *Solving Ordinary Differ-
            ential Equations I: Nonstiff Problems*. Springer-Verlag, Berlin.

[HW96]      Hairer E. and Wanner G. (1996) *Solving Ordinary Differential Equa-
            tions II: Stiff and Differential- Algebraic Problems, 2nd ed.* Springer-
            Verlag, Berlin.

[HY90]    Haug E. and Yen J. (1990) Implicit numerical integration of constrained
          equations of motion via generalized coordinate partitioning. In E.Haug
          and R.Deyo (eds) *NATO Advanced Research Workshop on Real – Time
          Integration Methods for Mechanical System Simulation.* Springer, Hei-
          delberg.

[JSD80]   Jackson K. and Sacks-Davis R. (1980) An alternative implementation
          of variable step-size multistep formulas for stiff odes. *ACM Trans.
          Math. Software* 6: 295–318.

[KL94]    Kortüm W. and Lugner P. (1994) *Systemdynamik und Regelung von
          Fahrzeugen.* Springer.

[Lam91]   Lambert J. (1991) *Numerical Methods for Ordinary Differential Sys-
          tems.* John Wiley.

[Leh87]   Lehmann E. (1987) Deckscontainer und Schiffsverformungen.    In
          *Jahrbuch der schiffbautechnischen Gesellschaft 1987.* Schiffbautechnis-
          che Gesellschaft.

[Lei88]   Leister G. (1988) Vergleichende Untersuchungen der Dynamik eines
          Schlaghammers. Master's thesis, Institut B für Mechanik der Univer-
          sität Stuttgart.

[LENP95]  Lubich C., Engstler C., Nowak U., and Pöhle U. (1995) Numerical
          integration of constrained mechanical systems using MEXX. *Mech.
          Struct. Mach.* 23: 473–495.

[Lub91]   Lubich C. (1991) Extrapolation integrators for constrained mechanical
          systems. *Impact Comput. Sc. Eng.* 3: 213–234.

[Lub93]   Lubich C. (1993) Integration of stiff mechanical systems by Runge-
          Kutta methods. *ZAMP* 44: 1022–1053.

[Man73]   Mannshardt R. (1973) Eine Darstellung von Gleitbewegungen längs
          Unstetigkeitsflächen von Differentialgleichungen mit Sprungfunktio-
          nen. *Z. Angew. Math. Mech.* 53: 659–665.

[Man78]   Mannshardt R. (1978) One-step methods of any order for ordinary dif-
          ferential equations with discontinuous right hand sides. *Numer. Math.*
          31(2): 131–152.

[MK75]    Mangoldt H. and Knopp K. (1975) *Einführung in die Höhere Mathe-
          matik,* volume 3. Hirzel-Verlag Stuttgart.

[Mod88]   Model R. (1988) Zur Integration über Unstetigkeiten in gewöhnlichen
          Differentialgleichungen. *Z. Angew. Math. Mech.* 68: 161–169.

[Mur97]    Murua A. (1997) Partitioned half-explicit Runge–Kutta methods for
           differential algebraic systems of index 2. *Computing* 59: 43–61.

[MvL78]    Moler C. and van Loan C. (1978) Nineteen dubious way to compute
           the exponential of a matrix. *SIAM Review* 20(4): 801–836.

[OR70]     Ortega J. and Rheinboldt W. (1970) *Iterative Solutions of Nonlinear
           Equations in Several Variables*. Academic Press, New York.

[Orl73]    Orlandea N. (1973) *Node-analogous, sparsity oriented methods for sim-
           ulation of mechanical dynamics systems*. PhD thesis, University of
           Michigan, Ann Arbor.

[OS98]     Olsson H. and Söderlind G. (1998) Stage value predictors and efficient
           Newton iterations in implicit Runge–Kutta methods. *SIAM J. Sci.
           Stat. Comput.*, to appear.

[Ost90]    Ostermann A. (1990) A half-explicit extrapolation method for
           differential-algebraic systems of index 3. *IMA J. Numer. Anal.* 10:
           171–180.

[Pfei84]   Pfeiffer F. (1984) Mechanische Systeme mit unstetigen Übergängen.
           *Ingenieurarchiv* 54: 232–240.

[PJY97]    Petzold L., Jay L., and Yen J. (1997) Numerical solution of highly
           oscillatory differential equations. *Acta Numerica* .

[PTVF92]   Press W. C., Teukolsky S. A., Vetterling W. T., and Flannery B. P.
           (1992) *Numerical Recipes*. Cambridge University Press.

[RS88]     Roberson R. and Schwertassek R. (1988) *Dynamics of Multibody Sys-
           tems*. Springer, Heidelberg.

[Rug84]    Ruggaber W. (1984) Numerische Berechnung des Frequenzganges aus
           der Zustandsdarstellung. *Regelungstechnik* 32(1): 801–836.

[Sac96]    Sachau D. (1996) *Berücksichtigung von flexiblen Körpern und
           Fügestellen in Mehrkörpersystemen zur Simulation aktiver Raumfahrt-
           strukturen*. PhD thesis, Universität Stuttgart, Institut A für Mechanik.

[SB93]     Stoer J. and Bulirsch R. (1993) *Introduction to Numerical Analysis*.
           Springer.

[SB95]     Schwerin von R. and Bock H. G. (1995) A Runge–Kutta starter for a
           multistep method for differential-algebraic systems with discontinuous
           effects. *Appl. Numer. Math.* 18: 337–350.

[Sch90]    Schiehlen W. (ed) (1990) *Multibody Handbook*. Springer, Heidelberg.

[Sey84]    Seydel R. (1984) A continuation algorithm with step control. In Küpper
           T. (ed) *Numerical Methods in Bifurcation problems*, volume 70 of
           *ISNM*, pages 480–494. Birkhäuser.

[Sey93]    Seyfferth W. (1993) *Modellierung unstetiger Montageprozesse mit
           Robotern*. PhD thesis, TU München.

[SFR91]    Simeon B., Führer C., and Rentrop P. (1991) Differential-algebraic
           equations in vehicle system dynamics. *Surv. Math. Ind.* 1: 1–37.

[SFR93]    Simeon B., Führer C., and Rentrop P. (1993) The Drazin inverse in
           multibody system dynamics. *Numer. Math.* 64: 521–539.

[SG75]     Shampine L. F. and Gordon M. K. (1975) *Computer Solution of Ordi-
           nary Differential Equations*. Freeman, San Francisco.

[SG84]     Shampine L. F. and Gordon M. K. (1984) *Computer Lösung
           gewöhnlicher Differentialgleichungen*. Vieweg.

[Sha86]    Shampine L. F. (1986) Conservation laws and the numerical solution
           of ODEs. *Comp. and Maths. with Appls., Part B*. 12.

[Sim88]    Simeon B. (1988) Homotopieverfahren zur Berechnung quasista-
           tionärer Lagen von Deskriptorformen in der Mechanik. Master's thesis,
           Technische Universität München.

[Sim95]    Simeon B. (1995) MBSPACK - Numerical integration software for con-
           strained mechanical motion. *Surv. Math. Ind.* 5: 169–202.

[Sim96]    Simeon B. (1996) Modelling a flexible slider crank mechanism by a
           mixed system of DAEs and PDEs. *Math. Modelling of Systems* 2: 1–
           18.

[Sjö96]    Sjö A. (1996) *Numerical Aspects in Contact Mechanics and Rolling
           Bearing Simulation*. Licensiate Thesis at the Dept. of Computer Sci-
           ence, Univ. Lund (Sweden).

[Söd92]    Söderlind G. (1992) Stability and error concepts in discretization meth-
           ods for initial value problems. Unpublished manuscript.

[Tau76]    Taubert K. (1976) Differenzenverfahren für Schwingungen mit trock-
           ener und zäher Reibung und für Regelungssysteme. *Numer. Math.* 26.

[Tho93]    Thomsen P. G. (June 1993) Event controlled simulation. Talk given at
           the Oberwolfach Conference "Differential-Algebraic Equations: The-
           ory and Applications in Technical Simulation".

[WA86]    Werner H. and Arndt H. (1986) *Gewöhnliche Differentialgleichungen.*
          Springer-Verlag.

[Wal89]   Wallrapp O. (1989) Entwicklung rechnergestützter Methoden der
          Mehrkörperdynamik in der Fahrzeugtechnik (Dissertation). Techni-
          cal Report DLR-FB 89-17, Deutsche Forschungsanstalt für Luft- und
          Raumfahrt (DLR).

[WWS94]   Wensch J., Weiner R., and Strehmel K. (1994) Stability investigations
          for index-2 systems. Technical report, Reports on Computer Science
          and Scientific Computing Universität Halle, Germany.

[Yen93]   Yen J. (1993) Constrained equations of motion in multibody dynamics
          as ODEs on manifolds. *SIAM J. Numer. Anal.* 30(2): 553–568.

# Index